CANADIAN AGRICULTURE *in the* 21ˢᵗ CENTURY

Change and Challenge

DR. MARVIN S. ANDERSON

◆ FriesenPress

Suite 300 - 990 Fort St
Victoria, BC, V8V 3K2
Canada

www.friesenpress.com

Copyright © 2020 by Dr. Marvin S. Anderson
First Edition — 2020

All rights reserved.

No part of this publication may be reproduced in any form, or by any means, electronic or mechanical, including photocopying, recording, or any information browsing, storage, or retrieval system, without permission in writing from FriesenPress.

ISBN
978-1-5255-5484-1 (Hardcover)
978-1-5255-5485-8 (Paperback)
978-1-5255-5486-5 (eBook)

1. BUSINESS & ECONOMICS, INDUSTRIES, AGRIBUSINESS

Distributed to the trade by The Ingram Book Company

Table of Contents

1. **INTRODUCTION** 1
2. **HISTORIC OVERVIEW** 7
   - 2.1 Beginnings 7
   - 2.2 Atlantic Provinces (PEI, NS, NB, NL) 8
   - 2.3 Quebec (Lower Canada) 9
   - 2.4 Ontario (Upper Canada) 11
   - 2.5 Prairies (Manitoba, Saskatchewan, Alberta) 13
   - 2.6 British Columbia 18
   - 2.7 Never Forget 18
3. **A FARM PROFILE** 19
   - 3.1 Number of Farmers & Average Farm Size 19
   - 3.2 Land Use & Land Ownership 21
   - 3.3 Farm Enterprise Composition 24
   - 3.4 Farm Organization – Legal Structure 25
   - 3.5 Demographic Characteristics 26
     - 3.5.1 Age of Farm Operators 26
     - 3.5.2 Gender 27
     - 3.5.3 Farm Education Levels 28
   - 3.6 Sources of Farm Revenue 29
   - 3.7 Farm Inputs & Costs 31
   - 3.8 Income & Off-Farm Work 33
   - 3.9 The Cost-Price Squeeze & Scale 36
   - 3.10 Commercial Farms Re-Visited 38
   - 3.11 Capital & Debt Structure 40
   - 3.12 Farmer's Status in Rural & Urban Canada 42
   - 3.13 Synopsis 43
4. **AGRICULTURAL PRODUCTION & MARKETING** 45
   - 4.1 Major Crops 45
     - 4.1.1 Canola 46
     - 4.1.2 Wheat 47
     - 4.1.3 Other Cereals 48
     - 4.1.4 Corn (Maize) 51
     - 4.1.5 Soybeans 52
     - 4.1.6 Pulses 54
     - 4.1.7 Other Crops 56
   - 4.2 Beef Production 56
     - 4.2.1 Cow-Calf Operations 56
     - 4.2.2 Feeder (Slaughter Cattle) Operations 57
   - 4.3 Pork Production 57
   - 4.4 Dairy Industry 59
   - 4.5 Poultry & Eggs 63
   - 4.6 Other Livestock & Poultry 66
   - 4.7 Synopsis 67
5. **INTERNATIONAL TRADE** 69
   - 5.1 Overview 69
   - 5.2 Major Agricultural Exports & Imports 70

		5.2.1	*Canola* 71
		5.2.2	*Wheat, Barley, and Oats* 71
		5.2.3	*Soybeans* 72
		5.2.4	*Pulses* 72
		5.2.5	*Beef* 72
		5.2.6	*Pork* 73
		5.2.7	*Dairy* 73
		5.2.8	*Poultry & Eggs* 74
	5.3	Agriculture & Agri-Food Trade Balance 75
	5.4	International Product Prices 77
	5.5	Foreign Exchange Rates 78
	5.6	Multinational Trading Rules 79
	5.7	National Levels of Agricultural Protection 80
	5.8	Trade Agreements 81
		5.8.1	*Bilateral Agreements* 81
		5.8.2	*Regional Trade Agreements* 82
	5.9	The Gains from Freer Trade 88
	5.10	Future Agricultural Trade Prospects 89

6. AGRI-INDUSTRY & VALUE-ADDED 93
	6.1	An Overview 93
	6.2	Forward Linkages – Food & Beverage Manufacturing 96
		6.2.1	*Grain and Oilseed Sector – Selected Profiles* 96
		6.2.2	*Livestock & Poultry Sectors – Selected Profiles* 104
	6.3	Backward Linkages – Farm Suppliers (Farm Inputs) 107
		6.3.1	*Grain & Oilseed Sector – Selected Profiles* 108
		6.3.2	*Livestock & Poultry Sectors – Selected Profiles* 114
		6.3.3	*Other Input Providers* 115
	6.4	The Total Impact 116

7. RURAL SOCIO-ECONOMIC PROFILE 119
	7.1	What is Rural? 119
	7.2	Historic Trends 120
	7.3	Human Capital in Rural Canada 124
	7.4	Related Challenges 124
		7.4.1	*Labor & Leadership Constraints* 125
		7.4.2	*Culturally Distinct Contiguous Communities* 125
		7.4.3	*Multiple Resource Use on Private Land* 126
		7.4.4	*Policing Low-Density Populations* 127
		7.4.5	*Adapting to Climate Change* 128
	7.5	Rural Consolidation & Growth 128
	7.6	Government & Community Initiatives 131
		7.6.1	*Top-Down* 132
		7.6.2	*Bottom-Up* 133

8. AGRICULTURAL ORGANIZATIONS 137
	8.1	Rural Cooperatives 137
	8.2	National Umbrella Organizations 139
	8.3	Commodity Groups 143
		8.3.1	*National Marketing Agencies* 145
		8.3.2	*Provincial Marketing Boards & Commissions* 147
		8.3.3	*Commodity-Group Research* 150
	8.4	Supplier & Processor Organizations 152
	8.5	Final Comments 152

9. ROLE OF GOVERNMENT 153
	9.1	Historical Overview, 1870–2019 153
		9.1.1	*The Decades 1870–1950* 154
		9.1.2	*The Decades 1950–1980* 155

		9.1.3	The Decades 1980–2000	157
		9.1.4	The 21ˢᵗ Century	159
	9.2	Organization of Agriculture and Agri-Food Canada (AAFC)		161
	9.3	The Regulatory Environment		163
	9.4	The Railways		164
	9.5	Agricultural Research		166
		9.5.1	Federal Government	166
		9.5.2	Provincial Governments	167
		9.5.3	Private Sector R&D	167
		9.5.4	Industry Groups	168
		9.5.5	Universities and Colleges	168
		9.5.6	NGOs and International Linkages	168
		9.5.7	International Comparisons	168
	9.6	Provincial Governments		169
		9.6.1	Ad Hoc Programs	169
		9.6.2	Resource-Based Programs	170
		9.6.3	Research – Extension	171
		9.6.4	Credit	171
	9.7	Support Measures for Agriculture		171
		9.7.1	Total Direct and Indirect Support	171
		9.7.2	Direct Budgetary Payments	172
	9.8	The Transition		173
10.	**RESOURCE MANAGMENT IN AGRICULTURE**			175
	10.1	Overview		175
	10.2	Agricultural Resource Profile		176
		10.2.1	Crop Land	176
		10.2.2	Pasture Land	177
		10.2.3	Woodlots and Wetlands	177
	10.3	Land Degradation		178
		10.3.1	Soil Salinization	179
		10.3.2	Soil Erosion	180
		10.3.3	Soil Acidification	182
		10.3.4	Organic Matter Loss	183
	10.4	Urbanization of Agricultural Land		184
		10.4.1	An Overview	184
		10.4.2	A Regional Perspective	185
		10.4.3	An On-Going Issue	187
	10.5	Regenerative Organic Agriculture (ROA)		187
	10.6	Public Land Management Initiatives		189
	10.7	Water Management		191
		10.7.1	Legal Framework	191
		10.7.2	Supply & Consumptive Uses	191
		10.7.3	Irrigation	192
		10.7.4	Drainage	199
		10.7.5	Flood Control	200
		10.7.6	Other Water-Related Government Initiatives	201
	10.8	The Value of Environmental Amenities		201
	10.9	Payment for Alternative Land Use Services (ALUS)		203
		10.9.1	European Union	203
		10.9.2	United States	203
		10.9.3	Canada	204
		10.9.4	The Future	205
11.	**CLIMATE CHANGE & AGRICULTURE**			207
	11.1	Overview		207
	11.2	A Global Perspective		208
	11.3	CO_2e Sources in Agriculture		210
	11.4	CO_2e Agriculture Sector Balance		213
	11.5	Role of Agriculture in CO_2e Mitigation		214

		11.5.1	Land Use Changes	215
		11.5.2	Land Management Changes	215
		11.5.3	Crop Mix & Yield Changes	215
	11.6	Climate Change & Canadian Agriculture		217
		11.6.1	Climate Change Impacts	217
		11.6.2	Potential for Agriculture Adaptation	219

12. RELATED POLLUTION ISSUES & MITIGATION MEASURES 221

	12.1	Sources & Impacts		221
		12.1.1	Synthetic Fertilizer	222
		12.1.2	Pesticides & Disease	222
		12.1.3	Irrigation	227
		12.1.4	Machinery Emissions & Discharges	228
		12.1.5	Livestock	228
		12.1.6	Agricultural Processing Wastes	228
	12.2	Mitigative Measures		228
		12.2.1	Nitrogen-Fixing Plants	229
		12.2.2	Increased Efficacy of Nitrogen-Based Chemical Fertilizers	230
		12.2.3	Biological Pest Control	230
		12.2.4	Genome Editing	230
		12.2.5	Alternative Fuels	231
		12.2.6	Lower Residual Feed Intake (RFI) in Beef	231
		12.2.7	Manure Treatment	231
	12.3	Sector Metamorphosis		232

13. ORGANIC CROPS, FUNCTIONAL FOODS, & GMO'S 233

	13.1	Certified Organic Production		233
		13.1.1	Legal Definition	233
		13.1.2	Regulation	234
		13.1.3	Market Profile	234
		13.1.4	Assessment	235
	13.2	Functional Foods		236
		13.2.1	Definition	237
		13.2.2	Regulation	237
		13.2.3	Market Profile	237
		13.2.4	Assessment	237
	13.3	Genetically Modified Organisms (GMO's)		238
		13.3.1	Definition	238
		13.3.2	Regulation	239
		13.3.3	Market Profile	239
		13.3.4	Assessment	240
	13.4	Implications for Farmers		241

14. THE FUTURE OF AGRICULTURE IN CANADA 243

	14.1	The Mega-Variables		243
		14.1.1	Internationalization	243
		14.1.2	Consumer-Driven	245
		14.1.3	Regulation	245
		14.1.4	Technological Change	246
		14.1.5	Water Availability	251
		14.1.6	Structure of Agri-Industry	252
	14.2	Farm Structure		253
		14.2.1	Heterogeneity	253
		14.2.2	Bi-Modality	253
		14.2.3	Increasing Commercial Scale	254
		14.2.4	Different Capitalization Mechanisms	254
		14.2.5	Vertical Integration	254
		14.2.6	Narrower Margins	254
		14.2.7	Increased Self-Sufficiency	255
		14.2.8	Environmental Sustainability & Enhancement	255
		14.2.9	Increased Competition for Land-Water Resources	256
	14.3	Agriculture is the Future		256

1. Introduction

There are over two million people directly or indirectly involved in Canadian agriculture. And there are millions more who also want to know and understand the dynamics of this exciting industry –the hot-button issues, the players, the passion, and the potential. So here it is—a comprehensive and contemporary profile of Canadian agriculture which can be easily understood and enjoyed by the average person. It can also serve as an orientation document for 1st year agricultural students at colleges and universities across Canada.

To adequately paint a panoramic portrait of this complex and ever-changing mosaic, however, is a humbling experience and from the onset we knew that we could not and would not fully succeed in reaching this lofty goal. The industry is complex, diverse, and ever-changing. The main types of agriculture vary across the country. Markets are both controlled domestically and subject to international forces. International and domestic trade and trade agreements play key roles in many sectors of the industry. Governments at many levels involve themselves in both supporting and regulating many parts of the industry. In addition, the farming community itself is a complex blend of many small to mid-sized producers and a relatively small number of large players, as well as many other rural residents. Employment and the vitality of rural communities are also very often inextricably linked to primary agriculture. Additionally, in a more global context, farmers and ranchers serve as custodians of a healthy and sustainable environment.

The resulting book is organized as follows:

2. **Historic Overview.** The history of agricultural development in Canada is fascinating. Initiated by Aboriginal communities, the influx of (largely) European immigrants brought with it the introduction of new seeds, plants, implements, animals, and farm technologies. First the Maritimes, then Quebec, then Ontario, and then the Prairies/BC. Much of the initial development in Quebec and Ontario took place in the 1700s and 1800s, and tended to reflect its European heritage, especially that of France and England/Scotland, respectively. The deluge of ethnically diverse and often penniless immigrants onto the prairie landscape began in the late 1800s and culminated in the early 1900s. Eastern agriculture is typically at least 100 years older than agricultural settlements on the prairies.

3. **Farm Profile.** There is no single "farm profile." Each and every farm/ranch in Canada is unique. Every enterprise is different; every region is different. And, yes, although statistics can be mind-numbing, in this case they can also paint a relatively comprehensive picture

of Canadian agriculture. Largely based on the 10-year **Census of Agriculture** conducted by Statistics Canada, we can determine the number, location, legal structure, capital and labor structure, production characteristics, land use, social characteristics, and income characteristics typical of various farm types throughout Canada. Fear not; aided by numerous figures and graphs, this is an invaluable primer and an easy read.

4. **Agricultural Production & Marketing.** Most of us don't really know where our food comes from; how it is so efficiently produced, processed, and marketed in such an abundant, timely, healthy, and environmentally sustainable manner. Here we provide a description of the production and **domestic** marketing characteristics of each major agricultural enterprise. The crops profiled include canola, wheat, barley, oats, corn, soybeans, pulses, and sugar beets. The livestock and poultry profiled include beef, pork, dairy, poultry, and other livestock and poultry. Regional data are also provided.

5. **International Trade.** Canadian agriculture lives or dies in the international marketplace. Here we provide a description of Canada's international trade in agricultural commodities and the variables affecting agricultural trading patterns. We identify our major agricultural exports and imports and our resulting agricultural trade balance. We also assess our international pricing power and the role of foreign exchange rates, as well as the rationale behind freer trade initiatives. The multinational trading rules, as well as national levels of agricultural protection are also detailed. Thereafter, bilateral trading agreements and our major regional trading agreements are described, followed by a synoptic assessment of our future trading prospects.

6. **Agri-Industry and Value-Added.** Primary agriculture (i.e., farms and ranches) is actually only a small part of the agricultural complex. Hundreds of thousands of people, the backbone of most rural communities (Chapter 7), also provide the farm inputs required, as well as the purchase, transportation, processing, domestic marketing, and international trade in agricultural commodities. Seventeen processing sectors and seven input providers are profiled, fully acknowledging that the on-going consolidation, increasingly global reach, and market power of many of these players may make some of these profiles outdated even before publication. The "ripple effect" of this economic activity—the structural linkages—throughout the Canadian economy is also highlighted.

7. **Rural Socio-Economic Profile.** Yes, thousands of rural communities are gradually dying. But public service consolidation, commercial consolidation/growth, evolving technologies (esp. communications), and urban out-migration are also making the remaining rural communities more sustainable, both socially and economically. This chapter details the prevailing socio-economic and demographic characteristics of rural Canada and identifies the major on-going issues and opportunities, including resource use conflicts, the difficulty of public service provision in sparsely populated areas, and the vital importance of local initiatives.

8. **Agricultural Organizations.** There are, literally, thousands of agriculture-based organizations in Canada, each reflecting their own particular interest or concern. This chapter looks at the historic and contemporary role of rural cooperatives, national umbrella

organizations, and specific commodity groups, as well as marketing boards and commissions. In particular, their respective funding sources and expenditure patterns are traced to suggest how money and political power seem to be intertwined. How they facilitate professional networking and commodity-based research is also highlighted.

9. **Role of Government.** Most rural Canadians would reluctantly concede that the historic role of the federal government in the development of Canadian agriculture has been pivotal. The particular structure, conduct, and performance of Canadian agriculture today is largely a reflection of federal government initiatives, however motivated or ill-conceived, during the last 150 years. This chapter details the extensive role the federal and provincial governments have played in the agricultural development process. It identifies the major federal initiatives for the periods 1870–1950, 1950–1980, 1980–2000, and post 2000. Also described is the organizational structure of Agriculture and Agri-Food Canada, provincial agricultural initiatives, the regulatory environment, and the historic role of the national railways. An overview of national agricultural research funding and performance is appended, and the federal government's funding for agriculture is also tabulated.

10. **Resource Management.** Laws change, people die, but the land remains the same (Abraham Lincoln). Canadians have a special affinity for the land and farmers/ranchers are the custodians of much of that land and water. It must remain pristine and productive. Yet there is a growing public perception that this trust is being violated. This chapter details the characteristics of our land and water resources and their required protection to maintain a sustainable ecosystem. We also look at the various land types, causes of land degradation, the urbanization of agricultural land, regenerative organic agriculture, public land management, water supply–demand and regulation, the role of reservoirs, drainage, and flood control, water and environmental values, and the extent and ramifications of public financial support to agriculture for protecting and enhancing environmental amenities.

11. **Climate Change & Canadian Agriculture.** This is a hot-button issue. Who isn't concerned about the ecological, social, and economic well-being of our planet? An army of publicly funded researchers further stoke the fire, sometimes generating more heat than light. We judiciously try to avoid all of this by, firstly, focusing exclusively on Canadian agriculture. And secondly, by separating the innumerable issues into three distinct elements: climate modeling and projections, anticipated adaptation, and possible public mitigation initiatives. Within this framework, we then provide a brief global perspective, a description of the CO_2e sources in agriculture (by region), as well as some international comparisons. Thereafter, we look at the CO_2e agricultural sector balance and the importance of agriculture as a major carbon sink. The anticipated climate change impacts on Canadian agriculture and the sector's capacity to successfully adapt is also examined.

12. **Related Pollution Issues and Mitigation Measures.** Mention the widespread use of glyphosate-based herbicides (e.g., Roundup©) in agriculture and the eco-warriors are in instant battle mode. And, eventually, public policy generally reflects public perception. Increasingly stringent environmental and health regulations on both primary agriculture and agri-industry are almost inevitable. This chapter adopts a broader perspective and

looks at six potential sources of agriculture-related pollution: synthetic fertilizer, pesticides and disease, irrigation, machinery emissions and discharges, livestock, and agricultural processing wastes. Thereafter, seven potential mitigation initiatives are also discussed.

13. **Organic Crops, Functional Foods, and GMOs.** In Canada and other relatively high-income countries, consumers are becoming increasingly health-conscious. They increasingly want "chemical"-free food, even if it costs more at the local supermarket. Food produced from Genetically Modified seeds (GMOs) are equally suspect. This chapter provides a legal definition of each, a brief description of the regulatory framework, an abbreviated market profile, and an objective third-party assessment. Regenerative (organic) agriculture (also see Chapter 10) may well be the sleeping giant of 21st century agriculture.

14. **The Future of Agriculture in Canada.** Historically, agriculture has undergone an industry-shaking metamorphosis every generation or so. Change is a constant, but we just don't know exactly what that will look like in the future. Nevertheless, this chapter fearlessly but somewhat foolishly purports to divine the future opportunities and challenges of the agricultural sector in Canada. We identify six mega-variables that we believe will largely drive how the structure, conduct, and performance of Canadian agriculture evolves in the immediate years ahead. And this, we think, will then have at least nine very important implications for primary agriculture, described thereafter.

The generous support of numerous friends and colleagues in the various agriculture-related disciplines has been invaluable in the preparation of this document. In particular, four long-time friends and associates must be singled out for their selfless contributions to more than one chapter of this document: Al Dooley, Mel Lerohl, Len Leskiw, and Don Macyk. The other very diligent and dedicated professionals involved included Jim Barlishen, Ray Bollman, Andrea Bravo, Terry Daynard, Tom Funk, Doug Hedley, Reynold Jaipul, Austin Leitch, Roy Larson, Sheilah Nolan, Mark Olson, Jim Phelps, Joe Rosario, Keri Sharpe, and Cliff Weber. This work simply could not have been completed without their personal and professional support.

At the same time, we readily acknowledge that some shortcomings persist. Some of the data may be incorrect or outdated. The book contains more than 300 tables, figures, and photos. Some errors are entirely possible. This work also has a western (i.e., prairie) and field crop bias. Greater emphasis on agriculture in other parts of Canada and on other farming types would be very useful in subsequent editions. Professional agronomists, in particular, may also be uncomfortable with the cursory examination allotted to some very complex issues facing Canadian agriculture. But an expert is, by definition, someone who knows more and more about less and less. We are more interested in the general contours of the land rather than the minute detail. Substantive factual errors, on the other hand, should not and cannot be tolerated.

All errors and omissions which remain, of course, are the sole responsibility of the author. I also fully acknowledge the numerous scholarly sources we have unabashedly lifted. Some of this was just cut-and-paste. But as one researcher whom we relied upon told us, "...if you are going to use my material, at least have the decency not to screw it up." We have diligently tried to acknowledge and be faithful to the original sources, and in any instances where we have failed to do so, we most humbly apologize. There may also have been an occasional and inadvertent copyright infringement without recognizing it as such. These, if any, were a benign oversight.

In any event, the preparation of this manuscript also has an emotional component. Recently, I was looking at my grandson's (age 6) latest art project and I commented, "You should always sign your art because one day you will be famous and it will be worth millions." And he matter-of-factly replied, "No, I will never be famous; I'm going to be a farmer." And, somehow, that loving exchange was incredibly sad.

Comments are encouraged and should be addressed to marvanderson360@gmail.com. All errors or omissions identified in this document will be conscientiously addressed in subsequent editions.

Thank you for caring, good luck, and straight ahead.

2. Historic Overview1

> The future of agriculture will be fueled by lessons from history... The past informs the present and the future.
>
> *Ontario Grain Farmer, 2-18*

Canadian agriculture experienced a markedly different evolution in every region of the country. Each region was settled at a different period in Canada's economic and political development and climate and geography also played a large part in making each region's agriculture unique. Both national and provincial governments played major roles in this maturation process.

Figure 2.1 Location of Canadian Agriculture

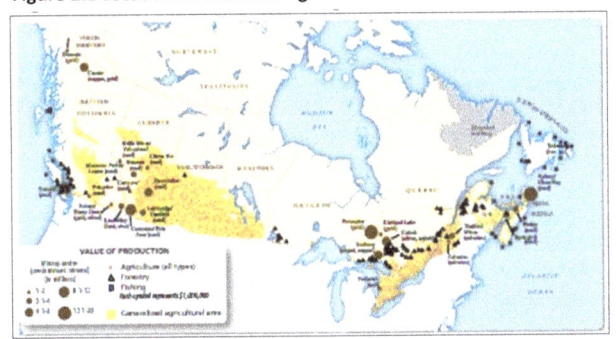

Source: Statistics Canada

Agriculture, and particularly the fur trade, initially dominated Canada's economic activity and in the latter part of the 19th century (the 1800s) it still generated about 30% of its gross national product (GNP). But in spite of the extensive settlement of the prairies in the 1891–1926 period, agriculture's role gradually diminished; indeed plummeted. Just one hundred years ago (in the early 1900s), over 50% of Canada's population was still farming/ranching. And even in 1931, one in three Canadians still lived on a farm or ranch. Yet today, only one in sixty Canadians (or about 1.6% of the total population) continues to live on a farm/ranch. What led to this evolution?

2.1 BEGINNINGS

For about 2.5 million years, hunters and gatherers "followed the food" that their local natural world provided. Agriculture, which "produced the food," only evolved during the last 10,000 years by members of different "hunter and gatherer" societies throughout the world, each society generally unaware of the other.[2]

It is now generally accepted that the first people to inhabit North America came from Siberia, by way of a land bridge, which existed until about 12,000 years ago. Archeological evidence of stone weapons, drills, knives, wedges, and scrapers indicates that these early hunter-gatherers first spread down the Pacific coast, then east across the Rockies and western prairies, and from there eventually to central and eastern Canada. Later, the Inuit moved across the Arctic in a series of west-to-east migrations.[3] Their original diets consisted mostly of the meat of larger animals and fish. Gradually, however, the dispersed clusters of Aboriginal peoples began farming the land as determined by their differing locales. They planted, tended, and harvested crops such as maize (known as Indian corn), potatoes, tobacco, beans, squash, pumpkins, sunflowers, and berries.[4]

By the time European explorers, traders, priests, and settlers arrived on the eastern shores of Canada, the transition from "follow the food"' to "produce the food" had substantially taken place in their own countries of origin. Consequently, for the most part, Europeans began farming in their own ways with the seeds, plants, implements, animals, and methodologies brought from the Old World, essentially ignoring what the Aboriginal peoples had already accomplished.

2.2 ATLANTIC PROVINCES (PEI, NS, NB, NL)

In 1497, three hundred years after a brief Viking presence in Newfoundland, the Venetian sailor Giovanni Caboto (John Cabot) landed his ship on the shore of Newfoundland (or possibly Labrador) and claimed it for England.[5] By the early 1500s, entrepreneurial fishermen from Britain and continental Europe came each summer to harvest the abundant and easily caught cod fish. They salted and/or air-dried the fish before shipping it home for sale. Soon, they would leave some fishing equipment and a few men over the winter. Eventually, some families joined them through the winters, leading to the establishment of settlements. In 1610, the Englishman John Guy established the town of Cupids on Conception Bay (Newfoundland), making it the oldest English colony in Canada. British and French settlements eventually drove the indigenous Beothuk inland and by the 19th century, the Beothuk disappeared completely.

In Newfoundland-Labrador, only a very small percentage of the land is suitable for horticulture or crop production because most is forest and tundra. But it was discovered that in the fertile soils of the Avalon Peninsula and other smaller areas, potatoes, turnips, and cabbage could be successfully grown. As well, grazing cattle and sheep could be raised on lands that served as natural pasture. With such exceptions, however, agriculture in Newfoundland was never more than marginally viable and only a supplemental food source for the few thousand people located in small fishing villages. (Even today, there are only about 400 relatively small farms.)

In Prince Edward Island, Nova Scotia, and New Brunswick, Europeans first encountered the indigenous MicMac (Mi'kmaq) who lived in coastal areas during the summer and inland during the winter.[6] As early as 1605, in the Annapolis basin (in Nova Scotia), French Acadians built dykes in saltwater marshlands, enabling them to grow wheat, flax, and vegetables, and create pasture for grazing animals.[7]

As farming by European immigrants gradually developed in the Maritimes, the most productive agricultural districts became Prince Edward Island, northern Nova Scotia, and eastern New Brunswick. Hay, potatoes, oats, and mixed grains (wheat, buckwheat, barley, and rye) were the principal crops, while turnips, apples, corn, and flax were produced in smaller quantities. In the Annapolis-Cornwallis Valley (Nova Scotia), potatoes and cattle were the commercial staples. Farms along the Northumberland Strait (the north coast of NS and NB) were the leading producers of grain. Land use generally reflected the importance of livestock in the mixed-farm economy, with the greater part of most clearings devoted to a combination of hay, oats, grasses (timothy, clover), and pasture. Farmers in Westmorland County, SE New Brunswick (Moncton area)—the longest continuously settled district in Atlantic Canada—regularly supplied hay and draft animals to Saint John and elsewhere.

Most full-time farmers concentrated on livestock raising, which required less manpower than did cereal growing. On the Atlantic coasts, small-scale subsistence farming generally prevailed. The roughness of the land and the labor-intensive character of the fishery limited most families to fewer than 3 acres (1.2 hectares) of improved land, enough, essentially, for a small kitchen garden and a cow or two. In contrast, in the more productive areas, most farms had 30 to 60 acres (12 to 24 hectares) cleared, and a few farms had more than 75 acres (30 hectares).

Prior to 1850, however, agriculture was still subordinate to the cod fishery in PEI, still subordinate to trade with the West Indies in Nova Scotia, and still secondary to the timber trade and shipbuilding in New Brunswick. Gradually, NS, NB, and PEI began to focus on dairy farming, along with livestock and mixed farming operations, as well as some fruit in Nova Scotia.

During the period 1880–1900, there was a large increase in the production of factory cheese and creamery butter and a rapid increase in the export of apples, particularly to Britain. After 1896, the boom associated with prairie settlement opened the Canadian market for Maritime fruit (especially apples) and potatoes.[8] Then, during the Great Depression of the 1930s, many farmers turned to more diversified self-sufficient agriculture, a change reflected in the increased dairy, poultry, and egg production.

In 1931, there were more than 86,000 farms in the Atlantic provinces.[9] Today, agriculture in the Maritimes is dominated by commercial potato production in PEI and New Brunswick and a prosperous fruit and vegetable industry in Nova Scotia. In 2016, there were still about 7,600 farms in the four Atlantic provinces and they were typically about 320 acres (130 ha) in size.

2.3　QUEBEC (LOWER CANADA)

In New France (Lower Canada), barley is said to have been grown as early as 1606 by Samuel de Champlain, explorer and founder of the French settlement that eventually became Quebec City. Between 1608 and 1610, Champlain also introduced cattle from Brittany and Normandy.

Louis Hebert, who came from France to Quebec in 1617, first used oxen to clear a small plot of land for cultivation in 1628. Small-scale clearing followed as other settlers followed his lead, planting cereal grains, peas, and corn. Hops, hemp, and other livestock were introduced in 1663. By 1667 there were 3,107 cattle in New France and, in 1721, they produced 4,584 bushels of barley.[10]

In the 1700s, land along the lower St. Lawrence River was initially conceded to *seigneuries* (lordships) within which all rural settlement took place. The first farm lots were long, narrow, almost rectangular plots along the river, very similar to the Métis lots along the Red and Assiniboine Rivers in the west. These long lots were easily and cheaply surveyed, gave all farmers frontage to the river, and allowed all of them to live close to neighbors. Most of these immigrants were of French origin.

Initially, farms were usually mixed farms that provided as much as possible for domestic consumption. Typically, there was a kitchen garden (vegetables, tobacco, and fruit trees), plowed fields (cereals, legumes, and fallow), and hay and pasture for foraging pigs, cattle, draft oxen, work horses, sheep, and poultry. At the end of the French regime (Treaty of Paris, 1763), about 60,000 people, some 85% of the non-indigenous Canadian population, lived in the countryside along the St. Lawrence River. And almost all of these people were farmers on their own long-lot farms.

After 1763, some Scottish, Irish, and American immigrants also purchased *seigneuries* while some New Englanders also settled in the Eastern Townships.

By the early 19th century (1800s) Quebec (Lower Canada) was one of the more mature settlement areas of North America. Numerous non-agricultural settlements already offered a wide range of services and production facilities, including grist mills, slaughter houses, breweries, tanneries, and textile manufacturing. And by the mid-1800s, the Montreal region, stretching southward from Lake Saint-Pierre and extending along the St Lawrence and Richelieu rivers, became the dominant agricultural area of the province. Still, by the 1830s, Quebec had already ceased to be self-sufficient in wheat and flour and increasingly began importing this from Ontario (Upper Canada).

Late 19th-century Quebec agriculture was marked by increases in the cultivated area and productivity and by a shift to dairying and stock raising. By the late 1800s, the number of farms in Quebec larger than 10 acres (4 ha) climbed to about 125,000. But whereas agricultural production in the more northerly settlement areas remained dispersed, agriculture production concentrated around the Montreal area became more specialized and more closely linked to commercial markets. Commercial dairies, cheese factories, and butteries developed around the towns and railways, but particularly in the Montreal plains and the Eastern Townships. During the 1880s and 1890s butter and cheese production increased exponentially and by 1900 dairying was already the leading agricultural sector in Quebec.

By the 1920s, however, agriculture in Quebec was already accounting for less than 1/3rd of Quebec's total output (GDP) and during the Dirty Thirties many farmers reverted to subsistence agriculture. The ***Union Catholique des Cultivateurs*** was founded in 1924. After World War II (1939–1945), and accelerated farm mechanization (as elsewhere), commercial farming did return, but then the number of farms dwindled and the remaining farms grew larger.

In 1931, there were almost 136,000 farmers in Quebec. Today, there are still about 28,900 farms, averaging some 280 acres (113 ha) per farm (2016 StatsCan data). And about 5,368 of these farms are still dairy farms.[11] The combination of rich and easily arable soils and relatively warm climate still makes the St. Lawrence River Valley Quebec's most prolific agricultural area. It remains a major producing area for dairy products, fruit, vegetables, *foie gras*, maple syrup (of which Quebec is the world's largest producer), fish, and livestock. Farm organizations, particularly the *L'Union des producteurs agricoles (UPA)* and the Dairy Farmers of Quebec, also still retain a strong political presence in Quebec.

2.4 ONTARIO (UPPER CANADA)[12]

By the early 17th century, in areas near the Great Lakes, some indigenous groups had already largely shifted from "following the food" to "producing the food." Champlain traveled extensively in both Lower and Upper Canada and when he visited Huronia (now Simcoe County) in 1615, he discovered an estimated 30,000 Hurons in eighteen villages already living off agriculture. They had cleared most of the land and were growing uncultivated wheat and hay. The Hurons also showed the French how to grow corn and make maple syrup. South and west of them, both the Tobacco/Allawandaron and Neutral/Petun indigenous peoples (both Iroquois-speaking) also farmed. Besides its ceremonial uses, tobacco served as a currency in trading.

The early settlers of Upper Canada soon developed three cash crops: timber, wheat, and potash. The potash was made by heating the ashes from land clearing in a large pot (hence, pot-ash). The British government bought all three while public markets were established in York and Kingston, where settler farmers could also sell their beef, pork, oats, and potatoes.

Increasing dissatisfaction with the rule and trading practices of Britain in the thirteen British American Colonies led to the American Revolutionary War and culminated with their Declaration of Independence in 1776. This resulted in a large wave of immigrants into Upper Canada. The British treated these "Loyalists" well, offering them free land for homesteading. Along with those of Scottish and English origin, the migration included some French Huguenots and many with a German background, as well as four of the Six Nations Indians. Basically, all of these groups already knew how to farm and would have no fear of Canadian winters. They were a gift to the strengthening of Upper Canada and its agricultural development, also making Upper Canada less likely to join the United States.

Governor Haldimand and Governor Simcoe in turn, followed by Col. Thomas Talbot and the Earl of Selkirk were significant figures in recruiting and settling immigrants on acreages suitable for farming. Most of these immigrants were of British origin, particularly from Scotland and England. In the early 1800s, agriculture in Upper Canada began to develop more quickly. Contributing factors were the availability of large expanses of good land, the opportunity to reach and serve a buoyant international market for wheat, accompanied by an influx of large numbers of land-hungry immigrants, many with capital to invest. Land clearing was rapid, a cash wheat economy was pervasive, and a service network of mills, merchants, and artisans, centered in small towns, began to appear.

The Crown land policy until 1825 rewarded United Empire Loyalists with large unencumbered grants in the thousands of acres. Thereafter, land was sold by auction, but by then large tracts of land were already owned by speculators. Farmers tried to purchase at least 50 acres of land and the average cultivated farm size was about 30 acres.

By the 1830s, most of the early settled lakeshore (especially along the north shore of Lake Ontario and Lake Erie) had about 1/3rd of the land cleared, and by the mid-1800s the last corners of the province, toward Georgian Bay and up the Ottawa valley, were also being surveyed for settlement. In one generation, from 1820 to 1850, the province's non-indigenous population increased almost fivefold, from about 200,000 to almost a million.

Between 1800 and 1860, in the central and western Upper Canada, wheat was the dominant crop because it was easy to plant, store, and transport. Wheat was exported to Britain and Quebec. Then the introduction of the Corn (cereal) Laws in Britain (which imposed tariffs and other restrictions on cereal imports into Britain from 1815 to 1846) effectively stopped Ontario wheat exports to British markets and precipitated a steep drop in wheat prices and land values. In the eastern third of the province, oats and potatoes remained popular while in the Niagara region corn (maize) was favored.

Agriculture in Ontario in the latter part of the 19th century experienced changes similar to those in Quebec earlier in the century. After 1850, Ontario's agriculture became increasingly diversified. While cereals (wheat, oats, and barley) remained prominent north of Lake Ontario, specialty crops such as corn, tobacco, peas, and beans became increasingly important in the southwest. Similarly, a unique fruit culture emerged in the Niagara Peninsula, Prince Edward, and Elgin counties; tobacco became important in Essex and Kent counties; while eastern Ontario typically adopted a hay-oats mix similar to that of Quebec. Dairying developed around urban areas.

Introduction of the cast-iron plow in 1815, the binder (or reaper) in the 1860s, sub-surface drains to reclaim swampy or bottom land, and improved crop rotations all increased crop production. The introduction by 1900 of the cream separator and refrigeration were catalysts to the growth of the dairy, beef, and pork industries.

Ontario agriculture then provided the essential raw materials for a number of industries, particularly butter and cheese manufacturing, flour milling, and meat packing.

The cheese industry had its greatest growth in the late 19th century. In 1864, factory cheese making was introduced and small plants proliferated across Ontario and a few districts of Quebec with about 80% of this production being exported. For a short period in the mid-1890s, cheese was actually Canada's leading single export commodity, and by 1900 Canadian cheddar cheese, largely from Ontario, had captured 60% of the English market. But by 1926, as processed cheese became more fashionable, the cheese industry had contracted and the butter output of creameries exceeded the output of cheese factories.

By the late 1800s, the flour-milling industry also blossomed. The development of the wheat-producing capacity of the Canadian prairies and the adoption of new milling technology re-established Canada as a major exporter of flour. Production came to be concentrated in a few very large mills located at break-points in the transportation network (Goderich, Midland, Port Colborne, and Montreal). The 1920s represented the heyday of Canadian flour milling as an export industry.

In the late 1800s, meat packing was largely confined to a few Toronto plants that processed hogs from the United States for export to Britain. But the industry grew rapidly as it developed local supplies of raw materials and also began to process beef in addition to pork. Vegetable canning (initially tomatoes and beans with pork) in relatively small plants clustered in southern Ontario also grew rapidly between 1890 and 1930.

At the same time, agricultural societies were formed early in the province's history as a way of encouraging mixed farming, particularly animal husbandry. Agricultural associations of dairy farmers, grain growers, fruit growers, and stockbreeders advocated for these specific enterprises, while the government-initiated Farmers' Institutes and Women's Institutes promoted the perspective and well-being of farm families.[13] The United Farmers of Ontario formed the provincial government in 1919 and to facilitate dairy modernization the Ontario Marketing Board was formed in 1931. By World War II, agricultural marketing boards and farmer-owned cooperatives were playing important roles in Ontario agriculture.

In 1931, there were about 192,000 farms in Ontario. Today, Ontario still has almost 50,000 farmers with a typical farm size of about 250 acres (2016 StatsCan data). There is extensive corn and soybean production, large acreages committed to intensive vegetable production, major grape and fruit-growing areas, and still a very strong dairy industry. And together they are still a potent political force.

2.5 PRAIRIES (MANITOBA, SASKATCHEWAN, ALBERTA)

Indigenous peoples were the first to introduce agriculture onto the prairies, but it was not a large part of their livelihood, tobacco being the most widely grown crop.[14] Some indigenous nations also harvested wild rice on a substantial scale. Of the traditional indigenous produce, maize was the best adapted to the prairies, although indigenous nations also successfully grew an abundance of turnips and potatoes, which they first acquired from Europeans.

American fur trader and map-maker Peter Pond came north and established garden plots at Lake

Athabasca in 1778. Mimicking Métis settlements along the Red and Assiniboine Rivers in Manitoba and the South Saskatchewan River in Saskatchewan (at Batoche), Scottish settlers also established river-lot agriculture at the Red River Colony (Manitoba) after their arrival in 1812. Missionaries there hoped they would set an example of farming for First Nations people, but at first the European settlers' wheat and barley crops regularly failed, leaving them dependent upon the indigenous people for food. By the 1870s, a new kind of settlement appeared in the west, replacing the old order based on fur-trading posts and river-lot farms. The new emphasis was on the use of land for commercial agriculture.

Large-scale ranching on leased land began in what is now southern Alberta and Saskatchewan in the 1870s and 1880s. In response to the growth of the cattle trade with Britain and the spectacular development of ranching on the American plains, Canadians, US cattlemen, and British investors established major ranches on land leased from the Crown after 1882. By 1891, there was a growing beef-cattle industry (largely American) in the foothills of southern Alberta and by 1900 ranches in the foothills dominated the cattle-shipping industry.[15] There, leased land had been replaced by deeded acreage and ranching methods had been intensified. For a time, extensive ranching techniques were still used on the short-grass areas of Saskatchewan, but with the advance of railways and subsequent agricultural settlement, the open range was greatly reduced. Moreover, the special position held by Canadian cattle in the British market was challenged by imports of good-quality chilled meat from the southern hemisphere, particularly Argentina. Prairie shippers turned instead to the expanding home market and, when tariff regulations allowed, to the lucrative Chicago market. First colonized by Latter-Day Saints in 1887, the southwestern corner of Alberta also became home to several Mormon communities. Small-scale irrigation began in the 1870s, followed by a pro-irrigation government policy in 1894.

The impetus for extensive crop cultivation across the prairies came through five major initiatives:

- A land survey of the prairies was initiated in 1871 and eventually covered 200 million acres, the largest survey grid laid down in a single integrated system in the world. This N-S and E-W checkerboard (see illustration) is laid out in terms of ranges, townships, sections, and quarters. Ranges are numbered east to west, 1 to 34. Townships are numbered from south (US border) to north, 1 to 129. Within this grid, there are thirty-six sections in one township, each section consisting of 640 acres. And, in turn, each section is divided into four quarters, each consisting of 160 acres. Thus, each quarter has a unique signature, e.g. SE1/4-34-57-15 W4 means this is the SE quarter of Section 34, Township 57, Range 15, west of the 4th Meridian. The land was then designated as either homestead land, Hudson Bay Company land (section 8, ¾ of section 26 and all of section 26 in every 5th township), railroad land (odd-numbered (red) sections except 11 and 29), and school land (sections 11 and 29).

- The Homestead Act (The Dominion Lands Act), which accompanied the land survey, was also initiated in 1871. This Act granted an applicant 160 acres (1/4 section) if he/she filed a claim, paid a fee of $10 for entry, and then maintained a three-year period of occupancy and cultivation before qualifying for a patent (fee simple ownership). Eventually, this led to the creation of more than 1.25 million homesteads on the prairies.

- Large-scale immigration was encouraged during the period 1895–1914. Spurred on by Clifford Sifton (Minister of the Interior under PM Sir Wilfred Laurier) Canada's total non-indigenous population during the first two decades of the 20th century skyrocketed from 5.4 M to 10.4 M and growth was most remarkable in the West. Potential farmers from the British Isles, Germany, France, Scandinavia, Ukraine, Gallacia (Ukraine-Poland), Poland, Hungary, USA, and Russia were actively recruited. In 1891, there were fewer than 100,000 non-indigenous people living between Manitoba and BC, while a quarter of a century later there were more than 1 million.

- Marquis wheat (in conjunction with summer fallowing to conserve soil moisture) was developed, along with more mechanized farming. In 1907, Charles Saunders developed Marquis hard red spring wheat, which was rust-resistant, had a shorter growing season, was relatively high-yielding, while the baking quality was excellent. This led to its rapid adoption and, by 1920, 90% of the entire prairie wheat crop was Marquis. At the same time, mechanization of the wheat economy with steam, gas tractors, gang plows, and threshing machines greatly facilitated further expanding the cultivated acreage.

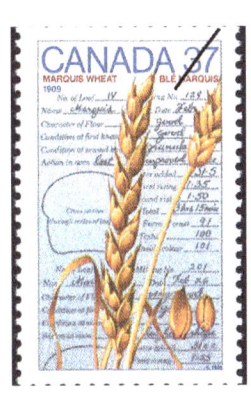

- The transcontinental (Canadian Pacific) railroad through the southern part of the Prairies was completed in 1885, followed by a second transcontinental railroad through the northern part of the prairies, completed in 1912. This second transcontinental railroad was created through the amalgamation of several already existing railroads, most prominently the Grand Trunk Pacific and the Canadian Northern which, combined, later became the Canadian National Railway (CNR). A greatly improved railway network reinforced the benefit of an earlier statutory reduction in rail transport costs for grains, enacted through the Crow's Nest Pass Agreement (1897).

In the early 1870s, almost all homesteaders took land as near Winnipeg as availability permitted, with the largest concentration on silty soils south of Lake Manitoba. By 1880, the homesteading frontier was just reaching the present western boundary of Manitoba, with the main westward thrust along proposed railway routes near the American border and north of the Assiniboine River. The greatest expansion occurred in the early 1880s, when settlement followed the proposed railway routes into Saskatchewan, particularly extending onto the prairie along the finally approved transcontinental route through Regina.

The expansion of the occupied area during the 1880s coincided with increases in crop and livestock production, setting the stage for the massive increases to follow. Before 1891, crop production was concentrated in Manitoba, the area first settled. The emphasis was on wheat as the export crop, with smaller acreages devoted to animal feeds such as hay and oats for horses and cattle for the potential export of beef and dairy products.

Wheat acreage continued to increase in the late 1880s, particularly in the heavily settled districts of Manitoba and along railway lines. Grain-elevator capacity to handle the increased production was another factor favoring wheat growing, and by 1891, an economy based on country elevators had spread throughout Manitoba and to the larger centers west along the transcontinental railway. The export of wheat in 1876 from the area around Winnipeg established the reputation of Manitoba wheat and began the process of making the prairies the bread basket of the Commonwealth. By 1891, significantly larger shipments of wheat were transported almost entirely through Winnipeg and eastward along the Canadian transcontinental railway to the lake head on Lake Superior.

Agricultural development followed the spread of railway branch lines and the accompanying wave of homesteaders. By 1931, 40 percent of all farmland was already producing crops, particularly wheat. But in Manitoba, southeast Saskatchewan, and the area around and south of Edmonton, oats and barley were important crops, along with the production of livestock and dairy products. Gradually, many farmers were also able to expand their operations beyond the initial 160-acre homestead plot. Still, in the arid southwest Saskatchewan and southern Alberta, while cash-crop wheat prevailed on the more fertile arable land, vast areas of range land continued to sustain extensive cow-calf operations.

And then came the Dirty Thirties. Newly erected tariff barriers and increased foreign production reduced the value of prairie wheat by almost one-third between 1929 and 1932. At the same time, a series of droughts between 1930 and 1937 brought about complete or partial crop failures over extensive areas. In Saskatchewan, per-capita income plunged by 72% between 1928 and 1933. Many prairie farmers, and especially those in Saskatchewan, simply abandoned their farms in order to survive. Eventually, however, various forms of mutual and external support did make a difference, particularly the following:

- Political populist movements were born, the most significant being the United Farmers of Alberta (UFA) and Social Credit in Alberta, and the Cooperative Commonwealth Federation (CCF) in Saskatchewan. The UFA was temporarily transformed into a political force, formed the government in Alberta in 1921, and remained in office until 1935. It focused on education and health services, as well as trying to meet the marketing needs of farmers. Thereafter, Social Credit, with its promise of monetary reform and populist oratory, held sway in Alberta from 1935 until 1972.[16] In Saskatchewan, the Cooperative Commonwealth Federation (CCF) came together as a political party in the 1930s and became the first socialist government in North America in 1944. In 1947, CCF introduced universal hospitalization and in 1959 they introduced a Medicare plan for Saskatchewan, the first in North America. CCF also created numerous Crown Corporations and emphasized industrial development to reduce the province's dependence on agriculture and to assist farmers who wished to move out of the drought areas to northern Crown lands and even outside the province. As the 1930s progressed, more assisted settlers chose destinations outside Saskatchewan, particularly northern Alberta.

- Cooperative grain movements gained traction. Farmers formed Grain Grower Associations in Saskatchewan-Alberta in 1901–02 and in Manitoba in 1903 and, subsequently, formed the Grain Growers' Grain Company, a cooperative, in 1906. Thereafter, in 1923–24, farmers organized compulsory grain pools: Alberta Wheat Pool, Saskatchewan Wheat Pool, and the Manitoba Wheat Pool, although these eventually collapsed during the Great Depression.

- The Prairie Farm Rehabilitation Act (PFRA) under the Ministry of Agriculture was established in April 1935 to implement various remedial programs for drought-prone areas, particularly the Palliser Triangle area. New agricultural practices, such as strip farming, less summer fallow, minimum-till, cover cropping, irrigation, and shelter belts were introduced and encouraged. Sub-marginal crop land was fenced, re-grassed, and permanently converted into community pastures to counteract soil drifting. Extensive water and pasture development enabled farmers to expand their livestock herds.

- The federal government established the Canadian Wheat Board (CWB) in 1935 to try to stabilize volatile grain markets. In 1943, the CWB was made compulsory for the marketing of western wheat, and in 1949 the board's authority was extended to western barley and oats. It then managed all farm grain deliveries to local elevators through a delivery quota system (based on crop acreage) and maintained pricing "pools" so all farmers received the same average annual price for their grain. The CWB also controlled all domestic sales to flour mills and was the sole exporter of wheat, barley, and oats abroad. The system was dismantled in 2012, although Ontario still has its own Ontario Wheat Board.

By the mid-1930s it appeared that large areas of the prairie would be lost to agriculture. Between 1931 and 1941 almost 200,000 demoralized people left the prairie provinces behind. Gradually, however, prairie agriculture recovered. And although the mechanization process essentially stopped during the 1930s, it dramatically resumed after WWII (1939–1945). Then, rubber-tired tractors and combines became commonplace, farm consolidation continued, and still more new technologies generated an on-going dynamic: hydraulics, synthetic fertilizers, pesticides, self-propelled machinery, and so on.

But after the Dirty Thirties and WWII, with increasing farm size, the number of farmers on the prairies went into a slow but persistent decline. In 1931, there were about 288,000 prairie farmers; today there are only about 90,000.

2.6 BRITISH COLUMBIA

Agriculture in BC was first established to provide for the fur trade, followed by a demand for agricultural products when the gold rushes began after 1858. Some early ranching operations were also established in the interior along the Thompson and Nicola Valleys. A further impetus to BC agriculture was provided by the railway camps during construction of the transcontinental Canadian Pacific Railway, completed in 1885. New Boundary-Kootenay mining industries in the 1890s, as well as the lumbering and fish packing industries, also provided a demand for locally produced agricultural products.

Gradually, mountainous BC gave way to some vibrant agricultural settlements. Large-scale farming continued in the Cariboo Similkameen and the northern Kootenay areas (predominately beef cattle). And somewhat later, a mixed farming area around Fort St. John (an extension of the Grand Prairie–Peace River area of Alberta) also developed. Smaller-scale specialized production evolved in the Okanagan Valley (fruits and grapes) and in the lower Fraser Valley near Vancouver (dairy and vegetables).

As early as 1913, the Okanagan fruit growers set up a cooperative marketing and distribution agency, followed by the 1938 establishment of the Tree Fruit Board as the sole agency for apple marketing.

In 1931, there were about 26,000 farms in BC. Today, there are still about 17,600 farms and the fruit farmers and wineries in the Okanagan Valley are relatively prosperous. Despite the existence of a provincially mandated Agricultural Land Reserve, urban encroachment around Vancouver has increasingly limited the extent of Fraser Valley agriculture.

2.7 NEVER FORGET

We are a relatively young country; in many ways, still an adolescent. But we must never forget that we stand on the shoulders of the hardy multi-cultural pioneers who preceded us; our bedrock: Aboriginals, English, French, Scots, Irish, Germans, Scandinavians, Finns, Ukrainians, Galicians, (Ukraine-Poland), Hungarians, Swiss-Dutch (Mennonites), Moravians (Hutterites), Americans, Chinese, Jews, and many, many others. They had incredible courage, perseverance, and civility, as well as an insatiable desire to make a better life for themselves, their community, and their offspring. They were the indestructible building blocks of today's vibrant agricultural communities throughout Canada. And rural communities retain that special affinity to the land, especially those who trace their heritage back to the poverty and persecution they often endured in the "old country."

3. A Farm Profile

Every farm is unique. Each and every farm has its own unique production and marketing characteristics, its own capital, labor, and management structure, and its own demographic characteristics. There are numerous socially, financially, and environmentally viable business models and, regionally, farms throughout Canada are also very diverse. The profile following details these and related structural attributes.

3.1 NUMBER OF FARMERS & AVERAGE FARM SIZE

The structure of agriculture has changed significantly over the last century with fewer but much larger farms today. While there were over 700,000 farms in the 1920s, 1930s and the early 1940s, the number plummeted to only about 194,000 by 2016.

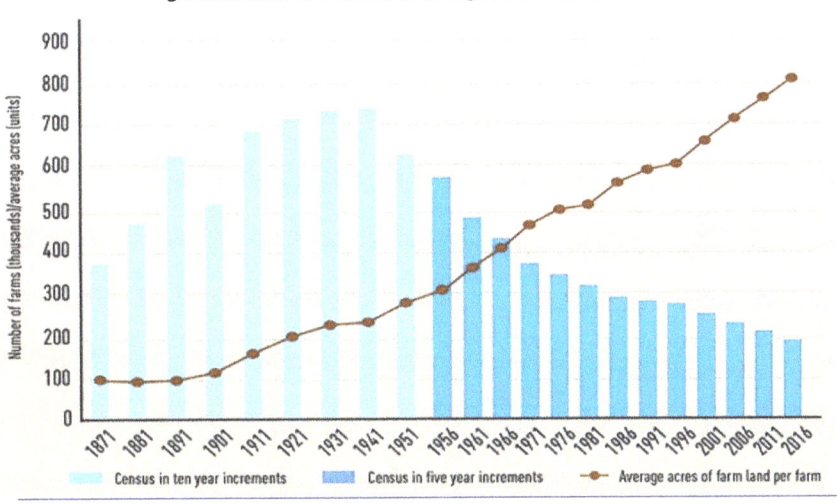

Figure 3.1 Number of Farms & Average Farm Size, 1871-2016

Source: Basic data from Statistics Canada.

By 2016, there were only about 7,500 farms in the Atlantic provinces, 28,900 in Quebec, 49,600 in Ontario, 89,900 in the prairies, and about 17,500 in British Columbia (Table 3.1). This is only about ¼ of the number of farms at the height of the Great Depression.

It is also important to note that the Statistics Canada **2016 Census of Agriculture** actually defines a "farm" as follows:

> ...any agricultural operation that produces at least one of the following products intended for sale: crops (hay, field crops, tree fruits or nuts, berries or grapes, vegetables, seed); livestock (cattle, pigs, sheep, horses, game animals, other livestock); poultry (hens, chickens, turkeys, chicks, game birds, other poultry); animal products (milk or cream, eggs, wool, fur, meat); or other agricultural products (Christmas trees, greenhouse or nursery products, mushrooms, sod, honey, maple syrup products).

In principle, if someone sells or intends to sell $10 worth of honey or blueberries at the local fresh market, that person or persons is classified a "farmer." The number of commercial farmers is much less (Section 3.10).

At the same time, there has been a gradual increase in average farm size. Just since 1991, the average farm area increased from 598 to 820 acres in 2016. This national average, however, still obscures the large differences in farm size that exist between various regions of the country. In much of Canada, farms are between 200 and 400 acres in size. In contrast, on the semi-arid prairies, with its focus on extensive dryland crop and beef production, the average size is about 1439 acres (Table 3.1 and Figure 3.2).

Table 3.1 Number of Farms and Average Farm Size, 2016

Province/Region	No. Farms (,000)	Average Farm Size (acres)*
MARITIME PROVINCES	7.6	320
PEI	1.4	425
NS	3.5	263
NB	2.3	370
NL	0.4	174
QUEBEC	**28.9**	**280**
ONTARIO	**49.6**	**249**
PRAIRIE PROVINCES	**89.9**	**1439**
Manitoba	14.8	1192
Saskatchewan	34.5	1784
Alberta	40.6	1237
BRITISH COLUMBIA	**17.5**	**365**
CANADA	**193.5**	**820**

* One hectare = 2.47 acres.

Source: Statistics Canada, *Census of Agriculture, 2016*

Figure 3.2 Average Farm Size, by Province, 2016

- N.L. (174 acres)
- B.C. (365 acres)
- Alta. (1,237 acres)
- Sask. (1,784 acres)
- Man. (1,192 acres)
- Ont. (249 acres)
- Que. (280 acres)
- P.E.I. (425 acres)
- N.S. (263 acres)
- N.B. (370 acres)

Source: Table 3.1

These averages also obscure commercial farm size. Dividing the total agricultural acreage by the nominal number of farmers greatly underestimates the typical operational size of a commercial farm (Section 3.10).

3.2 LAND USE & LAND OWNERSHIP

The total national agricultural land base and its composition is now fairly stable, as illustrated in Figure 3.3 and tabulated in detail in Table 3.2. This represents about 7 percent of all of the land in Canada.[17]

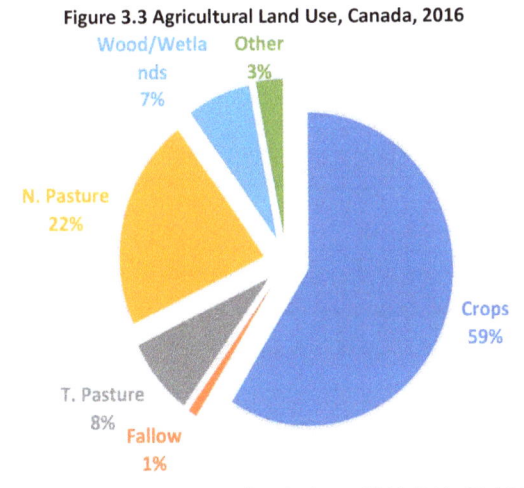

Figure 3.3 Agricultural Land Use, Canada, 2016

- Crops 59%
- N. Pasture 22%
- T. Pasture 8%
- Wood/Wetlands 7%
- Other 3%
- Fallow 1%

Source: Statistics Canada, Census of Agriculture, 2016, Table 32-10-0406-01

The one exception to this stability has been summer fallow. Summer fallow was used extensively throughout the early history of prairie agriculture as a means of conserving moisture and improving weed control. Many farmers, particularly in the driest areas, traditionally followed a 50/50 summer fallow/annual crop rotation, which meant that 50 percent of annual cropland was fallow each year. As late as 1976, there was still about 30 million acres of summer fallow on the Prairies. But since then, it rapidly declined, dropping to about 10 million acres by 2006 and, a decade later, to only 2.2 million acres (Table 3.2). The development of one-pass seeding tools, larger tractors, indiscriminate herbicides, and improved straw management all combined to make this possible. At the same time, labor costs declined and more soil moisture was retained at the time of seeding. This conversion to continuous cropping has resulted in an offsetting increase in the land annually cropped.

Irrigation is also very important to some regions of the country. Alberta accounts for about 70% of the national total (600,000 ha or about 1.5 million acres) with the largest irrigated field crop and irrigated forage areas. Most of this water is extracted from the St. Mary, Oldman, Bow, and South Saskatchewan Rivers (Figure 3.4). Potatoes, cereals for silage, and sugar beet production is pivotal to supply local demand (McCains, Lantic, numerous beef feedlots, etc.). A significant amount of irrigated production is also marketed elsewhere (e.g., beef to US slaughter plants). Saskatchewan also has about 150,000 acres of irrigation, and most of this water is extracted from Lake Diefenbaker on the South Saskatchewan River.

Figure 3.4 Irrigation Districts in Alberta

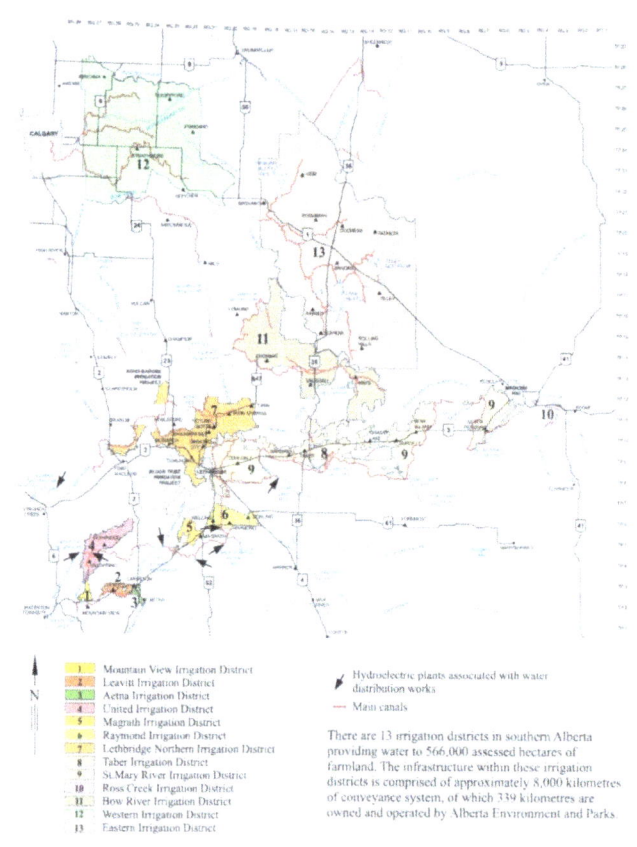

Source: Alberta Agriculture

Most of the vegetable irrigation is in Ontario, which has nearly one-half of the total irrigated vegetable area. Some irrigation is also required in the orchard-growing area of the Okanagan Valley in British Columbia. British Columbia accounts for about one-half of the total irrigated fruit area in Canada.

Table 3.2 Agricultural Land Use & Principal Crops, by Region, 2016 (,000 acres)

CATEGORY	MARITIMES	QUEBEC	ONTARIO	PRAIRIES	B.C.	CANADA	
CROPS:	1,031.9	4,613.0	9,021.3	77,281.2	1,435.2	93,382.6	59%
Wheat	53.5	227.4	1,202.3	21,846.2	107.1	23,436.5	15%
Barley	82.2	128.6	103.7	6324.3	57.2	6,696.1	4%
Other Cereals*	49.4	271.3	236.2	3,557.8	83.2	4,202.1	3%
Canola	2.4	34.2	39.5	20,434.9	95.2	20,606.8	13%
Other Oilseeds**	0.1	3.0	x	1,583.2	x	1,595.2	1%
Corn	66.5	1,144.6	2,457.7	850.3	44.2	4,563.2	3%
Soybeans	58.4	869.1	2,783.4	1,893.6	x	5,615.9	4%
Pulses***	0.5	x	x	10,309.3	x	10,518.3	7%
Potatoes	131.7	42.6	34.7	128.5	7.3	344.8	0%
Alfalfa, Hay, Seed	443.0	1,627.2	1,726.5	9,839.5	899.7	14,535.9	9%
All Other Crops	3.5	x	x	464.1	x	558.7	0%
SUMMERFALLOW	0.9	5.2	15.9	2,173.9	13.2	2,209.1	1%
TAME & SEEDED PASTURE	98.9	272.9	514.2	11,161.5	508.7	12,556.2	8%
NATURAL LAND FOR PASTURE	126.8	280.8	783.6	30,507.9	3,541.5	35,240.5	22%
CHRISTMAS TREES +	1,010.7	2,600.6	1,542.6	5,641.0	681.3	11,476.3	7%
ALL OTHER	128.0	330.7	470.9	2,708.1	220.6	3,858.4	2%
T OT A L	2,397.2	8,103.2	12,348.5	129,473.6	6,400.5	158,723.1	100%

* Includes barley, oats, rye & triticale.
** Includes flax and mustard.
*** Includes peas, beans, lentils, & chickpeas.
**** Includes grapes and sugarbeets.
***** Christmas trees = 58,800 acres
X Excluded to maintain confidentiality. Resulting summation errors.

Source: Statistics Canada, *Census of Agriculture, 2016*, Tables 32-10-0406-01 and 32-10-0416-01

Figure 3.5 Relative Crop Area, Canada, 2016

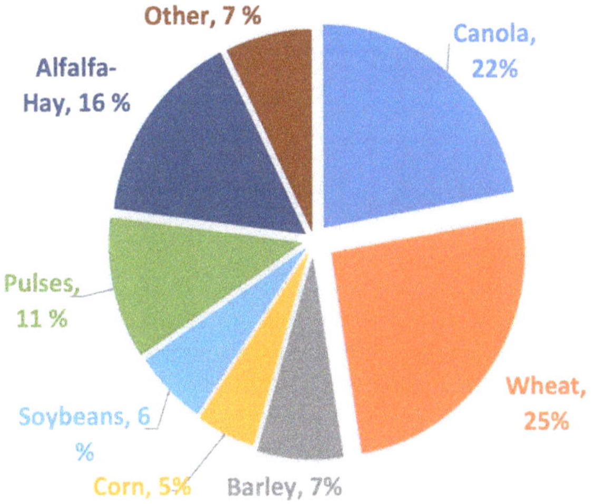

- Canola, 22%
- Wheat, 25%
- Barley, 7%
- Corn, 5%
- Soybeans, 6%
- Pulses, 11%
- Alfalfa-Hay, 16%
- Other, 7%

Source: Table 3.2

Throughout the history of Canadian agriculture, land tenure has always been overwhelmingly privately held *fee simple* ownership, which means the permanent and absolute **tenure** of a parcel of land with the freedom to dispose of it at will. This right, however, does not generally extend to sub-surface rights (other than gravel and groundwater) after the respective provinces entered Confederation or shortly thereafter. Most farmers today do not "own" the mineral rights. Mineral (including gas and oil) royalties generally accrue to the respective provincial governments.

At the same time, fewer and fewer farmers own all of the land that they actually farm. With the rapidly rising cost of land purchases, rentals, and leases are increasingly common. Now almost 40% of all farms with their own land base also rent some additional land in order to maintain a viable farming operation (Table 3.3). Nearly 40% of all agricultural land is rented: about 27% from, in most cases, neighbors, and another 13% from governments. Most government-leased land is pasture land (esp. community pastures) although some provinces (especially Saskatchewan) still own/lease some cropland.

Table 3.3 Farmland Area by Tenure, Canada, 2016

Tenure	Area Acres (,000)	Area Percent	Farms Number (,000)	Farms Percent
Own	99,631.2	63%	182.8	94%
Rent or Lease	67,316.1	42%	85.9	44%
Rented or Leased from Others*	44,604.2	28%		
Leased from Governments	21,209.2	13%		
Other Use Arrangements	1,502.7	1%		
Only Rent			10.7	5%
Own + Rent			75.2	39%
TOTAL FARM AREA**	158,723.1	100%	193.5	100%
Land Used by Others	8,224.1			

* Includes 4.5 M acres that is share-cropped.
** All land minus total area of land used by others.

Source: Basic data from Statistics Canada, *Census of Agriculture, 2016*, Table 32-10-0407-01

3.3 FARM ENTERPRISE COMPOSITION

There are nearly 64,000 grain/oilseed producers and 36,000 beef producers in Canada. Together they make up over 50% of all Canadian farmers. Additionally, forage is the principal crop for more than 20,000 producers (11%) while the supply managed sub-sectors, dairy, and poultry/eggs, only make up a modest 8% (or 15,000) of all producers. The relative importance of these various sub-sectors is illustrated in Figure 3.6 and the detailed statistics are provided in accompanying Table 3.4.

Figure 3.6 Total Number of Agricultural Operations, by Type, Canada, 2016

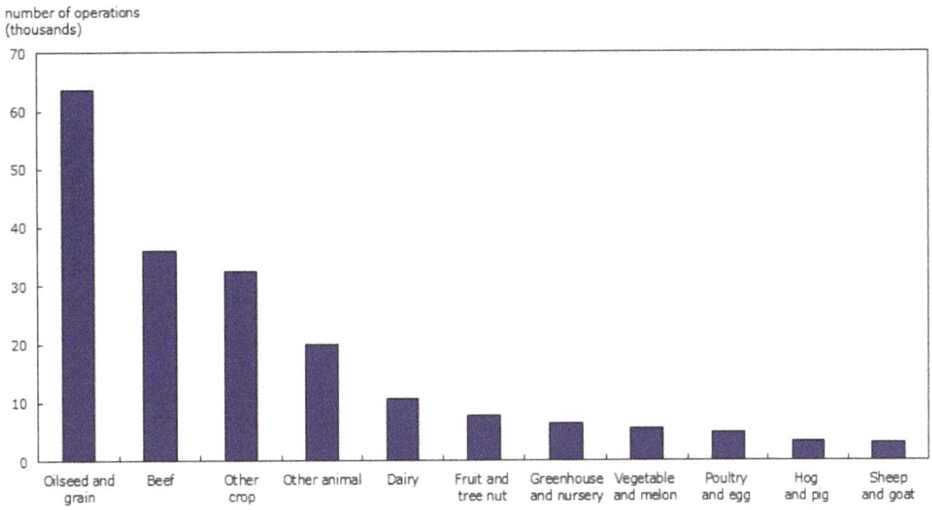

Source, Statistics Canada, Table 32-10-0403-01

Table 3.4 Number of Farms by Farm Type, by Region, 2016

FARM TYPE*	MARITIMES	QUEBEC	ONTARIO	PRAIRIES	B.C.	CANADA**	
Grain & Oilseeds	283	4506	16876	41659	304	63628	33%
Beef	1074	2474	6786	23317	2362	36013	19%
Dairy	615	5163	3439	791	517	10525	5%
Hogs	47	1463	1229	465	101	3305	2%
Poultry & Eggs	268	875	1816	724	1220	4903	3%
Other Animal/Combos	821	2093	6406	7861	3592	20773	11%
Apiculture	130	260	593	789	303	2075	1%
Vegetables	741	1172	1856	610	1135	5514	3%
Fruit	1506	1495	1362	302	3180	7845	4%
Greeenhouses-Nurseries	697	1247	2050	956	1499	6449	3%
Maple Syrup	160	4776	391	8	5	5340	3%
Hay	844	2508	4681	10043	2635	20711	11%
Other Crops/Combos	307	887	2115	2427	675	6411	3%
TOTAL	7493	28919	49600	89952	17528	193492	100%

* NAICS classification. Defined as the principal type of production; not exclusive.   **Excludes Yukon (142), Nunavut, & NWT (16).

Source: Statistics Canada, *Census of Agriculture, 2016*, Table 32-10-0403-01

3.4   FARM ORGANIZATION – LEGAL STRUCTURE

Most Canadian farms are still family owned. About 52% are sole proprietors; another 23% are family corporations, and yet another 23% are partnerships, typically father-son/daughter partnerships. Contrary to the perception of many, only about 5,100 farms in the whole of Canada are non-family corporations and a number of these are Hutterite colonies. In reality, there are only a very few "outside" investors, excluding family members who do not participate in the actual farming operation. And many of these so-called "outside" investors are, in fact, passive institutional investors who rent/lease their land back to active farmers, e.g. OMERS or OTPP."[18] (Figure 3.7)

Table 3.5 Organizational Structure of Canadian Farms, 2016

Type	Percent	Number
Sole Proprietorship	51.7%	100,061
Family Corporation	22.5%	43,457
Partnership	22.9%	44,237
Non-Family Corporation	2.7%	5,135
Other	0.3%	602
TOTAL	100.0%	193,492

Source: Statistics Canada, *Census of Agriculture, 2016*, Table 32-10-0158-01

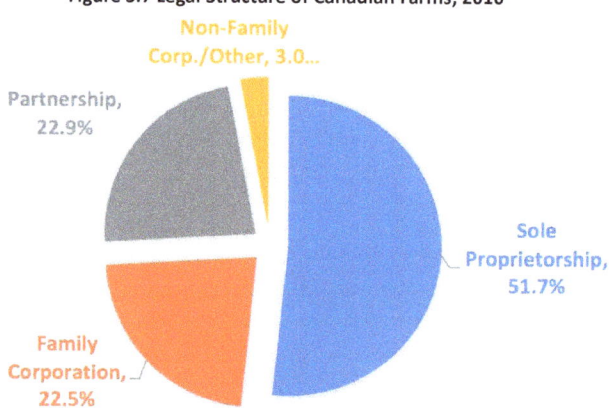

Figure 3.7 Legal Structure of Canadian Farms, 2016

- Non-Family Corp./Other, 3.0...
- Partnership, 22.9%
- Sole Proprietorship, 51.7%
- Family Corporation, 22.5%

Source: Table 3.5

Hutterites, a communal - religious brethren, have subsisted almost entirely on agriculture since migrating to North America in the late 19th century. Most colonies are located in Alberta (175), followed by Manitoba (110), and Saskatchewan (70); a total of about 355 colonies, or approximately 33,000 people and 4 million acres on the Canadian prairies.[19]

3.5 DEMOGRAPHIC CHARACTERISTICS

Defining demographic characteristics of Canadian farms and farmers include, in particular, age, gender, and education level.

3.5.1 AGE OF FARM OPERATORS

According to the **2016 Census of Agriculture,** the age of farm operators continues to increase, now averaging about 55 years, with the 55–59 age cohort accounting for the largest single share of farm operators (Figure 3.8). This is relatively old, giving rise to an increasing concern in the industry for more, better, and accelerated succession planning.

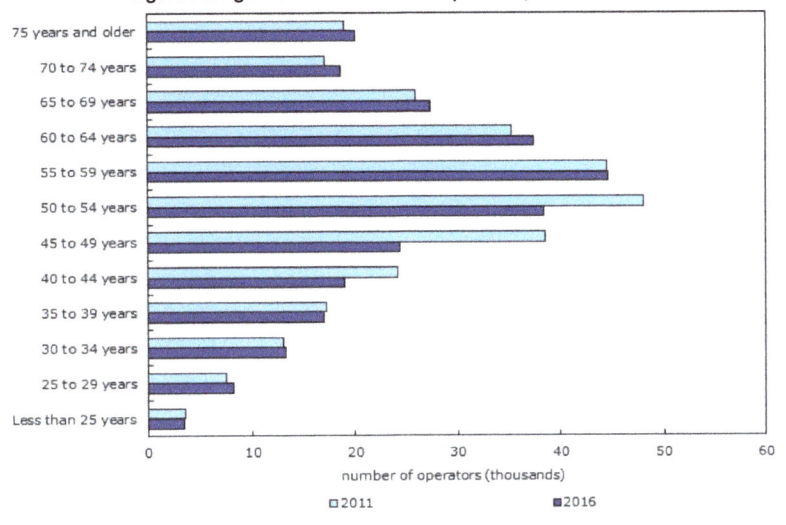

Figure 3.8 Age Distribution of Farm Operators, 2011 and 2016

Source: Statistics Canada, *Census of Agriculture*, 2011 and 2016, #3438,
From Statistics Canada, "A Portrait of a 21st century agricultural operation," May 17, 2017

At the same time, some analysts suggest that this may not really be a very accurate measure of the actual age of the ***de facto*** farm operator. With many inter-generational farm families, the parents still retain much of the capital base (particularly the land), while largely handing over the day-to-day operation to their sons or daughters. They retain this ownership for both cultural and legal reasons, but particularly the possible dissolution of a marriage by the younger generation, which might erode their land base. Since the "old guy," often the nominal head of the household, is also less active in the actual farm operation, he/she may also be more likely to accept responsibility for filling out the **Census of Agriculture** questionnaire.

Incorporating a family farm also facilitates succession. This allows for corporate shares to be allocated in any way desired rather than, for example, the division of land between siblings (both male and female) who are both active and inactive in the day-to-day operation of the farm. Ever-conscious of the fact that there is also only a $1 million capital gains exemption (i.e., ½ the increased value of a farm in one generation less $1 million is taxable) and that non-taxable transfers do not apply to off-farm siblings, other legal instruments (e.g., wills and life insurance) are also widely employed to help facilitate a smooth transition from one generation to the next. The imperative is always the need to maintain an economically viable farm unit.

3.5.2   GENDER

Historically and culturally, owner-operators in primary agriculture were almost all male. But this bubble has long been burst. In fact, women always played a vital role in the farm operation and, typically, a man and woman jointly owned the family farm. Increasingly, women have taken a lead role in the day-to-day management of the actual farm operation. In some sub-sectors (Figure 3.9) women are especially active, (e.g., equine, goat, and fruit/vegetable operations). Males still dominate the dairy and beef sectors, as well as the grains and oilseed sector, particularly on the prairies.

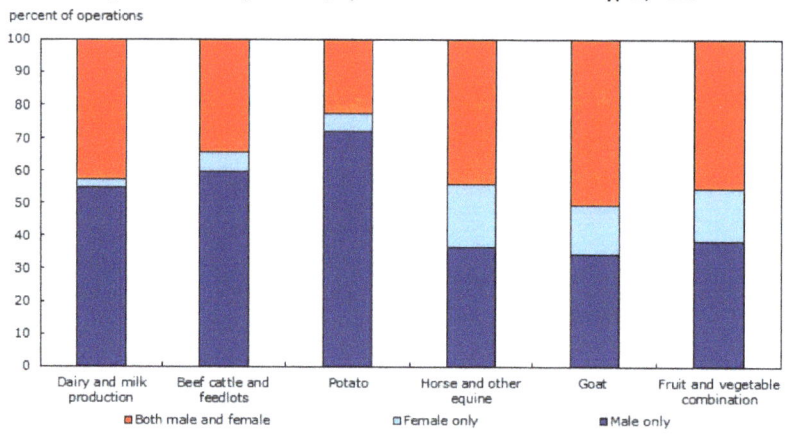

Figure 3.9 Farm Operations, by Gender and Selected Farm Types, 2016

Source: Statistics Canada, *Census of Agriculture*, 2016, 3438

This simply underlines how primary agriculture largely mimics the ever-changing social-economic milieu in Canada. It should also be noted that farm women are very often employed off-farm to supplement meager on-farm incomes. Employment as teachers, nurses, retailers, and related government services in nearby communities is extensive. Farm men, especially those who have smaller farm operations, also often work off the farm, especially during the off-season. (E.g., mechanics, welders, carpenters, truckers, miners, oil rig employees, oil/gas well maintenance employees, forestry employees, or as employees in some other (often) farm-related manufacturing operation or related retail trade.) The older generation often serve as "babysitters" to allow one or more members of the younger generation to pursue off-farm employment, again largely to supplement on-farm incomes for the extended family unit (Section 3.8).

3.5.3 FARM EDUCATION LEVELS

The educational level of Canadian farmers compares fairly favorably with other self-employed people in Canada (Figure 3.10). Highlights:

- In 2016, the majority of farm operators had completed at least a secondary school education (81.6%). The remaining 18.4% of farm operators versus 12.5% (in 2011) of the total self-employed labor force reported no certificate, diploma, or degree.

- A slightly larger proportion of farm operators than of the total self-employed labor force reported their highest level of educational attainment as a trades certificate, or college diploma (i.e., 35% versus 33 percent).

- Only about 18% of farm operators reported university credentials as their highest level of educational attainment, compared to about one-third (33%) of the total self-employed labor force.

- Among the farm operators, the proportion that attained a secondary school education or higher was larger for female operators (88.8%) than for male operators (78.6%).

A trades certificate (including apprenticeship) was the highest educational attainment for 16.7% of male operators and 7.4% of female operators. Almost 1/3rd, 28.5%, of all female operators reported a college diploma as their highest level of education compared to 17.9% of male operators. At 23.9%, the proportion of female operators with university credentials was larger than the 15.5% of male operators with university credentials.

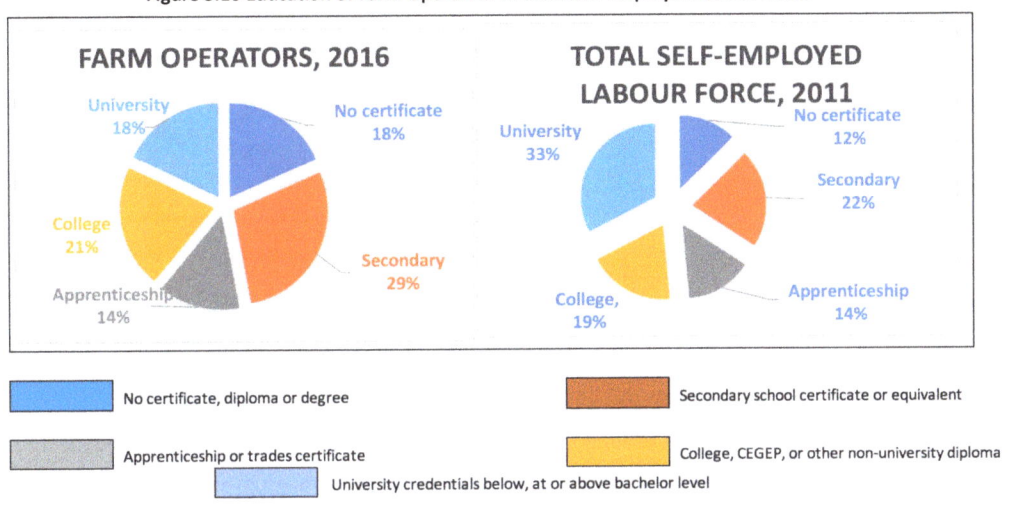

Figure 3.10 Education of Farm Operators vs Total Self-Employed Labour Force

FARM OPERATORS, 2016
- University 18%
- No certificate 18%
- Secondary 29%
- Apprenticeship 14%
- College 21%

TOTAL SELF-EMPLOYED LABOUR FORCE, 2011
- University 33%
- No certificate 12%
- Secondary 22%
- Apprenticeship 14%
- College 19%

Legend:
- No certificate, diploma or degree
- Secondary school certificate or equivalent
- Apprenticeship or trades certificate
- College, CEGEP, or other non-university diploma
- University credentials below, at or above bachelor level

Sources: Statistics Canada, Table 32-10-0024-01 and "Canadian farm operators: An educational portrait", November 27, 2018.

Farm operators still have a relatively low level of university training compared to the rest of the self-employed labor force. Often, however, they simply prefer a more practical hands-on college education. Still, the younger generation is giving ever-more emphasis to a higher education; something that was sometimes simply not available to older farmers. An increasingly digital technology in primary agriculture helps drive this trend.

3.6 SOURCES OF FARM REVENUE

Paralleling the farm enterprise structure in Canada (Table 3.3), an analysis of farm cash receipts in Canada provides us with additional insight into what crops and livestock are particularly important to primary agriculture. It also provides us with the information required to subsequently analyze on-farm incomes and gross margins.

Table 3.6 Farm Cash Receipts by Type, Canada, 2016

Category	Cash Receipts $ Million	Percent
CROPS:		
Wheat	5,680.8	9%
Other Cereals	1,048.3	2%
Canola	9,269.3	15%
Soybeans	2,983.9	5%
Corn	2,061.7	3%
Pulses**	3,765.9	6%
Potatoes	1,227.2	2%
Fruits & Vegetables	4,096.8	7%
Floriculture, Nursery & Sod	1,796.5	3%
Forages	558.4	1%
Maple Products	480.9	1%
Other Crops***	1,330.3	2%
TOTAL CROPS	**34,300.0**	**57%**
LIVESTOCK:		
Cattle & Calves	8,762.9	14%
Hogs	4,097.7	7%
Sheep & Lambs	183.5	0%
Dairy Products	6,174.2	10%
Poultry (Chicken & Turkey)	2,871.2	5%
Hatcheries & Eggs	1,106.9	2%
Honey	185.5	0%
Other Livestock	491.2	1%
TOTAL LIVESTOCK	**23,873.0**	**39%**
OTHER RECEIPTS:		
Crop Insurance	1,045.2	2%
Private Hail Insurance	276.4	0%
AgriInvest	297.3	0%
AgriStability	311.0	1%
Prov. Stabilization	330.0	1%
Other Payments	182.6	0%
TOTAL OTHER:	**2,442.0**	**4%**
GRAND TOTAL	**60,615.0**	**100%**

* Lentils, dry beans, dry peas, chickpeas.
** Incl. flaxseed, mustard seed, sunflower, canary seed.

Figure 3.11 Crop Revenue, by Source, Canada, 2016

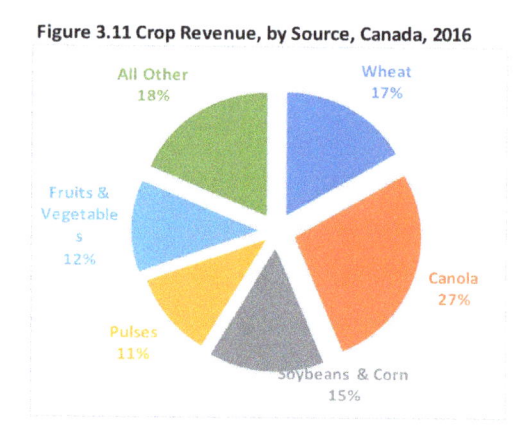

- All Other 18%
- Wheat 17%
- Canola 27%
- Soybeans & Corn 15%
- Pulses 11%
- Fruits & Vegetables 12%

Figure 3.12 Livestock Revenue, by Source, Canada, 2016

- All Other 8%
- Beef Cattle 37%
- Hogs 17%
- Dairy 26%
- Poultry & Eggs 12%

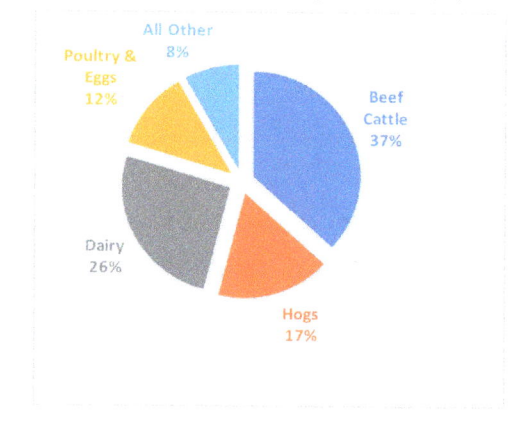

Figure 3.11 highlights how canola production now dominates crop production, displacing the historical king, wheat. Canola, which is a relatively profitable crop, now makes up almost 1/3rd of all crop receipts. But wheat is still a strong second (17%), followed by soybeans and corn (largely in Ontario and Quebec) at 15 percent. Concentrated in Ontario, fruits and vegetables also make up a healthy 12% of all farm receipts, followed by pulses (peas, beans, chickpeas, lentils, etc.), grown extensively in Saskatchewan (11%).

Livestock revenues largely comes from beef cattle (37%), dairy (26%), hogs (17%), and poultry/eggs (12%). Other livestock revenue largely comes from sheep and lambs, eggs, honey, and equine sales. (Figure 3.12)

In 2016 about one-half of "other receipts" (say $1 billion) came from (subsidized and government-funded) crop insurance. AgriInvest and AgriStability (also subsidized government programs) contributed another $600 million or so. Details regarding "Other Receipts" are provided in Chapter 9, Section 9.10.

Regional farm receipts in 2016 are indicated in Figure 3.13. Five provinces totally dominate the agricultural industry: the three prairie provinces (55%), Ontario (22%), and Quebec (15%). The other five provinces contributed a relatively meager 8 percent.

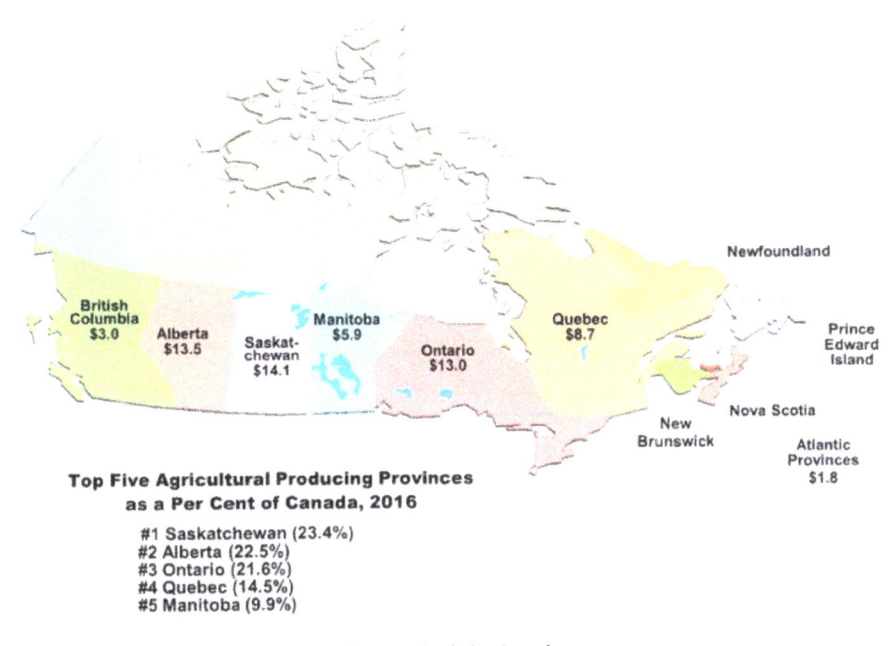

Figure 3.13 Regional Farm Cash Receipts, 2016—$ Billion

- British Columbia $3.0
- Alberta $13.5
- Saskatchewan $14.1
- Manitoba $5.9
- Ontario $13.0
- Quebec $8.7
- Atlantic Provinces $1.8

Top Five Agricultural Producing Provinces as a Per Cent of Canada, 2016
- #1 Saskatchewan (23.4%)
- #2 Alberta (22.5%)
- #3 Ontario (21.6%)
- #4 Quebec (14.5%)
- #5 Manitoba (9.9%)

Source: Statistics Canada

3.7 FARM INPUTS & COSTS

Although a composite listing of farm expenses does not reflect the cost of production of a particular enterprise, it is still very useful to look at this to see what costs are particularly important to Canadian agriculture (Table 3.7 and Figure 3.14).

Table 3.7 Farm Operation Expenses, Canada, 2016

Item	Expense $ Million	Percent
Land Rentals & Custom Work	4102.6	9%
Paid Wages	5444.3	12%
Interest Payments	2845.5	6%
Fuel (machinery & heating)	2650.3	6%
Machinery Repairs	3052.7	7%
Fertilizer & Lime	5407.3	12%
Herbicides, insecticides, fungicides	2897.3	6%
Seed & plant purchases	2582.4	6%
Commercial Feed Purchases	6467.9	14%
Livestock & Poultry Purchases	1831.3	4%
Other Livestock Expenses*	1506.0	3%
Building-Fence Repairs	1008.2	2%
Electricity, Telephone & Property Taxes	1932.8	4%
Other Operating Expenses**	4375.0	9%
TOTAL OPERATING EXPENSES*	**46103.6**	**100%**

* Includes AI, vet fees, twine, wire, containers.

** Includes business insurance, stabilization premiums, irrigation, crop & hail insurance, and legal/accounting fees.

*** Excludes depreciation = + $7.4 B and rebates = - $51.5.

Source: Statistics Canada, *Farm Operating Expenses and Depreciation Charges* (x1,000), Table 32-10-0049-01

Figure 3.14 Farm Operating Expenses, Canada, 2016

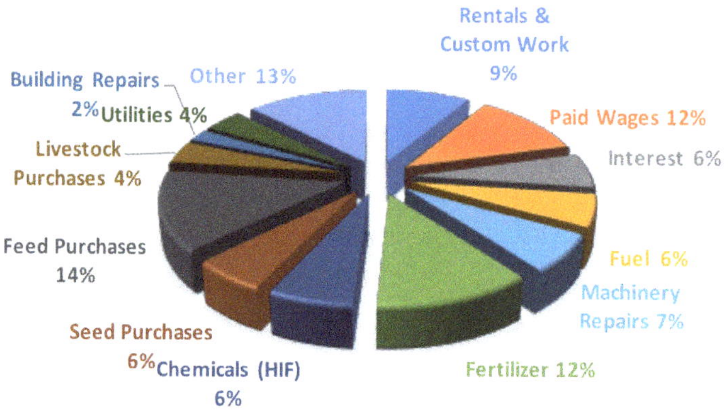

- Rentals & Custom Work 9%
- Paid Wages 12%
- Interest 6%
- Fuel 6%
- Machinery Repairs 7%
- Fertilizer 12%
- Chemicals (HIF) 6%
- Seed Purchases 6%
- Feed Purchases 14%
- Livestock Purchases 4%
- Utilities 4%
- Building Repairs 2%
- Other 13%

Source: Table 3.7

This cost structure is a composite of both crop and livestock producers, say 60–40 (Table 3.6). Thus, we see how four inputs dominate the cost structure of the crops sector: chemicals (herbicides, insecticides, fungicides), say 10%; synthetic fertilizer, about 20%; seed, approximately 10%; and fuel, around 10 percent. This totals perhaps one-half of their total annual cash costs. For the livestock sector, livestock and feed purchases alone make up about 35% and 10%, respectively, of total annual cash costs.

3.8 INCOME & OFF-FARM WORK

It is useful to initially clarify what "income" is in a self-employed enterprise. It is essentially a return to their own labor, management, and equity. The only return to capital that is considered a "cost" are the interest costs on borrowed capital. Nor is it a "profit," a widely abused concept that is sometimes used to refer to a return to capital or an income over-and-above what is payable to land, labor, capital, and management.

Gross Margin = Gross Farm Revenue - Cash Expenses
Realized Income = Gross Margin - Depreciation
Net Farm Income = Realized Income ± Inventory Changes
Disposable Income = Net Farm Income - Taxes

Note, as well, that simply having a positive gross margin does not imply long-term viability. To maintain a viable economic operation in the long-run, all fixed costs, including depreciation, must be paid. Depreciation is essentially a proxy for capital replacement although the nominal value of machinery and building replacement (with inflation) often climbs so that these don't exactly balance.

The distinction between **net farm income** and **disposable income** should also be made clear. The difference is taxes still payable for the year in question.

This, then, allows us to generate approximate regional farm income estimates for 2016, as summarized in Table 3.8.

Table 3.8 Calculation of Net Farm Income, by Region, 2016, $ millions

Item	MARITIMES	QUEBEC	ONTARIO	PRAIRIES	B.C.	CANADA
Total Farm Cash Receipts	1,795.1	8,775.5	13,261.3	33,664.0	3,119.1	60,615.0
Crops	816.1	3,227.9	6,837.8	21,794.9	1,623.1	34,299.8
Livestock	905.8	5,154.5	6,062.2	10,284.3	1,466.3	23,873.1
Program Payments/Adjustments	73.2	393.1	361.3	1,584.8	29.7	2,442.1
Total Operating Expenses	1,488.3	6,740.5	10,503.2	24,577.1	2,743.0	46,052.1
Gross Margin (Net Cash Income)	306.8	2,035.0	2,758.1	9,086.9	376.1	14,562.9
Plus Income-in-Kind less Depreciation	181.3	920.5	1,561.0	4,183.3	411.2	7,257.3
Realized Net Income	125.6	1,114.5	1,197.1	4,903.6	-35.1	7,305.7
Plus or minus Inventory Change	-0.3	-49.3	-138.6	1,352.7	38.3	1,202.8
TOTAL NET FARM INCOME	125.3	1,065.2	1,058.6	6,256.2	3.1	8,508.5
Expense/Revenue Ratio	0.93	0.87	0.90	0.89	1.02	0.90
Number of Farms (approx.)	7600	28900	49600	89900	17500	193500
Net Farm Income per Farm	$ 16,487	$ 36,859	$ 21,342	$ 69,591	$ 180	$ 43,971

Source: Basic data from: Statistics Canada, *Net farm income*, Table 32-10-0052-01; and *Farm cash receipts*, Table 32-10-0045-01

Table 3.8 highlights the following:

- Average net farm income for **all Census farms** for the whole of Canada in 2016 was a relatively modest $44,000.

- Average net farm income varies widely across Canada, ranging from basically zero in BC to $70,000/farm on the prairies. BC, on average, is highly dependent upon supplementary activities.

- Average net farm incomes in the Maritimes were about $16,500; clearly implying that farming is also a supplemental activity for many Maritime farmers.

- Average net farm incomes in Quebec were somewhat less than the national average; approximately $37,000 per farm.

- Average net farm incomes in Ontario were a relatively low $21,000/farm, perhaps somewhat surprising given the preponderance of supply-managed farms (i.e., dairy and poultry) and the vibrant fruit-vegetable-winery industry in the province.

- However measured (gross revenue, expenses, gross margin, realized net income, or net farm income), the dominance of the prairie provinces is also highlighted.

- Embedded in these estimates is an implicit expense/revenue ratio which, nationally, is determined, in 2016, to be about 0.90. That is, for every dollar a Canadian farmer earns, he/she pays out approximately 90 cents in operating expenses. This estimate varies by year and exactly how the calculation is made.

This is only an **average** net farm income for **all Census farms** derived **solely from farming operations**. It is generally not disposable farm family income.

In 2016, an estimated 44.4% of farm operators worked at least some time during the year at an off-farm job.[20] This has essentially remained much the same throughout the history of Canadian agriculture, as initially under-capitalized farm families have pursued a higher standard of living and/or continued capital acquisition. Additionally, with the social-cultural changes over time, a relatively large number of farm operating partners (usually the woman, often with a higher education) now further supplement incomes generated by the farm itself. Inter-generational farm members often generate even more off-farm income. Some farmers, either by choice or circumstance, are just life-long "part-time" farmers.

The net result is that the total income of farm families in Canada is typically (on average) much **higher** than the income generated solely from the farm (Table 3.9).

Table 3.9 Percentage Contribution of Income Components, Census Farm Families, Selected Years*

Income Source	1980	1990	2000	2009-2013** Income	2009-2013** Percent
Net Farm Income	31%	15%	18%	$ 25,649	23%
Wages & Salaries	47%	55%	57%	$ 57,016	52%
Net Non-Farm Self-Employment Income	6%	6%	4%	$ 1,020	1%
Net Government Program Payments	6%	10%	11%	$ 5,614	5%
Other Income Sources	11%	14%	11%	$ 20,036	18%
TOTAL INCOME	100%	100%	100%	$ 109,335	100%

* Average respondents in 2009-2013 = 106,900 farms.
** Average for 2009, 2011, and 2013. Pre-capital cost allowance (CAA) adjustment.

Sources: Data for 1980, 1990, and 2000 from Statistics Canada, Agriculture-Population data base. Data for 2009, 2011, and 2013 from Statistics Canada, Total and average off-farm income by source, Table 32-10-0057-01

Table 3.9 gives us a general indication of the importance of various income components on Canadian farms. As much as 75 percent of farm family income is probably derived from off-farm sources, with about 2/3rds of this coming from off-farm wages and salaries. The resulting average net disposable income (post accounting CCA adjustment) in 2009–2013 is estimated to have been about $92,000 per average farm family. This would generally be considered a good "middle class" income level for a Canadian family in 2016, especially for families who have some unique tax advantages, possibly substantial unrealized capital gains (Section 3.11), relatively low living costs vis-à-vis urban areas, and also have some perceived "life-style" benefits in a pristine rural environment.

Relying on American data, Figure 3.15 highlights how this varies across farm sizes (i.e. sales classes). Smaller American farms with annual sales of less than $350,000 rely almost exclusively on off-farm income sources. The situation in Canada is similar.

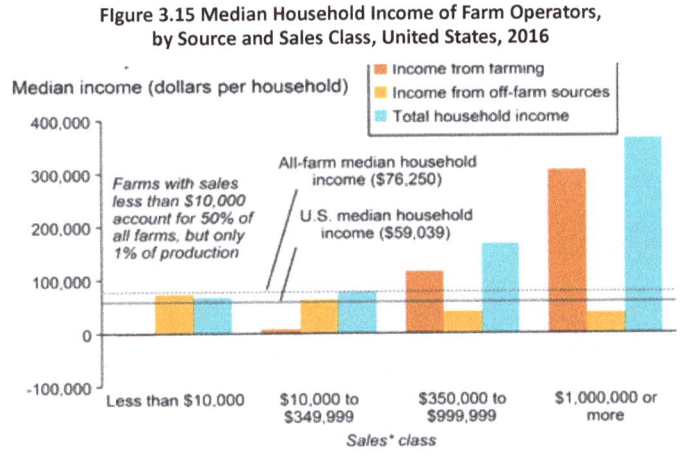

Figure 3.15 Median Household Income of Farm Operators, by Source and Sales Class, United States, 2016

*Sales = Annual gross cash farm income before expenses (the sum of the farm's crop and livestock sales, government payments, and other cash farm-related income).

Source: USDA, Economic Research Service and National Agricultural Statistics Service, Agricultural Resource Management Survey and U.S. Census Bureau, Current Population Reports. Data as of November 29, 2017

3.9 THE COST-PRICE SQUEEZE & SCALE

Generally speaking, a **cost-price squeeze** refers to ever-increasing (nominal) unit costs and the simultaneous slow increase, stable, or decreasing product prices. This is often attributed to the fact that individual farmers are generally price-takers, subject to input and product prices, which are largely determined by the national or international marketplace. Additionally, this is sometimes exacerbated by the oligopolistic pricing practices of both input suppliers and the buyers of farm products.

By-and-large, however, this is technologically driven and, thus, almost perpetually in motion. Farming has become increasingly capital-intensive and this capital has been biased in favor of larger-scale operations. Additionally, and sometimes for proprietary reasons, more and more farming operations are gradually being outsourced (e.g., custom spraying, machinery maintenance, digital installations and maintenance, fertilizer application, soil testing, trucking, etc.) and this also reduces gross margins per unit of output.

Basically, it is hypothesized that on-going supply shifts (largely due to technology-driven productivity increases in agriculture), subject to a more gradual increase in demand, will slowly, but mercilessly, push real (inflation-adjusted) product prices downward. Malthus be-damned. Yet there is a growing consensus that, historically at least, costs and prices have generally tracked each other fairly closely, as illustrated in Figure 3.16. Changes in the farm input price index (green line, 2003 = 100) have generally moved lock-step with changes in the farm product price index (purple line, 2003 = 100).

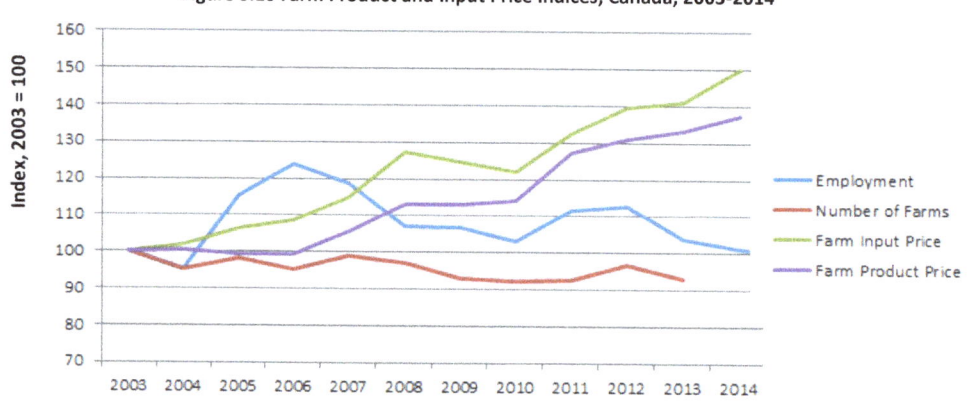

Figure 3.16 Farm Product and Input Price Indices, Canada, 2003-2014

Source: Statistics Canada, Table 18-10-0258-01 (input prices) and Table 32-10-0098-01 (product prices)

To counteract cost-price pressures, farmers have basically pursued two avenues: a) somehow reduce unit costs or b) somehow increase the value-added in their products, thus securing better prices.

The means of somehow increasing the value-added in farm products has been pursued in various ways, most notably through supply management, farmer-controlled input supply (e.g., UFA,

Cooperatives), farmer-controlled product purchases (e.g., grain cooperatives—the Pools, UGG), or various initiatives to, effectively, increase demand through local product processing, e.g. oilseed crushers or meat processors. Supply management (poultry and dairy) essentially locks in desired income levels by restricting supply and utilizing an elaborate cost-plus pricing policy. Both the federal and provincial governments have also regularly tried to moderate the impact of price and cost fluctuations and "stabilize" and/or improve farm incomes, most notable through a myriad of direct and indirect subsidies, R&D, and extension (Chapter 9, Sections 9.8–9.10).

It is the pursuit of **economies of scale** (that is, increasing size) that has had the most profound impact on-farm structure and farm numbers over the years, particularly since the rapid mechanization of the last half of the 20th century (Figure 3.17).

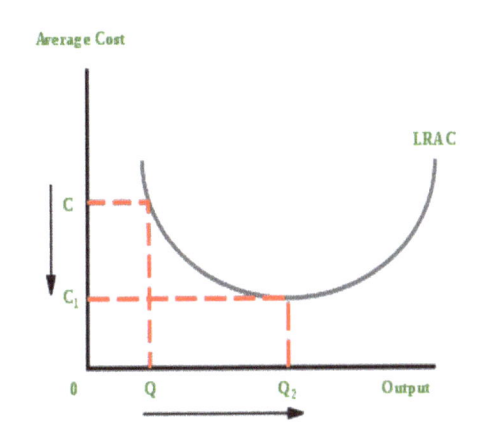

Figure 3.17 Economies of Scale in Production & Marketing

As the quantity of production/marketing increases from Q to Q2, the average cost of each unit decreases from C to C1. LRAC is the long-run average cost. Moreover, technological change generally keeps shifting this curve farther and farther to the right.

Virtually every sector in agriculture has been subject to this phenomenon. Thirty years ago, a cow-calf operator could "make a living" with fifty cows; now, he might need 300. Thirty years ago, a dryland grain farmer on the prairies needed, maybe, a section (640 acres) of crop land whereas now he might need four sections. And so on across virtually all of agriculture—swine, poultry, dairy, vegetables, etc.

Technology largely defines this optimal efficiency level at any point in time. For example, on a dryland grain/oilseed farm on the prairies today (2018), a new combine will thresh about 2,500 acres of crop, a new zero-till air-seeder/tractor unit might seed 5000 acres, and a new high-clearance sprayer might spray 5,000–10,000 acres. In this case, an economic unit might be 5,000 acres with two combines, one air-seeder, and (with multiple passes) one high-clearance sprayer. And then, labor and management permitting, this "module" can be endlessly cloned. This growth strategy is amenable to many other sub-sectors (e.g., hog barns).

3.10 COMMERCIAL FARMS RE-VISITED

The most descriptive characteristic of "farms" and "farmers" is that they are all different. They differ most profoundly in terms of their composition and their scale.

Table 3.10 and accompanying Figure 3.18 are instructive as they highlight the following:

- Farmers with sales in excess of $250,000 annually—about 28% of all farmers—account for 85% of sales (i.e., agricultural production) and at least 85% of farm income.

- Conversely, farmers with sales of less than $250,000 annually—some 72% of all farmers—account for a meager 15% of sales and probably less than 15% of farm income.

- Farmers with sales in excess of $100,000 annually—about 44% of all farmers—account for 94% of all farm income.

- The only sales classes that grew in the last 20 years were the top two: >$1 M and >$2 M. The "middle class," particularly the $100,000–$250,000 sales classes have been hollowed out. (This ignores the increasing nominal price effect.)

- Looking at imputed farm income levels for the various sales categories' paints approximately the same picture. Farms with annual sales between $250,000 and $500,000 might be expected to generate a net farm income of about 37,500 and, as such, might be considered "commercial" farms, as would all larger farmers.

Table 3.10 Gross Farm Receipts, by Sales Class, and Imputed Farm Income, Canada, 2016

Sales Class*	Number Farms	Percent of Farms	Sales Midpoint $*	Imputed Income**	Percent of Sales & Farm Income**
Under $10,000	34,156	18%	$5,000	$500	0%
$10,000 - $24,999	27,554	14%	$17,500	$1,750	1%
$25,000 - $49,999	23,519	12%	$37,500	$3,750	1%
$50,000 - $99,999	24,010	12%	$75,000	$7,500	3%
$100,000 - $249,999	30,721	16%	$175,000	$17,500	9%
$250,000 - $499,999	21,884	11%	$375,000	$37,500	14%
$500,000 - $999,999	16,907	9%	$750,000	$75,000	21%
$1,000,000 - $1,999,999	9,237	5%	$1,500,000	$150,000	23%
$2,000,000 and over	5,504	3%	$3,000,000	$300,000	28%
TOTAL	193,492	100%			100%

* Based on 2016 dollars. Approximate.
** Based on a Gross Margin of 10 percent. See Table 3.8.   0.10
Assumed applicable to all sales classes and all farm types. Very approximate.

Source: Basic data from Statistics Canada, *Census of Agriculture, 2016*, Table 32-10-0157-01

Utilizing a different methodology, different data sources, and a different time-frame, Figure 3.18 generally paints a similar picture.[21]

A FARM PROFILE 39

Figure 3.18 Average Net Operating Income by Revenue Class, 2012

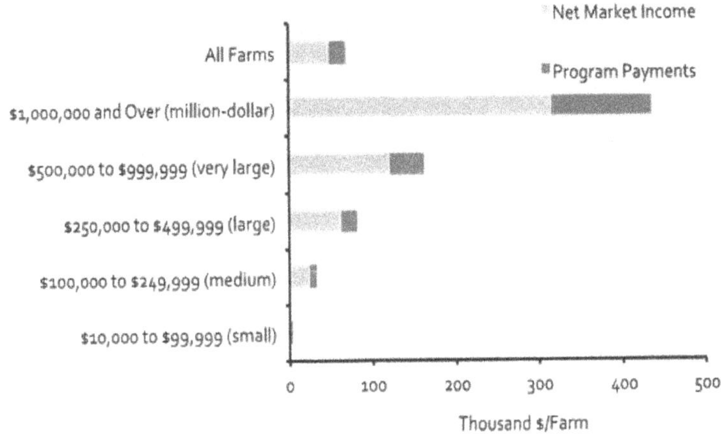

[Bar chart showing Net Market Income and Program Payments by revenue class, ranging from $10,000 to $99,999 (small) up to $1,000,000 and Over (million-dollar), and All Farms, measured in Thousand $/Farm from 0 to 500]

Source: AAFC, *An Overview of the Canadian Agriculture and Agri-Food System, 2015*, Chart E.1.8, p. 56

Thus, what we have is approximately the following for all of primary agriculture in Canada, as of 2016:

Commercial Farmers	53,532 (28%)
Marginal Farmers	30,721 (16%)
Hobby/Part-Time Farmers	109,239 (56%)
TOTAL CANADA	193,492

Over one-half of Census "farmers" are hobby or part-time farmers.[22] A more in-depth analysis of the financial status of truly "commercial" farmers engaged in various types of farming/ranching operation, all essentially full-time farmers/ranchers, is provided in Table 3.11 and Figure 3.19.

Table 3.11 Financial Profile, Commercial Farms, by Farm Type, 2017*

Farm Type	Number***	Revenue $,000	Expenses** $,000	Margin** $,000	E/R Ratio	NET FARM INCOME****
CROP FARMS:						
Grain & Oilseed	48216	557.7	445.3	112.4	0.80	$ 56,630
Potato	791	1784.1	1405.6	378.5	0.79	$ 200,090
Other Vegetables/Melon	1606	917.7	755.5	162.2	0.82	$ 70,430
Fruit & Nuts	3121	379.8	325.3	54.5	0.86	$ 16,520
Greenhouse, Nursery, & Floriculture	2963	1148.6	990.3	158.3	0.86	$ 43,440
Other Crops	4805	223.0	176.5	46.5	0.79	$ 24,200
LIVESTOCK FARMS:						
Beef Cattle	28012	390.6	345.7	44.9	0.89	$ 5,840
Dairy Cattle	10062	770.4	620.4	150.0	0.81	$ 72,960
Hogs	2180	1289.1	1054.7	234.4	0.82	$ 105,490
Poultry & Eggs	2996	1369.1	1104.7	264.4	0.81	$ 127,490
Other Animals	4838	277.3	245.8	31.5	0.89	$ 3,770
ALL COMMERCIAL FARMS:	109590	569.2	468.3	100.9	0.82	$ 43,980

* "Commercial" farms, i.e. gross sales > $25,000/year.
** Before (-) depreciation and (+) income-in-kind.
*** Number participating in survey.   **** Authors estimate after adjusting for depreciation and income-in-kind (Table 3

Source: Statistics Canada, *Financial Structure by farm type*, Table 32-10-0102-01

Figure 3.19 Average Net Operating Income, by Farm Type, 2012

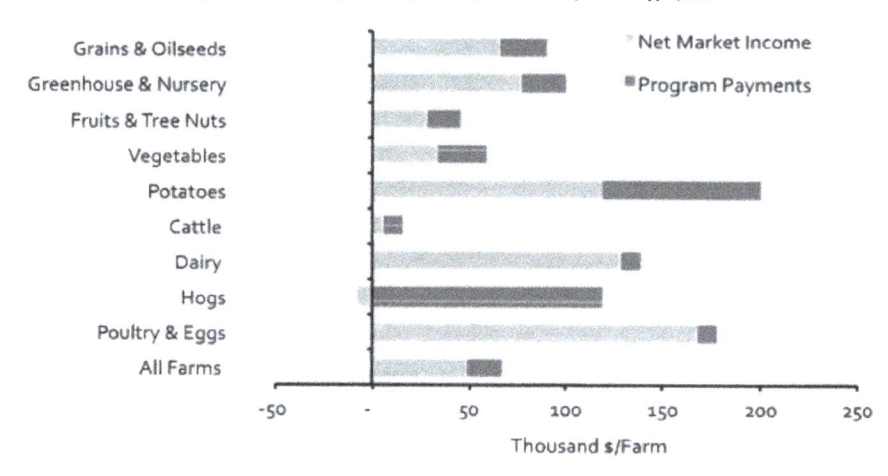

Source: AAFC, *An Overview of the Canadian Agriculture and Agri-Food System, 2015*, Ottawa, 2016, Chart 3.1.9, p. 56

Table 3.11 highlights both typical net farm incomes for the various types of commercial farming operations and provides some insight into the average size of various farm types. On average, potato farmers, hog farmers, and poultry/egg farmers tend to have relatively large operations (in terms of sales) and relatively high net farm incomes. In contrast, the average beef operation tends to be relatively small and generate relatively low net farm incomes.

Figure 3.20 Gini Coefficient = A/(A+B)

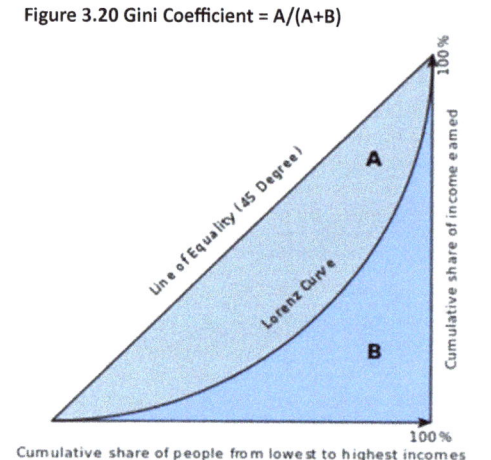

Utilizing a different methodology, different data sources, and a different time-frame, Figure 3.19 generally paints a somewhat similar picture.[23]

For the agricultural sector as a whole, our estimate of the Gini Coefficient is 0.46, implying a very unequal farm income distribution within the sector.[24] The national average for all sectors is about 0.32; much more equal.

3.11 CAPITAL & DEBT STRUCTURE

The old adage that farmers' "live poor and die rich" is still true. Capital values, particularly with respect to agricultural land, have steadily climbed. Capital values per farm during the last 20 years have increased approximately five times (Figure 3.21). Farmland and buildings appreciated at an average rate of 11.7% per year during 2011–2016. Relatively good crop/livestock conditions, relatively strong prices, historically low interest rates, the continued urbanization of farmland, and an infusion of some external funds have all coalesced to accelerate agricultural land price increases.

Figure 3.21 Average Total Capital Value per Farm, Canada, 1996-2016*

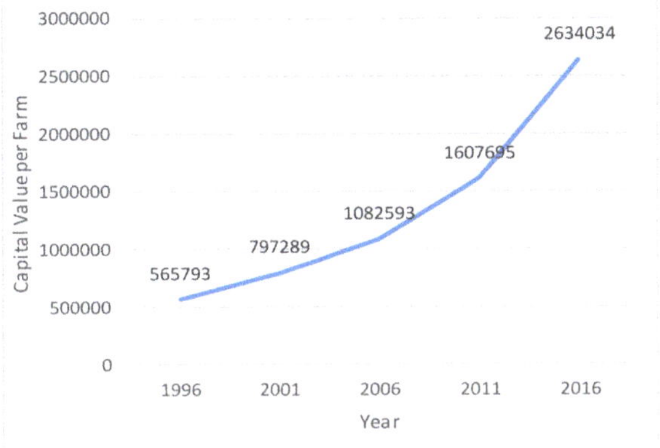

```
1996: 565793
2001: 797289
2006: 1082593
2011: 1607695
2016: 2634034
```

Includes land, buildings, machinery, and other (e.g. quotas)

Source: Statistics Canada

A closer examination of the data (Table 3.12) reveals that about 84% of all farm capital is now embedded in land and buildings.

Table 3.12 Regional Agricultral Land Values, Other Capital, & Outstanding Debt, Canada, 2016

Region	No. Farms	Acres (,000)	$/Acre*	Total Value $,000	Average Land-Bldg. Value/Farm $,000
LAND & BUILDINGS:					
Maritimes	7493	2397.2	$ 2,141	$ 5,132,455	$ 685.0
Quebec	28919	8103.2	$ 5,199	$ 42,126,266	$ 1,456.7
Ontario	49600	12348.5	$ 10,006	$ 123,561,422	$ 2,491.2
Prairies	89952	129473.6	$ 1,714	$ 221,924,670	$ 2,467.1
BC	17528	6400.5	$ 5,487	$ 35,119,287	$ 2,003.6
CANADA	193492	158723	$ 2,696	$ 427,864,100	$ 2,211.3

LAND & BUILDINGS ($M)	$ 427,864.1	84%
Machinery & Equipment ($M)	$ 53,869.2	11%
Livestock & Poultry ($M)	$ 27,931.3	5%
TOTAL ($M)	$ 509,664.6	100%
DEBT ($M)	$ 95,993.9	
Debt/Equity Ratio		18.8%

Adjusted.

Sources: Statistics Canada, *Census of Agriculture, 2016,* Tables 004-0021, 004-0234, 002-0003, and 002-0008

The 2016 value of agricultural land and buildings varied across the country, but, nationally, it averaged some $2.2 million per farm. Ontario and the prairie farmers led the way, averaging about $2.5 M. per farm.

At the same time, farm debt levels have climbed at approximately the same rate. In 2016, farm debt reached about $96 billion, implying a debt-asset ratio of about 18.8 percent. In 2015, this stood at about 15.5% while the fifteen-year average (2001–2015) was 16.7 percent. The source of these funds is as follows:

SOURCE OF FUNDS	$ BILLIONS	PERCENT
Chartered Banks	36.1	38%
Federal Government Agencies (FCC)	26.9	28%
Credit Unions	14.5	15%
Private Individuals/Supply Companies	10.0	10%
Provincial Government Agencies	5.8	6%
Other	2.7	3%
TOTAL	96.0	100%

Finally, and not surprisingly, with rapidly inflating asset values, the rate of return on these assets may gradually be declining. The rate of return was a very modest 2.3% in 2015, down from 2.7% during the previous 5 years and even less than the 15-year (2001–2015) average of 2.6 percent.

3.12 FARMER'S STATUS IN RURAL & URBAN CANADA

Solely in terms of numbers, farms and farm families are now greatly outnumbered by the much larger rural community and the much, much larger rural-urban population. Primary agriculture now contributes less than 2 percent to the national gross domestic product (GDP), a measure of the value-added (land, labor, capital, and management) contributed by the respective sectors (Figure 3.22).

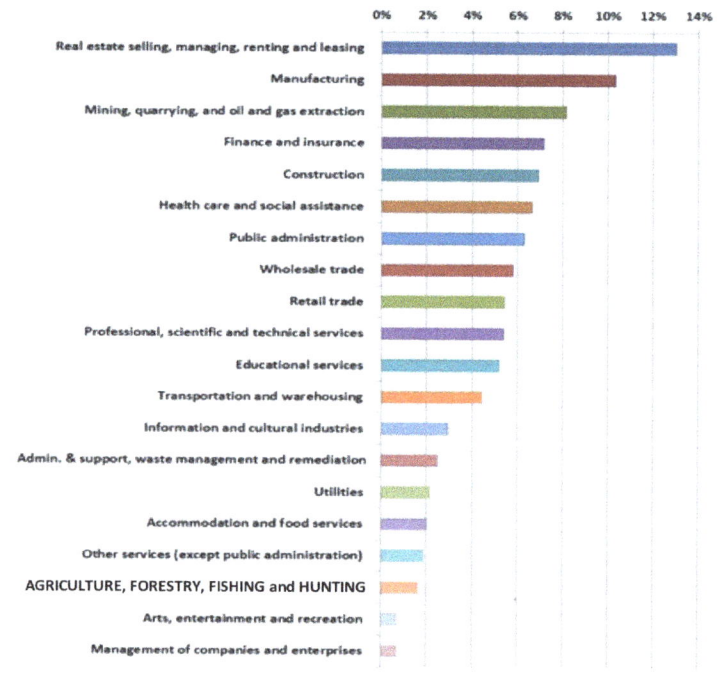

Figure 3.22 Canada Gross Domestic Product, by Industry, $2007

Basic source: Statistics Canada

This is not surprising. Outside of the prairies, farm families now make up no more than 1–3 percent of the rural population and no more than 1–2 percent of the total population. On the prairies, however, farmers **per se** still make up about 8 percent of the rural population and about 4 percent of the total population (Table 3.13). But everywhere this translates into a diminished role in the social-political arena.

Table 3.13 Farm Population, Rural Population, & Total Population, 2011

Region	No. Farms	Farm Populaton	Rural Population Number	Rural Population Farm/Rural	Total Population Number	Total Population Rural/Total	Total Population Farm/Total
Maritimes	8521	24155	1027600	2%	2327638	44%	1%
Quebec	29437	90735	1511525	6%	7903001	19%	1%
Ontario	51950	163435	1775670	9%	12851821	14%	1%
Prairies	96063	257015	1236765	21%	5886906	21%	4%
B.C.	19759	49840	600305	8%	4400057	14%	1%
CANADA*	205730	585180	6151865	10%	33369423	18%	2%

*Excludes the NWT, Nunavut, and Yukon.

Source: Statistics Canada, Table 32-10-0012-01

Further details are provided in Chapter 7, Rural Socio-Economic Profile.

3.13 SYNOPSIS

There is no single "farm profile." Each and every farm/ranch in Canada is unique. Every enterprise is different and every region is different. Nevertheless, a composite national profile highlights the following:

- There are surprisingly few primary producers (farmers/ranchers) left in Canada. Numbering only about 194,000 (in 2016), this total, nationally, is only about 3 percent of the rural population and 2 percent of the total population. They are now only a major social/economic and political force on the prairies.

- The liberal definition of a "farmer" actually artificially inflates bona fide commercial farm numbers. We estimate that less than 1/3rd of **Census of Agriculture** "farmers" (i.e., about 54,000) are actually full-fledged, full-time farmers securing most of their annual income from farming/ranching operations. The remaining 140,000 or so are, essentially, marginal, part-time, or hobby farmers.

- The farm "middle class" has been and is being gradually hollowed out. The only sales classes that have grown in the last 20 years are the top two: > $1M and > $2 million. To remain economically viable, most commercial farmers have had to either continue to become larger and larger or develop a different business model—off-farm work, intensive niche markets, non—farm investments, or some other income source.

- Twenty percent of the farms generate approximately 80% of agricultural production while the remaining 80% generate approximately 20% of the agricultural production. The net farm income distribution and farm capital (including land) concentrations are even more skewed in favor of larger farming operations. Farm input providers, processors, financial institutions, and even research and development (R&D) initiatives are, generally, similarly focused on providing their services to larger farming operations.

- There is a greater reliance on net farm incomes on the prairies, in Ontario, and in Quebec. Net farm incomes in the Maritimes and BC are, on average, almost negligible. The "average" farm makes more money off the farm (primarily wages and salaries) than it does on the farm. Extensive off-farm work has always been a trademark of Canadian agriculture; often a way to gradually accumulate enough capital to become a "full-time" farmer.

- The capital value of an "average" Canadian farm/ranch now exceeds $2.6 million and over 80 percent of this is embedded in land and buildings. About 80% of this is equity; approximately 20% is debt. The price of land and other assets (including production quotas) has increased four or five times in the last 25 years, essentially locking beginning farmers out of the market. (There are parallels in the housing market.)

- The social and cultural fabric embodied in the "family farm," typically a mixed farm with a few chickens, pigs, cows, and a garden, is becoming frayed. Increasing farm specialization and increasing scale, accompanied by ever-encroaching urbanization, has irreversibly changed family farms, rural communities, and the rural landscape.

4. Agricultural Production & Marketing

This chapter describes the current production and domestic marketing characteristics of canola, wheat, barley, oats, corn, soybeans, pulses, beef, pork, dairy, eggs, and broilers. The international marketing characteristics of each are described in more detail in Chapter 5 immediately following.

4.1 MAJOR CROPS

The major crops briefly profiled following are canola, wheat, other cereals (barley & oats), corn, soybeans, and pulses. Their relative importance, in terms of domestic production, is indicated in Table 4.1 while their respective acreages in Canada for the 1997–2017 period are traced in Figure 4.1. Table 4.2 highlights Canada's status in a global context.

Table 4.1 Production of Principal Crops, Canada, 2013-2017 (million tonnes)

	2013	2014	2015	2016	2017*
Wheat, all	37.5	29.4	27.6	31.7	29.9
Wheat, excl durum	31.0	24.2	22.2	24.0	25.0
Wheat, durum	6.5	5.2	5.4	7.8	4.9
Canola	18.5	16.4	18.3	19.6	21.3
Soybeans	5.3	6.0	6.4	6.5	7.7
Barley	10.2	7.1	8.2	8.8	7.8
Pulses	6.6	6.3	6.1	8.4	7.1
Oats	3.9	3.0	3.4	3.2	3.7

*Estimated as of November 2017.
Source: Statistics Canada

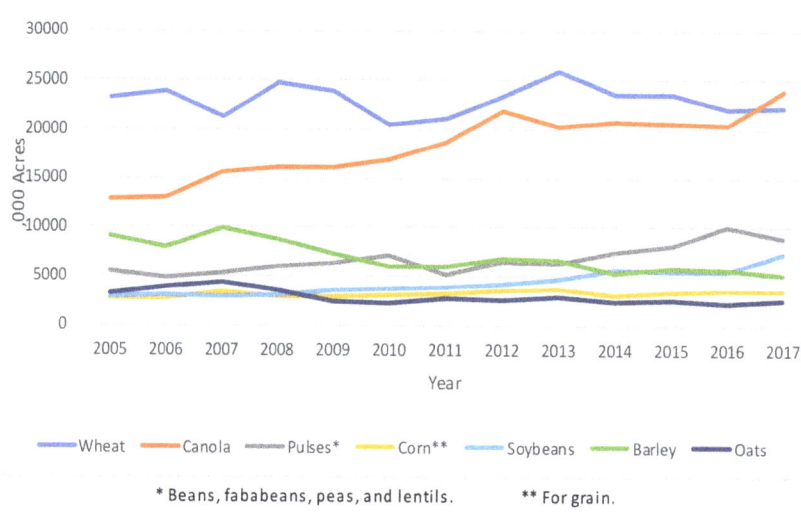

Figure 4.1 Area in Principal Crops, Canada, 2005-2017

* Beans, fababeans, peas, and lentils.   ** For grain.

Source: Basic data from Statistics Canada, Table 001-0017

Table 4.2 Share of World Production and World Ranking, Canada, 2011

Commodity	Share of World Production (%)	World Ranking
Canola	22.6	1
Wheat	3.6	7
Barley	5.8	7
Oats	13.2	2
Rye	1.5	10
Maize (Corn)	1.2	11
Soybeans	1.6	7

Source: FAO/UN

4.1.1 CANOLA

Canola is now (2019) the single most important crop in Canada, finally surpassing "king" wheat in terms of planted acreage and profitability. It currently generates about one-quarter of all crop revenue in Canadian agriculture (Figure 3.11). About 43,000 farmers now grow canola, mostly in Saskatchewan, Alberta, and Manitoba, but BC, Ontario, and Quebec also grow a substantial amount of this so-called Cinderella crop.

Canola was developed through conventional plant breeding from rapeseed, an oilseed plant already used in ancient civilizations as a fuel. It has a lower erucic acid and glucosinolate content and is low in saturated fats.[25] Later, in 1995, a higher-yielding genetically modified (GM) canola was developed and most (>90%) canola grown in Canada today is GM. (See

Chapter 13, Section 13.3.3.) Canada is also now the largest hybrid canola seed producer in the world.

Canola is a relatively easy plant to grow and harvest. Most canola on the prairies is now planted with a large no-till drill in mid-May and then harvested in mid-September. Typically, fairly heavy manufactured fertilizer applications accompany this seeding operation, especially nitrogen. Thereafter, it is generally sprayed with an herbicide and, if necessary, an insecticide. On the prairies, the crop is sometimes subject to flea beetle and grasshopper infestations. Clubroot, blackleg, and *Sclerotinia* can also reduce yields. Spraying with a fungicide for *Sclerotinia* (white mold), in particular, is very common under irrigation and in areas that have ample to excess moisture. Historically, it has been swathed prior to threshing, although straight-cutting is becoming increasingly popular. A "good" yield now exceeds 50 bushels per acre.

From on-farm storage (at <10% moisture) it is then transported to either a local canola crushing plant or a high-throughput elevator. (For details, see Chapter 6.)

Canola crushers (Chapter 6, Section 6.2.1) transform the harvested seed into oil for human consumption and meal for livestock feed (about 44:56 oil/meal). The oil can also be used as a biofuel and, like soybean oil, is often used interchangeably with non-renewable petroleum-based oils in products such as industrial lubricants, candles, lipsticks, and newspaper inks. Other emerging industrial uses include plastics, protein isolates, adhesives, and sealants.

4.1.2 WHEAT

Cultivated forms of wheat evolved from the natural crossings of wild species. Wheat was first introduced into North America in the 15th century. Most cultivars are common (hard & soft, spring & winter) and durum wheats. The common cultivar Red Fife was developed in Ontario in the 1870's and Marquis was developed on the prairies in the early 1900s. These provided the basis for all the Canadian wheat varieties developed thereafter. There is no authorized GM wheat in Canada.

The principal classes of wheat are:

- CWRS, Canadian western red spring, a hard spring wheat with a high protein content that is highly regarded for its superior milling and baking quality; about 60% of annual production;
- CPSR, Canada prairie spring red with a medium protein content;
- CWAD, Canada western amber durum, especially good for pastas; and

- CWSWS, Canada western soft white spring, a favorite of the ethanol industry (Chapter 6) because of its low protein-high starch content.

The wheat acreage was only surpassed by canola in 2017. But even in 2017, wheat was still seeded to some 22 M acres (9 M ha.), although this is down by over 6 million acres (2.5M ha.) from 20 years ago (Figure 4.1). Canada is now the world's seventh largest producer (Table 4.2) and in 2017 production (incl. durum) was more than 27 million tonnes (almost 1 billion bushels). Spring wheat production is concentrated in Saskatchewan (41%), Alberta (39%), Manitoba (18%), and Ontario (2%). Virtually all Canadian durum production takes place in Saskatchewan (83%) and Alberta (17%).

Wheat is one of the easier crops to plant, maintain, and harvest. Most wheat on the prairies is now planted with a large no-till drill by mid-May and then harvested in early-September. Manufactured fertilizer is generally applied at the time of the seeding operation. Thereafter, it is typically sprayed with an herbicide and, increasingly, with a subsequent fungicide. Fungicides are employed to reduce the damage from, especially, fusarium head blight and leaf rust. Historically, it was generally swathed (windrowed) prior to threshing, but straight-cutting has now become widespread. A "good" HRS yield now exceeds 60 bushels per acre.

After on-farm storage (<14.5% moisture), it is generally delivered/sold to a high through-put elevator (Chapter 6, Section 6.2.1) or, in some cases, a feed mill. Poorer grades also sometimes get sold directly to beef feedlots (Chapter 6, Section 6.2.2).

Wheat has several uses, including flour for bread and pasta. As such, it is the principal carbohydrate of the Western world. But it is also utilized as a feed for livestock (particularly poultry) and as a basic ingredient in some beers, vodka, and biofuel. Still, its magic is that it contains gluten protein, which allows bread dough to rise and produce a light bread. Canada western red spring is recognized worldwide as a premium wheat for bread production. Importers of Canadian wheat often blend it with weaker (and less expensive) wheats before using it for bread-making. Durum wheat is prized by pasta manufacturers in both Canada and abroad.

Since the demise of the Canadian Wheat Board (in 2012), Prairie farmers can sell their wheat anywhere and anytime. Ontario still retains the Ontario Wheat Board. Domestically, flour mills (which purchase their wheat from intermediaries) are generally located in Ontario and Quebec. (See Chapter 6.)

4.1.3   OTHER CEREALS

The other cereals of particular importance are **barley** and **oats**. Rye and triticale are yet other cereals but because of their relatively small planted area, they are not discussed further.

Barley:[26]

Although its origin is not known, barley has been cultivated for thousands of years in the Middle East, Far East, North Africa, and east-central Africa. It is thought to be the first domesticated grain to be grown by humans. In Canada, it is the third most important cereal crop grown, after wheat and corn (Table 4.1), and production ranks 7th in the world (Table 4.2).

There are two types of barley: 2-row and 6-row, and it is planted and harvested in much the same manner as wheat (Section 4.1.2). It prefers a well-drained loam or clay loam and is strongly negatively affected by acid soils (< 6.0 pH). Most barley is now planted with a large no-till drill in mid-May and then harvested in early-September. Manufactured fertilizer is generally applied at the time of the seeding operation at about one-half the rate utilized for wheat. Thereafter, it is typically sprayed with an herbicide and, increasingly, with a subsequent fungicide. Fungicides are employed to reduce the damage from, especially, fusarium head wilt and leaf blight, smut, and leaf rust. It is then generally swathed (windrowed) prior to threshing. A "good" barley yield now exceeds 90 bushels per acre.

Barley is a relatively short-season crop (80–90 days), is fairly hardy, and can tolerate more saline soils than wheat or oilseeds. Negative characteristics are that it has a weak (hollow, except at nodes) straw, which often allows for lodging (plant collapse), quality deterioration if subject to too much rain prior to harvest, and pre-harvest head shattering and loss.

At the same time, barley is a versatile crop with numerous end-uses: a) livestock feed (grain and silage); b) malt; and c) food. It is used as a feed for livestock, including cattle, hogs, poultry, and sheep, and is the principal grain used by Canadian beef feedlots. Malt barley varieties are processed into malt for beer or malt-enriched food products while pearled barley can be used in soups, flour for flatbread, or as a porridge.

Canada produces about 8 million tonnes (370 M bushels) of barley annually on about 5.5 M acres, of which about one-third is utilized by maltsters (domestic and international, Section 6.2.1) while most of the remaining two-thirds is utilized by domestic livestock operations. About one-half of the grain is marketed through commercial channels. Most Canadian barley is grown in Alberta (50%) and Saskatchewan (40%). With numerous beef feedlots in central and southern Alberta, up to 80 percent of total annual production in Alberta is used as a feed.

Canadian barley production has dropped about 35% in the last ten years and this can largely be traced to the following:

- Competition from other feed ingredients in the domestic livestock feed market—distillers dried grain, corn, feed wheat, etc.
- Growth of the malt market has been limited by a flattening of the domestic beer market

and more robust competition in international markets because of lower-priced ingredients and recipes used by some brewing segments and beer styles

- Selection and storage risks for malting barley do not exist for most other grains
- Persistent genetic and agronomic constraints (i.e., disease, lodging, etc.)
- On-farm profitability is typically relatively low, especially for feed barley (a price of say, $4.00/bus. for feed barley vs. $6/bus. for malt barley). If a barley is not selected for malting, the feed barley market is the default destination. In recent years, the profitability of barley has also typically been less than that of CPS wheat and field peas.

Oats:[27]

Oats were first utilized as both a grain for human consumption and as a forage for animals in Western Europe as early as 1000 AD. In Canada today, it is grown on about 2.5 M acres, mostly in the northern prairies, and Canadian production ranks 2nd in the world (Table 4.2).

Oats, like wheat and barley, is generally planted with a large no-till drill in mid-May and then harvested in late September. Oats usually require 100–103 days to mature. Manufactured fertilizer is usually applied at the time of seeding and, thereafter, it is typically sprayed with a broadleaf herbicide. (Wild oat herbicides will also kill tame oats.) It is then generally swathed (windrowed) prior to combining. To minimize shattering, swathing should start when the kernel moisture level is about 35 percent. A "good" dryland oat yield now exceeds 120 bushels per acre.

Oats prefer higher moisture conditions than most other cereals. It is less drought-resistant than wheat, barley, and rye. It is susceptible to damage by hot, dry weather, especially in the reproductive phase. Thus, oat yields are generally higher in the more northerly dark brown and black soil zones. If mature oats are left standing too long, weather may cause the stems to break down and thus lodge. Heavy rain, wind, or hail will cause grain shattering.

There are basically three types of oats to serve two distinct markets: a) milling oats for human consumption, as well as (especially) horses; and b) feed and hull-less oats for livestock.

Milling oats are common in cereals (e.g., Quaker Oats), soups, cookies, granola bars, and pet foods. For horses, this good-quality oat is also called a **performance oat** or a **pony oat**.

As a livestock feed, oats can be utilized in a variety of ways: grain, green feed, silage, or pasture.

For green feed and silage, oats are cut at the late milk stage. When used as a forage, oats are cut and conditioned in the same manner as regular hay—wilt the crop to 40–60% moisture and then bale it at about a 15 percent moisture content. Pasture (or forage) oats are easy to seed, establish quickly, and can provide pasture in 8–10 weeks after seeding or be swath-grazed later.

The world production of oats has been slowly declining because of its yield and comparative feed value relative to barley and corn as an animal feed. At the same time, there is an increasing demand for oat-based food products . which have a health benefit. Their high beta-glucan content can decrease blood levels of LDL cholesterol and, thus, reduce the risk of cardiovascular diseases. Oat flour is also being studied as a replacement for individuals who suffer from celiac disease.

4.1.4   CORN (MAIZE)

Corn evolved in the Americas and was initially grown by Native Americans. Now, especially in Latin America and the USA, grain corn is king. Tortillas in Latin America (and, increasingly, in the USA) are piled high on every plate, just as rice is piled high on every plate in Asia.

Even in Canada, in terms of farm cash receipts, corn is the fourth most important crop, after canola, wheat, and soybeans (Table 3.6). Corn is grown extensively in Ontario and Quebec, with some short-season varieties also becoming increasingly important on the prairies, especially in Manitoba. Ontario accounts for about 60% of the seeded area, Quebec 30%, and Manitoba 10 percent. Some corn is also grown in Alberta, especially in the irrigated area of southern Alberta. Corn still ranks as the second most important crop in Ontario in terms of both production and farm cash receipts. Total Canadian production ranks 11th in the world (Table 4.2).

Over the years, farmers and corn breeders have developed multiple varieties suited to particular uses and adapted to distinct environments. Most seed is now GM, already 70–75% in 2011, and estimated at >90 percent today. (See Chapter 13, Section 13.3.3.) In Canada, three broad types of corn dominate: **corn for grain**, **corn for silage**, and **sweet corn**. There is no popcorn. In 2011, the number of Canadian farmers producing each was:

>   Corn for Grain . . . . 23,472 farmers
>   Corn for Silage 13,184
>   Sweet Corn 2,997

About 80% of the corn area is harvested as grain corn, and almost all of the remainder as silage.

Corn is typically planted in May and, depending upon its end-use, harvested during September–November. Because corn is planted in discrete rows and row-spacing, a special corn planter is required. To obtain a good yield, corn requires more nutrient additions from manufactured fertilizers, manure, and other soil amendments compared to other crops to prevent soil depletion. Corn grown in rotation with a nitrogen-fixing crop (often soybeans), cereal, or hay improves soil fertility and structure, reduces fertilizer requirements, helps with weed control, and also breaks disease and pest cycles. Harvesting grain corn requires a special straight-cut header and yields are usually in the 150—200 bushels/acre range.

Grain Corn:

Most grain corn is utilized on-farm, an important part of the ration for many dairy and beef herds. This is ground or cracked before feeding. Some grain corn is also sold to manufacturers to produce corn grits, hominy, corn meal, or masa (utilized for tortillas). Still other grain corn is first processed into bran, gluten, starch ,and germ before being converted into flaked corn cereal, corn starch, etc. In the United States, considerable grain corn is also processed into **biofuel**.

Silage Corn:

Corn for silage is harvested as a whole plant—the cob, grain, stem, and leaves—when it is still green. This requires the use of a very powerful silage cutter. Silage is then stored in bunkers or silos on dairy or beef farms/feedlots for use over the winter and the following spring/summer before another crop can be harvested. No other crop produces more nutritional tonnage/acre and a certain amount of fermentation enhances its nutritional value even more. Corn can also be left standing to allow cattle to forage on it over the winter.

Sweet Corn:

Often grown in your own garden or purchased on a road-side, sweet corn is what most consumers devour as fresh corn-on-the-cob. It is a sugar-enriched corn, which maintains these sugar levels over time. Sweet corn is also sold directly to canning plants, (e.g. Leamington [Ontario] and Lethbridge [Alberta])[28]. In Alberta, "Taber corn," produced under irrigation, is also a very well-known sweet corn.

4.1.5   SOYBEANS

The soybean is a dietary staple which originated in China and, later, spread throughout Asia. It reached North America in about 1765. It only became a commercial oilseed crop in Canada when a crushing facility was built in southern Ontario in the 1920s. Soybeans continued to largely be restricted to southern Ontario until the mid-1970s because the plant requires a relatively warm climate. But plant breeders have now gradually developed different varieties of soybeans for different uses, shorter growing seasons, and colder climates. Most (>90%) soybeans are now GM. (See Chapter 13, Section 13.3.3.)

An important characteristic of soybeans, from a farmer's perspective, is that it is a legume and like all legumes, "fixes" its own nitrogen requirements from the air. This means that soybeans require little in the way of manufactured nitrogen fertilizer.[29]

Soybeans are now the fourth most important field crop in Canada (Table 3.6). Production is still concentrated in Ontario, where it is now the top crop in terms of farm cash receipts (surpassing corn). But it is also gradually creeping westward, first to Manitoba and, more gradually, into Saskatchewan and Alberta. Total Canadian production ranks 7th in the world (Table 4.2).

Aside from their relative on-farm profitability, soybeans are attractive because they are extremely versatile. The beans can be converted into numerous food and industrial products, as well as an animal feed (Figure 4.2). Particularly noteworthy is its growing use as a food additive, meat alternative, or meat extender; a protein supplement. (The bean is 18% oil and 35% protein.) In the USA, soybean oil is also utilized extensively in the biofuel industry.

Soybean oil dominants the international oilseeds complex, effectively competing with Canadian canola oil, palm oil, coconut oil (mostly grown in Asia), and others. Soybean oil accounts for about 20 percent of global vegetable oil production, while high-protein soybean meal accounts for over 60 percent of world vegetable and animal meal production (FAO). Soybean meal is the most valuable component obtained from processing soybeans, ranging from 50–75 percent of its value. Soybean oil generally makes a smaller contribution to soybean value, as it constitutes just 18–19% of soybean weight. Canola prices in Canada generally track US soybean prices. About two-thirds of the soybeans grown in Canada are exported, either as raw soybeans or as processed products. (For details, see Chapter 5.)

Figure 4.2 Soybean Uses

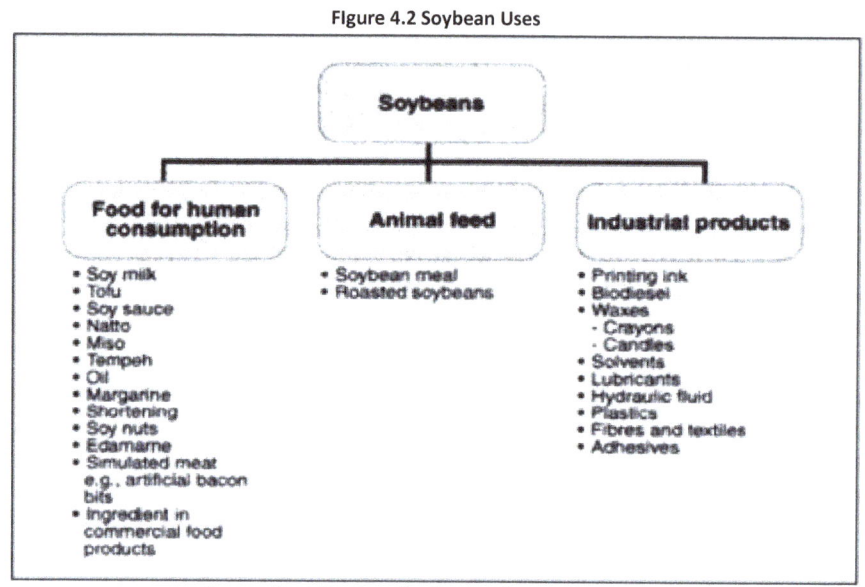

```
                        Soybeans
        ┌──────────────────┼──────────────────┐
   Food for human      Animal feed      Industrial products
   consumption
   • Soy milk          • Soybean meal    • Printing ink
   • Tofu              • Roasted soybeans • Biodiesel
   • Soy sauce                            • Waxes
   • Natto                                   - Crayons
   • Miso                                    - Candles
   • Tempeh                               • Solvents
   • Oil                                  • Lubricants
   • Margarine                            • Hydraulic fluid
   • Shortening                           • Plastics
   • Soy nuts                             • Fibres and textiles
   • Edamame                              • Adhesives
   • Simulated meat
     e.g., artificial bacon
     bits
   • Ingredient in
     commercial food
     products
```

Source: Statistics Canada, *Census of Agriculture, 2007*

4.1.6   PULSES

The term "pulse" is reserved for crops harvested solely for the dry seed. This excludes green beans and green peas, which are considered vegetable crops.

A pulse is an edible seed harvested from the pod of a variety of annual leguminous plants. They are cool-season crops and include hundreds of varieties of dry pea, dry bean, chickpea, lentil, lupin, mung bean, faba bean, cow pea, and numerous others. Pulses are especially attractive because they are high in protein and fiber, and low in fat. As such, they are increasingly popular as a protein substitute for North American consumers. Pulses have double the amount of protein compared to other cereal crops (including rice) and they also have a low glycemic index to help manage blood sugar levels.

Since the 1980s, pulse crops have become increasingly important in Canada, largely due to their highly successful development in Saskatchewan, which now has about 75% of the total pulse area in Canada. Alberta led development of the field pea. Prairie soils and climate favor pulse crops. The principal pulse crops in Canada are dry peas, chick peas, lentils, and dry beans (Figure 4.3). The development and expansion of the pulse industry has been closely tied to its profitability, and to research into new varieties that are taller, resist lodging, or have a shorter growing season, in conjunction with the growth of local processing facilities.

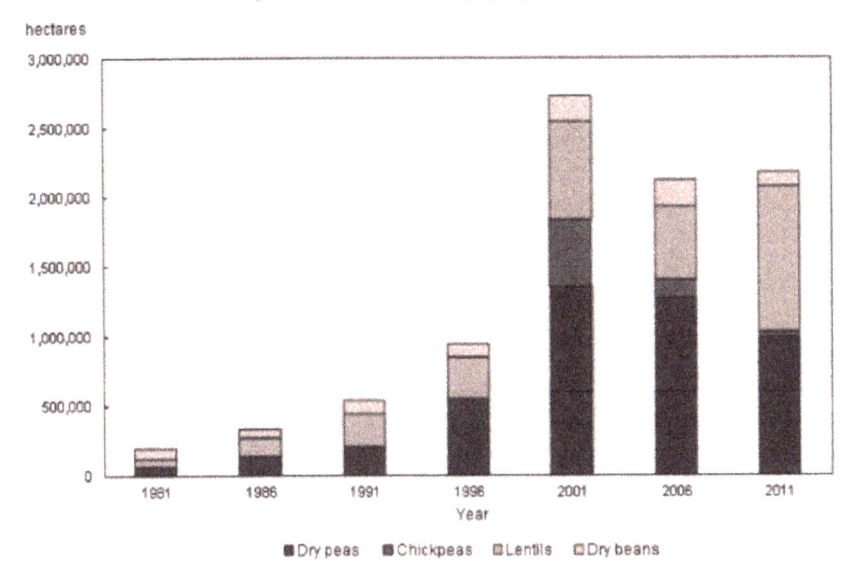

Figure 4.3 Canadian Pulses, by Type, 1991 - 2011

Source: Statistics Canada, *Census of Agriculture*, 1981-2011

The favored dry peas are yellow and green, while lentils are usually classified as yellow, red, or green. Chickpeas are usually ***kabuli*** or ***desi,*** while dry beans are usually white or colored. Bean varieties include kidney beans, navy beans, Great Northern beans, pinto beans, and black beans.

Zero-till or minimum-till seeding and the harvesting of pulses, with large equipment, is now commonplace on large dryland farms on the prairies. Aside from their profitability, growing pulses, with their nitrogen-fixing capability, is also desired to reduce manufactured nitrogen requirements in the current crop as well as the follow-up crop. Growing pulses in rotation also helps to disrupt disease and insect cycles.

In 2017, the largest area of white and colored dry beans was in Manitoba (41%), Ontario (41%), and Alberta (15%), with the balance in Quebec.[30] Canada produces more than 1/3rd of global lentil production and about 20% of global dry pea production.

The main domestic use of pulses is for livestock feed—particularly low-quality production—with dry peas for hogs being the most common. Relatively small quantities are consumed by Canadian households; much different than countries where pulses are a dietary staple. The export of dry peas, lentils, and chickpeas typically accounts for about 40 percent of annual production (Chapter 5).

Primary processing of pulses includes cleaning and quality sorting. Additionally, fractionation can separate pulses into concentrates, or proteins, starches, and fibers used in various food and industrial applications. (See Chapter 6.)

Further expansion of pulse production in Canada is anticipated in response to:

- Rapidly growing national and international market for plant protein
- Recognized health benefits of lentils, beans, and other pulses
- Crop's lower environmental impact, regarding lower CO_2 emissions
- General compatibility with tenets of regenerative/sustainable/diversified cropping practices.

4.1.7 OTHER CROPS

One other crop of particular interest is **sugar beets**. In Alberta, the current contract with processors (Lantic Sugar Inc.) called for 28,000 acres of (irrigated) sugar beets to be planted and a minimum of 21,000 aces in 2019 and 2020. Sugar beet growers and the Taber sugar factory produce more than 100,000 tonnes of sugar annually, which is sold in both the domestic and international markets. Products include refined sugar, molasses, icing sugar, and livestock feed.

4.2 BEEF PRODUCTION

Beef production is the single largest agricultural sub-sector in Canada, currently accounting for over 40% of all farm cash receipts in Canada (Table 3.6). There are about 36,000 beef producers; about 19% of all farmers (Table 3.4). The beef cattle herd exceeds 11 million, while natural and improved pasture land makes up about 30 percent of all farm land in Canada. Canada now processes approximately 1.3 million tonnes of beef annually.[31]

Beef cattle production is concentrated in Alberta (41%), Saskatchewan (30%), Manitoba (11%), Ontario (7%), and BC (5%). Correspondingly, the average beef herd per farm is 255 in Alberta, 191 in Saskatchewan, 167 in Manitoba, 74 in Ontario, and 117 in BC.

Beef production has two fairly distinct components: (1) cow-calf operations; and (2) feeder (slaughter cattle) operations.

4.2.1 COW-CALF OPERATIONS

The cow-calf producer maintains the feeder breeding stock, raising cattle from gestation to feeder weight, now usually 600–700 pounds. Spring calving, summer forage, weaning, and fall marketing is the general production cycle. These cow-calf operations are often complementary to related crop production and generally they depend fairly heavily on an extensive land base, particularly if the land is marginal with a relatively low opportunity cost. Scavenging on crop stubble before and after the crop season is also commonplace.

These cow-calf herds, are handled and fed in a

way not too different from a generation ago. But such things as mechanized forage handling, automatic watering systems, electronic monitoring, etc., have made it possible for a farmer or rancher to now look after about three times as many cows than he/she could 60 years ago (averaging 169 in 2016 versus 56 in 1961). At the same time, there are still numerous relatively small cattle farms and, nationally, 39 percent of beef farms still have less than 47 cows. However, periodic droughts, impending retirement, labor requirements, and variable market conditions increasingly discourage maintaining a cow-calf operation.

4.2.2 FEEDER (SLAUGHTER CATTLE) OPERATIONS

When the calves are weaned, some remain with their owners, but large numbers are marketed to other cattlemen or feedlots to grow and finish.

Calves can be directly put into feedlots and fed high-energy rations to finish out for slaughter at 13–16 months of age at 1200–1300 pounds (550–600 kg) live weight while others can be grown and finished entirely on pastures and harvested forages. More frequently, however, calves are fed on forages, grown more slowly, and then finished in feedlots at 18–24 months with high-energy feeding periods. This two-stage process is referred to as ***backgrounding*** and ***finishing***. Backgrounding is often done by the same farmers/ranchers who also have a breeding herd.

As of 2017, there were 158 feedlots on the Canadian Cattlemen's Association's CANFAX list with a minimum capacity of 1000 head, and some of these are relatively large. Feedlots with capacities of 30,000-40,000 head are not uncommon. Canadian feedlots are concentrated in the irrigated Lethbridge area of southern Alberta. Alberta has bunk space for an estimated 1,328,000 head, 93% of the western Canadian total. Major slaughter plants exist at Brooks (JBS) and High River (Cargill). (Also see Chapter 6.)

4.3 PORK PRODUCTION

Over time, farming operations have generally become more intensive and specialized and the hog sector is no exception to this trend. As of 2016, there were still about 8,402 farms reporting some hog production; about 3,305 specialized hog farms (Table 3.4 and Figure 4.4). Together, in 2016, they produced about 7% of all farm cash receipts in Canadian agriculture, the fourth largest agricultural sector.

Virtually all of commercial hog production in Canada now takes place in a controlled environment where, at all times of the year, the animals are kept in buildings specialized in the farrowing, growing, and finishing stages of raising market hogs.

Figure 4.4 Number of Hogs and Hog Farms, 1921 - 2016

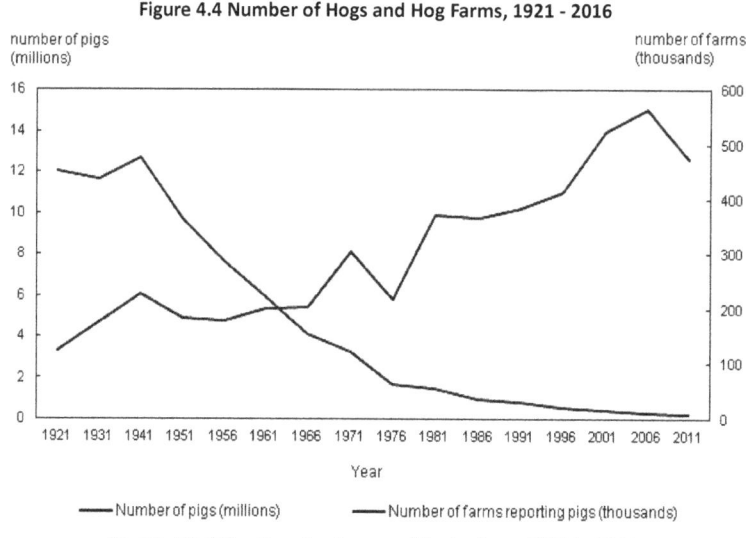

Source: Statistics Canada, *Census of Agriculture*, 1921 to 2011

There are basically three types of commercial hog operation, farrowing nurseries, finishing, and farrow-finish, as profiled in Table 4.3.

Table 4.3 Types of Hog Operations, Canada, 2011

Type	Number of Farms		Number of Hogs	
	Number	%	Number (M)	%
Farrowing Nurseries	2150	29%	1.9	15%
Finishing	3052	41%	3.5	28%
Farrow - Finish	2169	29%	7.3	57%
TOTAL	7371	100%	12.7	100%

Source: Statistics Canada, *Census of Agriculture*, 2011

Farrowing nurseries (about 30% of all hog farms) focus exclusively on the farrowing and raising of piglets. Finishing operations are only involved in finishing piglets to bring them up to slaughter weight and these operations represents the largest proportion of all hog farms (about 40%). Farrow-finish operations (also about 30% of all hog farms) raise hogs throughout all stages of development. Most of the hogs, however, are raised by farrow-to-finish operations (about 60%) and most of the larger commercial hog operations (about 3,300) are either farrow-to finish, or just finishing operations (Table 4.3).

This translates into the marketing of about 27 million head per year, or (at 220 lb or 100 kg per head and net of live exports), about 1.5 million tonnes of pork per year.

These operations are fairly heavily concentrated in Quebec (34%), Ontario (25%), and Manitoba (22%) (Figure 4.5). Annual fluctuations in hog numbers are usually attributed to either relatively high feed prices or the outbreak of a particular disease. Since feed is the largest expense item in hog production, and since hogs are monogastric animals primarily fed on meal composed of grains (typically barley or corn), any feed price increase has a particularly strong impact on the profitability of hog farms.

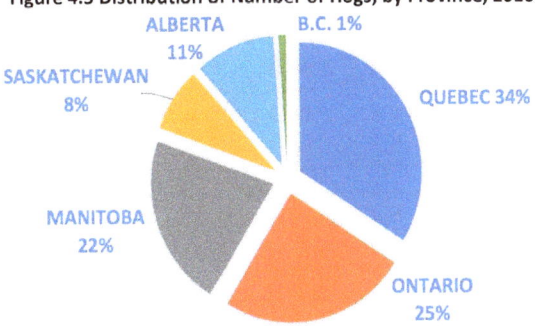

Figure 4.5 Distribution of Number of Hogs, by Province, 2016

- QUEBEC 34%
- ONTARIO 25%
- MANITOBA 22%
- ALBERTA 11%
- SASKATCHEWAN 8%
- B.C. 1%

Source: Basic data from Statistics Canada

Hogs are usually marketed directly to hog slaughter plants located where hog production is most concentrated: Quebec (13), Ontario (6), Manitoba-Saskatchewan (4), and Alberta-BC (6) (Chapter 6).

About one-third of the pork produced in Canada is consumed domestically. Exports account for the other two-thirds. In addition, the industry typically exports about 6 million live hogs/year to the United States (Chapter 5).

4.4 DAIRY INDUSTRY[32]

Dairy farmers are the second largest (after beef) livestock-based industry in Canada, typically generating about 11% of total farm cash receipts (Table 3.6). Dairy farmers total about 10,600 active shippers, about 5% of all farmers in Canada (Table 4.4).

There are almost a million dairy cows in Canada, mostly Holstein (93%). This translates into an average of about 92 cows per farm, but many commercial dairy farmers are much larger. Milk production now totals about 8.5 billion liters/year; about 1/3rd destined for the fluid milk market and 2/3rds for industrial processing (Table 4.4). About 75% of the barns are tie-stall, and 25% free stall. About 634 robotic milking systems are already (2016) being utilized.

Table 4.4 Profile of Dairy Farmers, Canada, 2016-2018

Category	Quantity
Number of Farmers	12895
2018 Farmers with Milk Shipments	10593
2018 Cows	969700
Average Cows/Farm	92
2018 Heifers (Replacements)	435500
Milk Production (billion litres)	8.47
Approx. litres/cow (litres)	8963
Fluid Milk (billion litres)	2.77 (33%)
Industrial Milk (billion litres)	5.68 (67%)
Organic Milk Production (billion litres)	0.11

Source: Canadian Dairy Information Centre, 2019

Compared to the United States, Canada still has a relatively large number of dairy farmers with considerably fewer cows per average farm. The average size of a US dairy is 900 cows (about ten times Canada) and a mere 3,300 US dairy farms are responsible for over 50 percent of all US dairy production for a domestic market, also ten times as large as the Canadian market. These mega-dairy farms average more than 1,400 dairy cows per farm.[33] According to US research, larger and more modern dairy farms can produce milk for almost one-third less cost.[34] In Canada, can quota-leasing and giant milking parlors with robotic milking systems be far behind?[35]

Regionally, the Canadian dairy industry is very heavily concentrated in Quebec and Ontario. Quebec has almost one-half of all the dairy cows in Canada, only averages about 70 cows/farm, and generates only about 36% of total national dairy receipts. In contrast, Ontario has about 1/3rd of all the dairy cows in Canada but averages about 90 cows/farm and generates a corresponding 32 percent of total national dairy receipts (Figures 4.6 and 4.7).

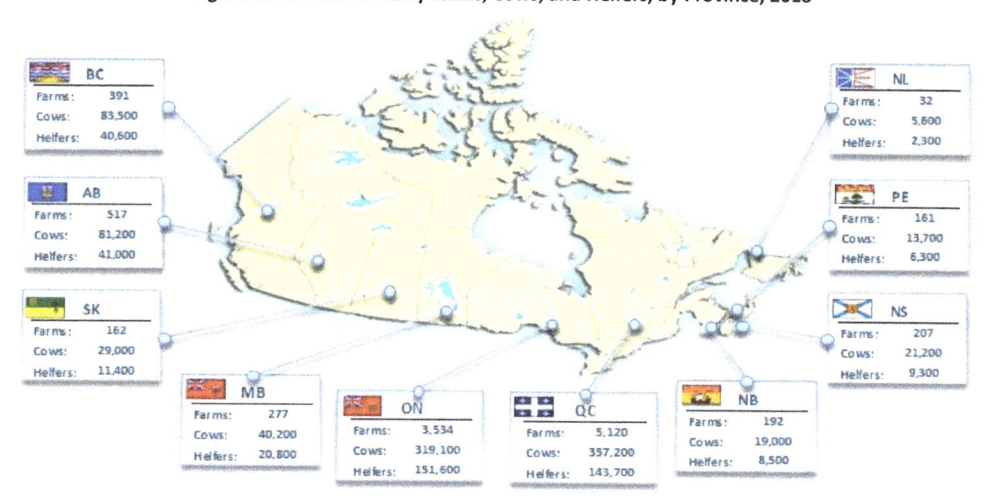

Figure 4.6 Number of Dairy Farms, Cows, and Heifers, by Province, 2018

Province	Farms	Cows	Heifers
BC	391	83,500	40,600
AB	517	81,200	41,000
SK	162	29,000	11,400
MB	277	40,200	20,800
ON	3,534	319,100	151,600
QC	5,120	357,200	143,700
NB	192	19,000	8,500
NS	207	21,200	9,300
PE	161	13,700	6,300
NL	32	5,600	2,300

Source: Canadian Dairy Information Centre, 2019

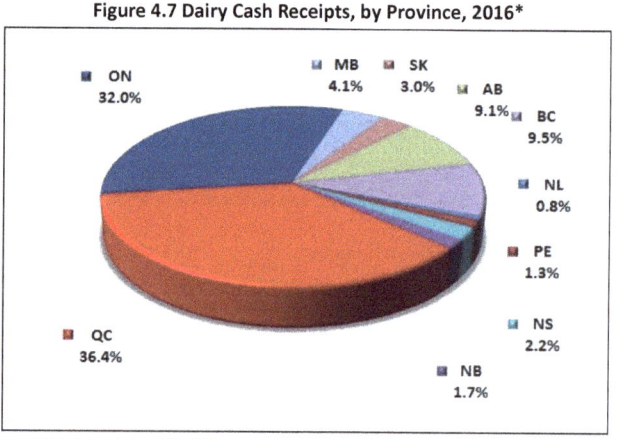

Figure 4.7 Dairy Cash Receipts, by Province, 2016*

- ON 32.0%
- QC 36.4%
- MB 4.1%
- SK 3.0%
- AB 9.1%
- BC 9.5%
- NL 0.8%
- PE 1.3%
- NS 2.2%
- NB 1.7%

* Total receipts = 6.2 billion.

Source: Canadian Dairy Information Centre

Virtually every aspect of the dairy industry is managed by the Canadian Dairy Commission (CDC, a Crown Corporation established in 1966) and the Canadian Milk Supply Management Committee (CMSMC), which is made up of provincial dairy boards and provincial officials. The CDC chairs the CMSMC, but its authority is only over industrial milk (processed, Classes 2 through 5), historically deemed to be under federal jurisdiction. Fluid milk prices and volumes (Class 1) are set by the individual provinces within the "western pool" and the P5 (eastern provinces), with coordination between the two pool areas.

The CDC and the CMSMC manage the industry principally through three market mechanisms: a) production controls through milk quotas to equate domestic supply to domestic demand; b) guaranteed minimum cost-plus pricing, and c) import controls. The CMSMC applies the terms of the National Milk Marketing Plan (NMMP) to establish the provincial shares of the quota. The national quota is allocated to provinces approximately in proportion to their respective populations and then these provincial quotas are sold to producers within each province. The quotas for both fluid milk and industrial milk are based on the butterfat content. Farmer-owned quotas are marketable at their current market price, which effectively restricts entry. Resulting production and sales (Figure 4.9) almost exactly mirror provincial quota allocations.

Production Controls:

To prevent surpluses and shortages, commercial dairy farmers must hold quota, a kind of license authorizing them to produce a given volume. Quotas in the diary industry are transacted in terms of the daily kilograms of butterfat produced, the equivalent of one cow's production. Each year the CMSMC sets the national industrial milk production target. This is adjusted to reflect changes in demand for industrial milk and dairy products, such as butter and cheese, as measured in terms of butterfat.

Quota values greatly affect existing dairy farm values and deter new entrants. The quota value for a single cow now ranges from $25,000 in Ontario and Quebec (legally capped) to $42,500 in BC.[36] It is estimated that dairy quota values now total at least $25 billion, about $2.4 million for an average dairy farm with 92 cows[37] (Figure 4.8). Two-thirds of this total quota value is located in Ontario and Quebec. This does not include investments in other assets, such as other livestock, land, buildings and machinery. For the larger dairy farms (and there are many), this capital gain (or "entry fee") is proportionately greater.

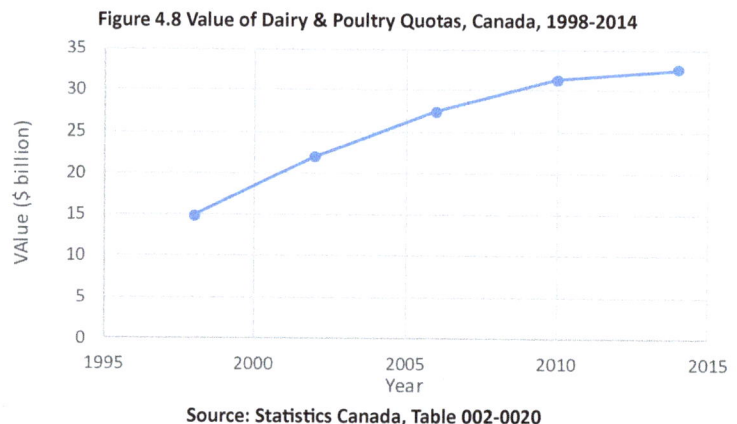

Figure 4.8 Value of Dairy & Poultry Quotas, Canada, 1998-2014

Source: Statistics Canada, Table 002-0020

Minimum Prices:

In addition to production controls, dairy farmers are guaranteed a minimum price for their products. Through their provincial marketing boards, dairy farmers collectively negotiate minimum farm-gate prices with processors. This minimum price is based on production costs and market conditions, such as consumer demand, inventory available on the market, and the price of competing products.

Established farm prices translate into consumer prices further down the food chain and it is well documented that these inflated prices likely result in an implicit consumer-to-producer financial transfer, in effect a hidden tax on consumers and a hidden subsidy to producers.[38] During the period 2002–2013, this averaged about $523 million (or ½ billion dollars) per annum.[39] This translates into an average "subsidy" of about $40,000 per dairy farm/year. but only a relatively modest "tax" on consumers of approximately $37 per household per year.

The annual per-capita consumption of milk and milk products in Canada is approximately as follows:

Fluid Milk	69.5 L
Cheese	13.4 kg
Cream	10.1 L
Yogurt	10.6 L
Ice Cream	4.3 L
Butter	3.2 kg

Import Controls:

Finally, in addition to relying heavily on production controls and price-setting, the dairy supply management system also relies on import controls.

In accordance with various trade agreements (Chapter 5), Canada restricts imports by setting tariff-rate quotas (TRQs). This means that it grants trading partners a "minimum level of access" to imports and imposes a high customs tariff on imports over a certain quantity to prevent foreign products from flooding the Canadian market. For example, the import quota for yogurt is currently set at 332,000 kg.[40] Imports within these quotas are not subject to customs tariffs or, if they are, the tariffs are low. Imports in excess of these amounts, however, are subject to very high tariffs, (e.g., 300% in the case of butter).

4.5 POULTRY & EGGS

The poultry industry is essentially made up of the broiler, turkey, and egg sub-sectors and together they contribute about 7% to total farm cash receipts in Canada (eggs = 2%) (see Table 3.6). The number of commercial producers, however, only makes up about 3 percent of all Canadian farmers (Table 3.4). The location of these approximately 4,430 producers generally mirrors population and consumption patterns across Canada (Table 4.5).

Table 4.5 Number of Commercial Chicken, Egg, and Turkey Producers, by Region, Canada, 2016

Region	Broilers	Turkeys	Eggs	TOTAL Number	Percent
MARITIMES	138	38	50	226	5%
QUEBEC	748	152	117	1017	23%
ONTARIO	1177	179	345	1701	38%
PRAIRIES	429	115	409	953	22%
B.C.	325	67	137	529	12%
TOTAL	2817	551	1062	4430	100%

Note: The total (commercial & home-use) numbers are much larger: broilers 7249; turkeys 2690; eggs 18664. (**Census, 2016**, #004-0225.)

Source: Chicken Farmers of Canada, Turkey Farmers of Canada, and Egg Farmers of Canada

In 2016, Canada produced 1.1 billion kilograms of chicken, 60% of which was produced in Quebec and Ontario.

Broilers are raised in confinement and the space required is about 0.8 square feet per bird. A typical broiler operation has over 30,000 birds. They generally grow to about 2.2 kg in 38 days with a feed conversion of 1.8 kg feed/kg meat. Broiler birds are produced from crosses that originally involved White Plymouth Rock and/or New Hampshire on the mother's side and Cornish on the father's side. The commercial broiler is white-feathered, fast growing, vigorous, and well-fleshed. Most broilers are marketed in approximately 35–39 days, leaving the remainder of the 8-week cycle for clean-ups and preparation for placement of the next flock. Revenue is generated over 6.5 cycles per year.

Turkeys are also raised in confinement and require about 4 square feet per bird. A typical turkey operation has about 15,000 birds. Various crossbreds have been developed, which are generally smaller and less broody. When marketed as broilers, these birds are usually slaughtered at 10–12 weeks of age.

Egg-laying hens are usually confined to an individual cage and are generally a White Leghorn. A typical laying egg operation has 15,000–20,000 hens. The commercial white egg layer is white-feathered and weighs about 1.8 kg. Brown egg layers (produced from hybrids) are colored and usually weigh slightly more. Laying hens produce 300 to 340 eggs during the 12 to 13 months they lay eggs.

The Farm Products Marketing Agencies Act (1972), in conjunction with provincial legislation, enabled producers of poultry products to establish national marketing agencies to manage the supplies of poultry products being marketed by Canadian producers. The Canadian Egg Marketing Agency (CEMA), the Canadian Chicken Marketing Agency (CCMA), and the Canadian Turkey Marketing Agency (CTMA) were all established in the 1970s.

Like the dairy industry (Section 4.4), the respective organizations (CEMA, CCMA, and CTMA) principally manage the industry through three mechanisms: a) chicken, turkey, and egg production quotas to equate domestic supply to domestic demand, b) minimum on-farm cost-plus pricing, and c) import controls.

Production Controls:

Similar to the dairy industry, commercial poultry and egg producers must hold quota, a kind of license authorizing them to produce a given volume. For poultry, quota is sold by the number of units produced or square meters of floor space. In Quebec, for example, one square meter of chicken production is equivalent to the production of 7 to 10 birds, while in Manitoba one production unit is equivalent to the production of one chicken.

Some jurisdictions allocate quota on the basis of animal units. For broiler chickens, 200 broiler chickens equals one animal unit. For layers, the conversion is 100 birds per animal unit, while 50 turkeys are equal to one animal unit.

For eggs, chickens, and turkeys, a national council sets national production quota allocations, as well as producer or marketing levies. The respective national agencies (CEMA, CCMA, and CTMA) set both national and provincial quotas which, in turn, are allocated by the provincial boards to existing or new producers in their respective provinces. The national egg and turkey agencies also have the power to purchase and remove surplus supplies.

Generally, quota units can also be purchased or leased from another licensed producer. Quota utilization percentages change from cycle to cycle in order to respond to market demand. At the same time, quota values greatly affect existing farm values and deter new entrants. Our best estimates of the respective quota values are tabulated in Table 4.6.[41]

Table 4.6 Estimated Quota Values in the Poultry Industry

Industry	Total $ Billion	Average Farm Value $ Million
Broilers	3.7	1.3
Turkeys	1.5	2.7
Eggs	6.0	5.6
TOTAL	11.2	

This is a farm capital value aside from the cost of land, buildings, equipment, and so on. It is either an accrued value (much of it a capital gain) for existing producers or an "entry fee" imposed on new producers.

The combined poultry/egg quota value is approximately $11 billion, while the dairy quota value is approximately $25 billion (Section 4.4), together amounting to about $36 billion (Figure 4.8). Estimates by other researchers vary.

Minimum Prices:

The national egg agency (CEMA) has a cost-of-production formula to set prices, but neither the national chicken nor turkey agencies set prices. This is done by provincial chicken and turkey boards. A national council sets the interprovincial price of the regulated commodity.

Through their provincial marketing boards, commercial chicken and turkey producers collectively negotiate minimum farm-gate prices with processors. This minimum price is based on production costs and market conditions, such as consumer demand, inventory available on the market, and the price of competing products.

Import Controls:

The international trade in poultry and eggs is also controlled by utilizing tariff-rate quotas (TRQs), which set a limit on the amount of various poultry products allowed into Canada with little or no tariff. For example, the import quota for chicken is currently 39,900,000 kg or 7.5 percent of domestic production, whichever is greater.[42] Imports within these quotas are not subject to customs tariffs or, if they are, the tariffs are low. But imports in excess of these amounts are subject to very high tariffs (e.g. 240%, in the case of chicken). (For further details, see Chapter 5.)

4.6   OTHER LIVESTOCK & POULTRY

Other livestock and poultry here includes sheep, goats, horses, other fur-bearing animals,[43] ducks, geese, and other types of poultry.

In 2016, there were an estimated 2,862 farms producing **duck**, primarily in Ontario and Quebec (Table 4.7). The top ten farms reporting ducks in Canada accounted for two-thirds of all reported ducks. The total per-capita consumption of ducks, geese, and related fowl is about 2 kilograms/year, miniscule compared to chicken (31 kg) or even turkey (4 kg). Production exceeding domestic demand is primarily exported to the United States, Japan, and Mexico.

Additionally, there are approximately 1 million **sheep** in Canada on about 11,000 farms, generally concentrated in Ontario, Quebec, and Alberta.[44] The main market for sheep in Canada is for mutton and lamb. The per-capita consumption of mutton and lamb in Canada is about 1 kilogram/annum, miniscule compared to that of chicken, beef, pork, and fish (Figure 4.9). About 60% of domestic requirements are imported from Australia and New Zealand. Wool is also marketed and some milk is utilized to make specialized cheeses.

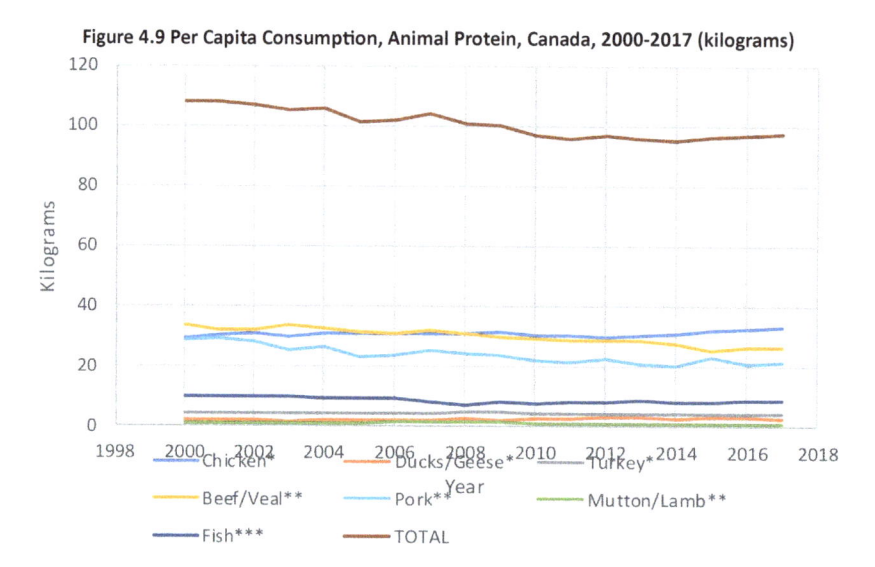

Figure 4.9 Per Capita Consumption, Animal Protein, Canada, 2000-2017 (kilograms)

Source: AAFC, *Canadian Agriculture at a Glance*, Ottawa, 2017

Three-quarters of the global population also eat **goat** meat; about 10% of worldwide meat consumption and 60% of all red meat. In Canada, however, production (on about 867 goat farms) and consumption of goat meat and milk is very limited. Still, in southern Ontario, the goat milk industry is reportedly one of the fastest growing livestock industries in the region.[45] Goat milk is typically collected by milk brokers and sent to processors to be made into cheese, yogurt, ice cream, or powdered milk. Angora and Cashmere goats also produce mohair and cashmere fiber, a niche market.[46]

Table 4.7 Total Census Farms, Specialty Census Farms, and Livestock/Poultry Numbers

Type	Total Farms,	Total	Total	Total	Total Animals,
Ducks/Geese*			2,862		1,700,000
Sheep**			11,000		1,065,400
Goats	6,725	177,698		867	
Horses & D/M***	54,169	453,965	47,996	10,507	300,393
Rabbit/Mink/Fox	327	1,916,327		284	
Bison (Buffalo)	1,898	195,728			
Llamas & Alpacas	4,302	31,708			
Deer & Elk	1,525	115,916			
Wild Boars	256	20,898			
ALL OTHER				6,926	
TOTAL			45,925	18,584	

* Data is only for ducks. In 2016, there were a total of 5,223 agricultural operations with other poultry.

** Approximate. Source: Alberta Lamb Producers. Number of animals is for 2018 (Stats. Canada, 32-10-0129-01)

*** 2010 grand total = 963,500 on 145,000 properties. Source: Equine Canada.

Basic Source: Statistics Canada, *2006 and 2016 Census of Agriculture*, 95-629-XWE and 3438, respectively

About one-quarter of Census farms also have one or more **horses**, particularly beef or feedlot operations. Some of these are work horses. Most horses (i.e., > 600,000), however, are actually found on non-Census farms, primarily rural acreages. Thirty-four percent of Canada's horses reside in Alberta and 21 percent reside in Ontario.[47] Most of these are recreational horses, breeding stock, race horses, or show horses. A relatively small number of **donkeys** and **mules** are also kept, primarily intended to protect flocks of sheep and, occasionally, herds of cattle.

4.7 SYNOPSIS

What are the take-aways from this overview of the prevailing production and marketing characteristics of Canadian agriculture? Some are obvious; some are less so.

- The vastness of Canadian agriculture is breathtaking, especially on the prairies. Fly from Edmonton to Winnipeg in July and below you will see an almost endless checkerboard of green and yellow quarter-sections (160 acres = 65 ha) of what is probably a cereal and canola. Fly from Calgary to Regina and rangelands for beef cattle sometimes reach to the horizon. Fly from Lethbridge to Saskatoon, and see the endless circles that trace the path of center pivots on more than a million acres (400,000 ha.) of irrigated land, methodically watering the parched soil. In total, there are about 130 million acres (50 million ha; 500,000 km^2), an area larger than the entire land mass of 70% of other countries in the world.

- Most of Canadian agriculture has a temperate climate, which limits crop production to one crop per year, and even that crop is sometimes subject to drought, hail, frost, wind, snow, or other adverse weather-related events. In more northerly areas, the abbreviated frost-free growing season and the limited heat-units available also narrows the production alternatives. For livestock producers, the harsh snow-bound winters similarly constrain production and inflate costs. But this is not all bad. The harsh winters also limit pest survival rates and, thus, reduce disease and insect infestations endemic to tropical agriculture.

- Canadian agriculture depends very heavily on only a few major crops—think a two-year rotation of canola and wheat in the West; a two-year rotation of corn and soybeans in the East. Long-term, monoculture agriculture potentially has some very negative ecological implications: a) fewer biological controls; b) extensive use of synthetic fertilizers, herbicides, insecticides, and fungicides; c) reduced organism resistance; d) soil degradation; e) water resource depletion; and, f) high fossil fuel usage. Is this ecologically sustainable? It is definitely not consistent with regenerative (organic) agriculture's dictates.

- As major agricultural exporters, a large segment of Canadian agriculture is at the mercy of prices in and access to the international marketplace. Most farmers (excluding dairy and poultry farmers) are price-takers with respect to both input purchases and product sales, and this is only exacerbated by the growing presence of just a few oligopolistic input providers and sales outlets. This is particularly true with respect to cereals, oilseeds, beef, and pork.

- New profit-driven technological developments largely dictate the current production and marketing systems adopted. This has, by design or happenstance, generally favored capital-intensive, large-scale monoculture.

- Consumers, especially in the domestic market, are becoming increasing concerned about food-related health issues, food-related environmental issues, and food-related social issues. We expect this to eventually have a profound impact on the present structure, conduct, and performance of Canadian agriculture.

5. International Trade

This chapter highlights the many facets of agricultural commodity trade in the international marketplace. This includes a brief commodity overview, followed by a description of international price characteristics, multinational trading rules, national levels of agricultural protection, existing trading agreements, the anticipated gains from freer trade, and a brief assessment of future agricultural trading opportunities.

5.1 OVERVIEW

> And the princes said unto them, let them live;
> but let them be hewers of wood and drawers
> of water unto all the congregation.
>
> *Joshua 9:21 KJV*

Trade has always been central to Canada's prosperity. Canada's economic development historically depended on the export of large volumes of raw materials, particularly fish, fur, grain, and timber.

Figure 5.1 Canada, International Trade as a Percentage of Gross Domestic Product, 1960 - 2017

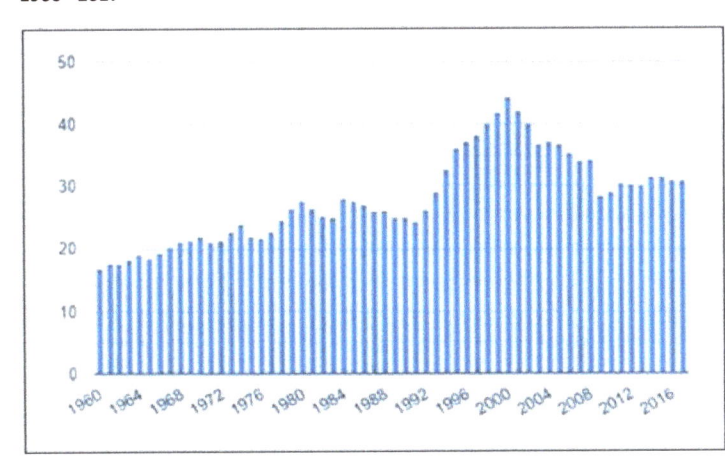

Source: World Bank, 2018

Although the composition of trade has changed over time, Canada remains very heavily dependent upon international trade to maintain its standard of living. During the period 1960–2017, the export of goods and services as a percent of gross domestic product (GDP) was about 28 percent, with a minimum of 17 percent in 1960 and a maximum of 44 percent in 2000 (Figure 5.1).

The top ten **net** exports from Canada in 2017 were still **all** resource-based and three were generated by the agricultural sector (Table 5.1).

Table 5.1 Canada: Top Ten Net Exports, 2017*

Commodity	Net Value US$B	% of Top 10
1. Mineral fuels, incl. oil	$54.9 B	50%
2. Wood (timber)	$11.1 B	10%
3. Precious metals	$8.8 B	8%
4. Oil Seeds	**$6.9 B**	6%
5. Woodpulp	$6.0 B	5%
6. Aluminum	$5.9 B	5%
7. Cereals	**$5.6 B**	5%
8. Ores, slag, ash	$3.8 B	3%
9. Fertilizers	$3.3 B	3%
10. Meat	**$3.0 B**	3%

* Net of off-setting imports.
Source: CIA, *The World Factbook: Country Profiles,* 2018

The destination of Canada's exports and the source of its imports is equally revealing. Three-quarters of Canada's exports go to the United States and two-thirds of Canada's imports come from the United States. Canada's second-largest trading partner is China (Figure 5.2).

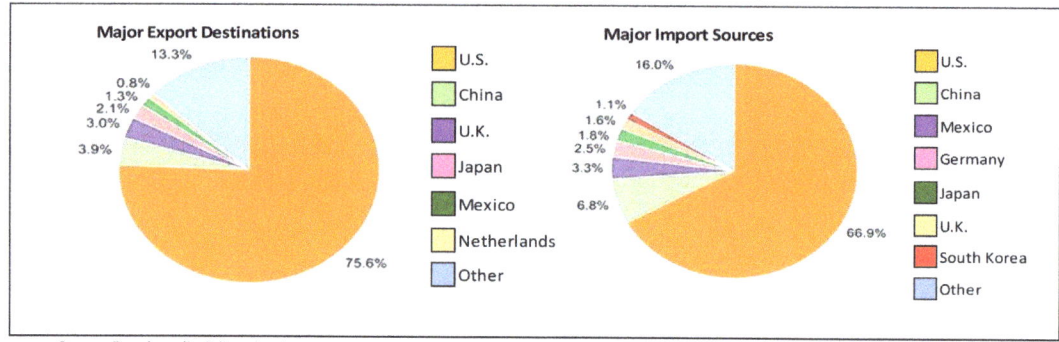

Figure 5.2 Canada: Major Export and Import Destinations, 2014

Major Export Destinations: U.S. 75.6%, China 13.3%, U.K. 3.9%, Japan 3.0%, Mexico 2.1%, Netherlands 1.3%, Other 0.8%

Major Import Sources: U.S. 66.9%, China 16.0%, Mexico 6.8%, Germany 3.3%, Japan 2.5%, U.K. 1.8%, South Korea 1.6%, Other 1.1%

Source: Encyclopedia Britannica, Inc.

5.2 MAJOR AGRICULTURAL EXPORTS & IMPORTS

Canada is one of the few areas of the world that produces more food than its population can consume. Canada grows 1.5 percent of the world's food with just 0.5 percent of the world's population. As the fifth largest agricultural exporter in the world, we take great pride in considering ourselves the "breadbasket of the world." But Table 5.2 also highlights how dependent our major agricultural enterprises are on successfully exporting our oilseeds, wheat, pulses, beef, and hogs to the rest of the world. The world is also our breadbasket.

Table 5.2 Trade Balances for Major Agricultural Products, Canada, 2016, $M

Product	Exports	Imports	Balance	Balance/Production*
Oilseeds (except soybean)	6343.7	247.8	6095.9	66%
Soybeans**	2500.0			87%
Wheat	5972.8	51.6	5921.2	104%
Pulses (peas/beans)	4127.9	232.7	3895.2	160%
Lentils**	2100.0			
Other Grains (barley/corn/oats)	1031.3	96.6	934.7	29%
Beef	1373.9	37.0	1336.9	15%
Hogs	437.1	1.9	435.2	11%
Dairy (milk)	56.9	0	56.9	1%
Eggs	60.3	97.1	-36.8	-3%
Poultry Hatcheries	56.3	70.6	-14.3	
Broilers & Turkeys	14.7	7.2	7.5	0%
TOTAL	24074.9	842.5	18632.4	39%

* Approximate. Production data from Farm Cash Receipts, Table 3.6.
** AAFC, **Canada, At A Glance**, 2017.
Basic Source: Trade data from Statistics Canada & US Census Bureau

Canada is the world's largest exporter of **canola, mustard, flaxseed, pulses** (beans, lentils and peas), and **durum wheat**, and Canada also produces approximately 80% of the world's maple syrup.

5.2.1 CANOLA

Canada exports 90% of its canola as seed, oil, or meal to about fifty markets around the world.[48] It's biggest buyer of canola oil and meal has traditionally been the United States, accounting for about 65% of oil exports and 82% of meal exports (2016). For raw seed, the most important destinations are China, Japan and Mexico.

5.2.2 WHEAT, BARLEY, AND OATS

Figure 5.3 Wheat Exports, 2000-2018 (M. tonnes)

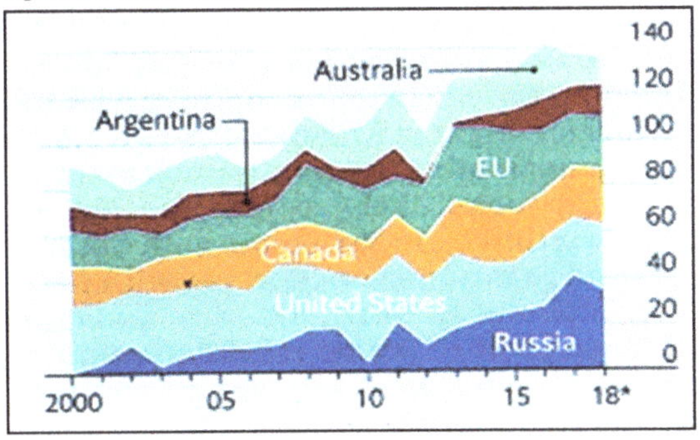

Source: USDA.   *Forecast

Canada is still a major **wheat** exporter (similar to the USA, EU, Australia, and Argentina), typically exporting about 15 million tonnes (1/2 billion bushels). A resurgent Russia, however, has captured most of the recent export growth (Figure 5.3). In the crop year 2016–2017, Canadian wheat was exported to eighty-seven countries. Principal markets are Indonesia, Japan, USA, Peru, Colombia, Mexico, Nigeria, and Bangladesh.

Barley exports generally amount to about 20% of total production and approximately 80% of this now goes to China. Additionally, Canada is the world's largest **oat** exporter, with 95% of this going to the United States.

5.2.3 SOYBEANS

About three-quarters of the soybeans grown in Canada are exported, either as raw soybeans or as processed products.[49] Principal buyers are Japan, Malaysia (with lots of its own palm oil), the Netherlands, and Iran.

5.2.4 PULSES

Canada produces more than one-third of global lentil production and about 20 percent of global dry pea production. The export of dry peas, lentils, and chickpeas typically accounts for about 40 percent of annual production. Most of the global trade in lentils is in red lentils, the most commonly consumed lentil worldwide. Historically, the principal export destinations for pulses have been Turkey (for lentils and chickpeas), India and China (for dry peas), and the USA (for dry beans). (See Chapter 6.)

5.2.5 BEEF

During the 2011–2013 period, about 46% of cattle and beef products were exported.[50] In terms of world beef exports, Canada has about 5 percent of the international beef trade. In 2015, this amounted to about 322,000 tonnes valued at about $2.2 billion, almost three-quarters of it to the United States. Hong Kong/China (7%), Japan (6%), Mexico (6%), and South Korea (2%) imported most of the remainder.

At the same time, live beef cattle exports to the United States (Figure 5.4) have typically amounted to about 1 million beef animals (say 300,000 tonnes, carcass weight) per year, less beef imports (say 100,000 tonnes). Only the BSE outbreak in 2003 (when exports were prohibited) and US Country of Origin Labelling (COOL) in 2008 severely interrupted this movement south.

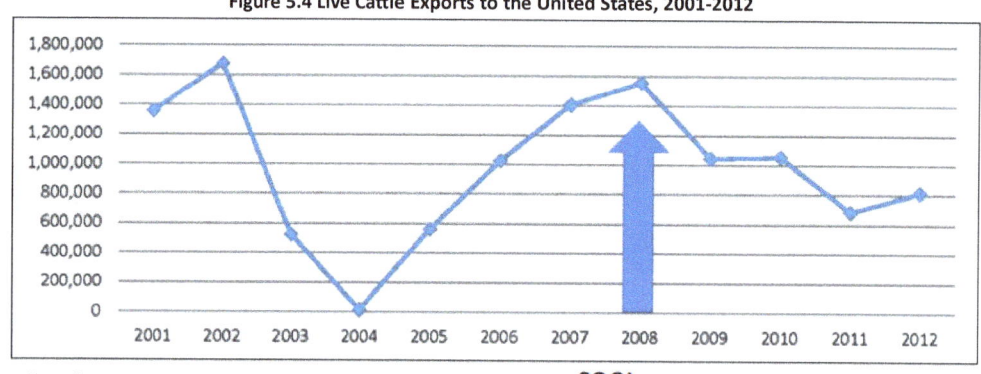

Figure 5.4 Live Cattle Exports to the United States, 2001-2012

Note: Country of origin labelling (COOL).   **COOL**

Source: Statistics Canada, CANSIM Table 003-0088

In terms of world beef exports, however, Canada ranks a distant sixth. This market is dominated by India, Australia, Brazil, and the USA. The total world beef trade in 2017 was about 9.6 M metric tonnes.

5.2.6 PORK

Exports absorb about two-thirds of the pork produced in Canada.[51] In 2015, pork exports reached 1,171,000 tonnes, valued at about $3.4 billion (excluding live animal exports) and this easily surpasses the export contribution of the beef industry. Canadian pork is exported to more than eighty countries, but principally the United States, Japan, Russia, China, and South Korea. In addition, the industry typically exports about 6 million live hogs/year to the United States, down (after the introduction of COOL) from a high of about 10 million in 2007 (Figure 5.5). COOL wasn't repealed by the United States until December 2015.

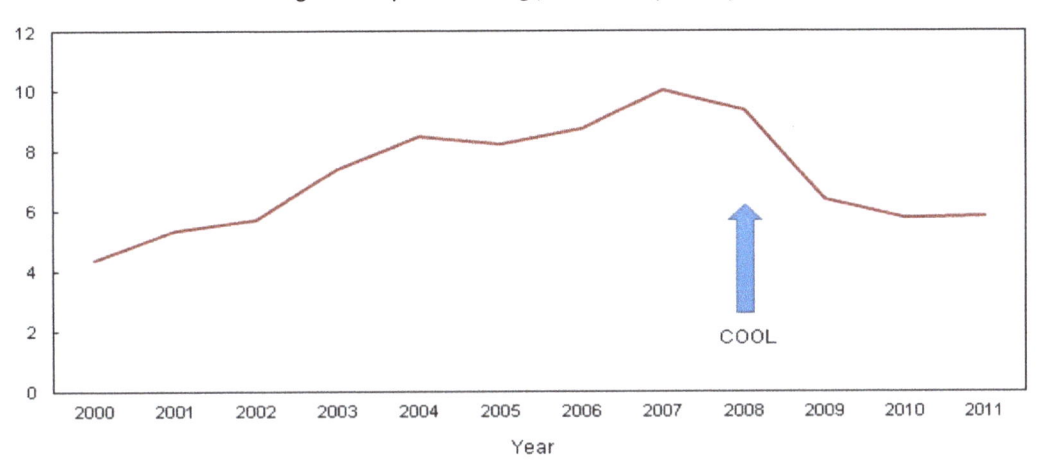

Figure 5.5 Export of Live Hogs, 2000 - 2011 (millions)

Note: Country of origin labelling (COOL).

Source: Statistics Canada, CANSIM Table 003-0088

5.2.7 DAIRY

In the dairy industry, trade is limited by both quotas and import tariffs. Quota levels are set to primarily meet domestic requirements. Nevertheless, dairy exports include cheese, skim milk powder, whey products, and products consisting of natural milk constituents. Major destinations are North America (52%), Egypt, the Philippines, Brazil, and Saudi Arabia.[52] Offsetting imports are somewhat larger but are still severely limited by employing extremely high WTO over-quota tariffs (tariff-rate quotas =TRQs) at their point of entry. This tariff ranges from 202% for skim milk to 298% for butter, with cheese, yogurt, ice cream, and regular milk within that range (Figure 5.6). Most imports originate from North America (58%), or Europe (31%). Ignoring possible trans-shipments, this trade balance for the last decade is tracked in Figure 5.7. Recent trade agreements (CUSMA, CPTPP, and CETA, discussed further in Section 5.8.2) will allow more imports of processed milk products into Canada and, thus, potentially displace some domestic production. This, in turn, could ultimately reduce domestic quota requirements in the coming years (Section 5.7).

Figure 5.6 Customs Tariffs on Selected Over-Quota Dairy Products, Canada, 2015

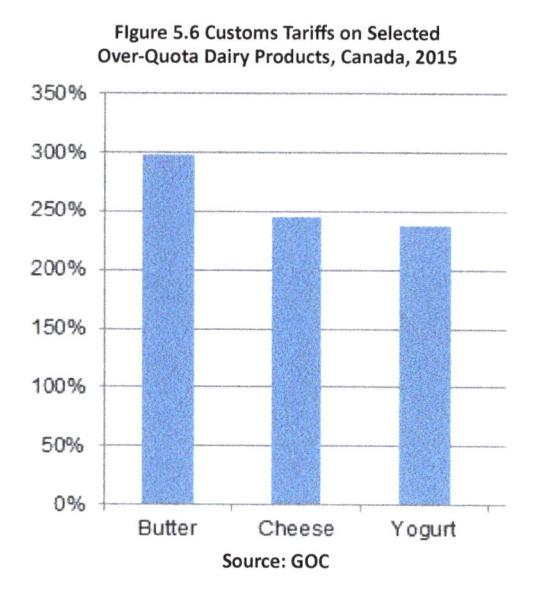

Source: GOC

Figure 5.7 Canadian Dairy Trade Balance, 2007-2016 ($million)

Year	Exports	Imports	Trade Balance
2007	283	622	-338
2008	255	679	-424
2009	230	573	-343
2010	227	610	-383
2011	252	670	-418
2012	237	677	-440
2013	262	751	-489
2014	281	899	-618
2015	211	900	-689
2016	235	969	-734

Source: Canadain Dairy Information Centre, 2018

5.2.8 POULTRY & EGGS

In 2016, Canada exported 14.4 million chicks and poults (young turkeys), 35.8 million hatching eggs, and 165 million kilograms of poultry meat and edible byproducts (fresh, chilled and frozen) to a dozen or so countries around the world (esp. Russia and the USA). This had a combined value of over $600 million.

Imports, on the other hand, are very limited. This is done utilizing tariff-rate quotas (TRQs), which set a limit on the amount of various poultry products allowed into Canada with little or no tariff. Under this WTO Agricultural Agreement, TRQ imports are subject to low "within access commitment" rates of duty up to a predetermined limit, e.g. x percent of the previous years production. Unrestricted imports in excess of this limit are subject to significantly higher "over access commitment rates of duty." These are almost 250% for chicken and about 150% for turkey and eggs (Figure 5.8).

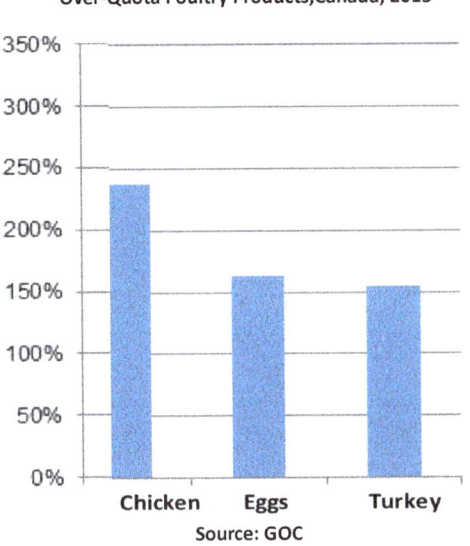

Figure 5.8 Customs Tariffs on Selected Over-Quota Poultry Products, Canada, 2015

Source: GOC

5.3 AGRICULTURE & AGRI-FOOD TRADE BALANCE

Total export sales of both agriculture **and** agri-food products in 2013 was $46 B, about 3.5% of the world total. Offsetting imports amounted to $34.3 B, about 2.9% of the world total. As such, Canada was the world's fifth largest exporter of agriculture and agri-food products in the world (behind, the EU, the US, Brazil, and China). At the same time, Canada was also the sixth-largest importer of agriculture and agri-food products (behind the EU, China, the US, Japan, and Russia).[53]

Table 5.3 provides us with a total agricultural trade balance for 2016 and these data again highlight Canadian agriculture's dependence on the export of bulk agricultural commodities. In 2016, canola seed, wheat seed, soybeans, and lentils made up about 28% of all agri-food exports from Canada. Beef, pork, and other grains do not make the short-list, even though exports are still very important to these particular agricultural industries. Conversely, our agri-food imports tend to be products that our climate does not fully support, such as some year-round fruits and vegetables, foreign wines (#1,) and coffee (#5).

Table 5.3 Canadian Agri-Food Trade, Top Five Exports & Imports
and Agr-Food Trade Balance, 2014 - 2016 (Cd$ billions)

Category	2014	2015	2016	AVERAGE	% TOTAL
EXPORTS: $B					
Canola Seed	5.2	5.0	5.6	5.3	10%
Wheat Seed (excl. durum)	5.8	5.9	4.5	5.4	10%
Soybeans	1.9	2.2	2.5	2.2	4%
Lentils, dried, shelled	1.5	2.5	2.1	2.0	4%
Bakery Products	1.2	1.6	1.9	1.6	3%
OTHER	36	38.4	39.4	37.9	70%
TOTAL AGRI-FOOD EXPORTS	51.6	55.6	56.00	54.4	100%
IMPORTS: $B					
Grape Wines <2 litres	1.9	2.0	2.0	2.0	5%
Food Preparations	1.4	1.6	1.7	1.6	4%
Bakery Products	1.2	1.4	1.4	1.3	3%
Dog & Cat Food	0.7	0.8	0.8	0.8	2%
Coffee, unroasted	0.7	0.8	0.8	0.8	2%
OTHER	33.6	36.9	37.8	36.1	85%
TOTAL AGRI-FOOD IMPORTS	39.5	43.5	44.5	42.5	100%
AGRI-FOOD TRADE BALANCE $B	12.1	12.1	11.5	11.9	

Source: AAFC, *Canada - At a Glance*, 2017

The food processing industries, particularly international branch plants, are especially challenged by the on-going globalization phenomena and other countries comparatively lax social policies (e.g. labor laws), environmental regulations, and tax structure. The Kraft-Heinz closure of their Leamington tomato processing factory (2014) to move it abroad is indicative. Yet another constraint is the shortage of skilled and, especially, unskilled labor, made even more acute by restrictions on the hiring of foreign labor. Meat processing plants, for example, find it very difficult to find required laborers and, thus, incur productivity losses while deferring additional investment and expansion. The increasingly acute food & beverage trade deficit is traced in Figure 5.9.

Figure 5.9 Net Agricultural Trade, Canada, 1990 - 2014

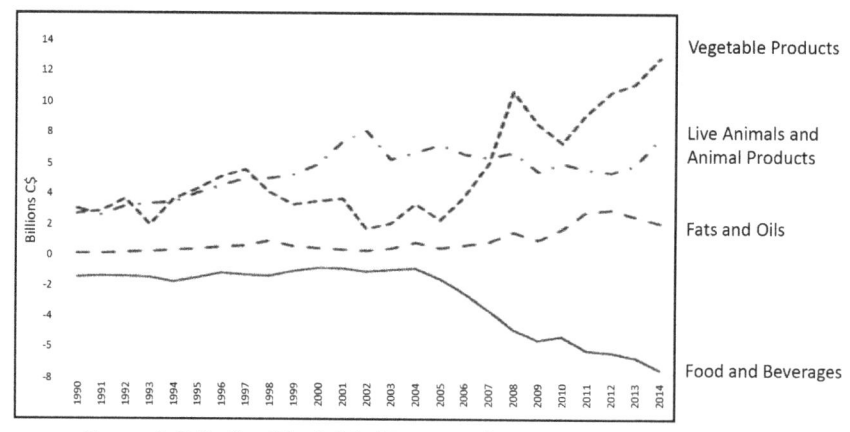

Source: D. D. Hedley, "The Political Economy of Agricultural Policy in Canada",
in: *Handbook of International Food and Agricultural Policies*, forthcoming

The bottom line is that we have a net trade surplus in agri-food products that is surprisingly small and uncertain. With agri-food exports of, say, $55 B and agri-food imports of, say, $43 B, the difference is a relatively meager $12 B. To put this in perspective, crude oil usually generates $50–$100 B in annual export revenue. Nevertheless, agri-food exports of $55 B remain a very significant component of total Canadian sales abroad.

5.4 INTERNATIONAL PRODUCT PRICES

Generally consistent with our international production and export ranking, our principal agricultural export commodities are essentially all **price-takers**. Aside from qualitative differences or unique niche markets, the international prices for oilseeds, grains, beef, and pork largely establish our domestic and export prices, and, in turn, prices at the farm gate:

- Since soybeans dominate the international oilseeds market, and are grown extensively in the contiguous USA, soybean prices largely establish international oil and meal prices. Canadian canola prices, therefore, generally track USA soybean prices fairly closely.

- The dominant world feed grain is corn, again led by the USA. Thus, Canadian barley prices also tend to track USA corn prices.

- Since Canada only accounts for about 15% (and decreasing) of the world's wheat exports, the price of wheat is effectively set by the private grain traders who market most of the USA and world's grain (see Chapter 6). Again, we are price-takers.

- The well-integrated North American beef industry also means that the price of Canadian beef is chiefly determined by the much larger (10X) American beef industry, although the grading system in Canada does tend to neutralize some of the effects of the US market.[54] (See discussion of COOL, above.) Low tariffs on imported beef and meat products further strengthen this linkage.

- Similarly, with low tariffs and the relatively free movement of fresh and processed pork across the Canada–US border, Canadian pork prices generally track American pork prices fairly closely.

Managed trade by the major trading nations (e.g., USA and China) can also have a major impact on particular markets (e.g., soybeans and canola).

Thus, given this general lack of pricing power in the international marketplace, the relationship between Canadian and international costs-of-production must also be maintained. And this, in turn, means that if comparative on-farm costs in Canada increase because, for example, the Canadian tax or regulatory regime becomes relatively more punitive, this will almost inevitably reduce on-farm margins in Canada and have corresponding impacts across other sectors of Canadian society.

5.5 FOREIGN EXCHANGE RATES

The exchange rate is the rate at which one currency can be exchanged for another. It is also regarded as the value of one country's currency in relation to another currency. This is usually expressed in terms of the US$, the currency most often used in international transactions. Thus, if the official rate-of-exchange is, say, 1.33, this means that we must pay CD$1.33 to buy one US$. Or, conversely, we can say that the CD$ is a given fraction of a US$. Today this would be 1/1.33 = 0.75. Seventy-five cents. This relationship for the past 64 years is tracked in Figure 5.10.

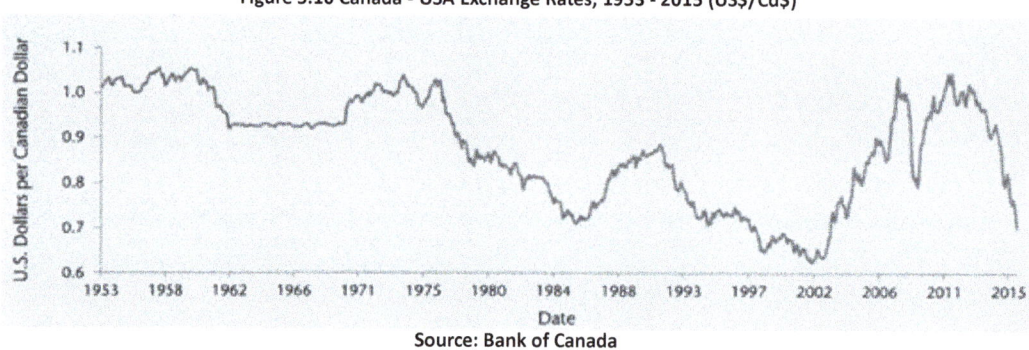

Figure 5.10 Canada - USA Exchange Rates, 1953 - 2015 (US$/Cd$)

Source: Bank of Canada

This rate can weaken because the US$ gets stronger and/or the Canadian economy gets weaker. The CD$ tends to follow the strength or weakness of international commodity prices (esp. extractive product prices) with a long-term equilibrium rate of perhaps 0.85; essentially reflecting embedded productivity differences. These exchange rate fluctuations impact both physical trade flows and capital movements.

In any event, with 60–70 percent of our agricultural trade going to or coming from the USA, the prevailing CD$/US$ exchange rate is critical to the balance sheet of the agricultural sector in Canada:

- When the exchange rate with the US$ climbs, Canadian agricultural exports are depressed. (Canadian imports into other countries cost more in terms of their local currency vis-à-vis the US$.) On the other hand, farm imports of various inputs effectively become less expensive in terms of US$. Thus, farm revenue goes down but costs effectively go down as well.

- Conversely, when the exchange rate with the US$ drops, Canadian agricultural exports are stimulated and the price of imported farm inputs climb. So then farm revenue goes up but so does the imported price of farm inputs, especially machinery.

At the same time, when the US$ also strengthens relative to other major currencies, especially the Euro and the Yen, as well as most emerging market currencies, this CD$/US$ disparity loses some of its potency. Because 30–40 percent of our exports (and substantial capital movements) involve the rest of the world and these countries often experience a similar change in their exchange rate vis-à-vis the US$, the Canadian terms-of-trade with these countries often remain largely unchanged.

Canadian net farm income is generally enhanced by a decline in the value of the Canadian dollar vis-à-vis the US dollar.

5.6 MULTINATIONAL TRADING RULES

Given the importance of trade to Canadian agriculture, the international rules and regulations strongly influence our agricultural policies, which, in turn, affect the structure, conduct, and performance of the sector.

The **World Trade Organization** (WTO), 1994, and its predecessor, the **General Agreement on Tariffs and Trade** (GATT), 1947, are by far the most influential multilateral agreements governing agricultural trade. The Uruguay Round (1994) made substantial progress toward agricultural trade liberalization and this led to a commitment to continue (e.g., Nairobi Declaration, 2015) with further efforts toward even more agricultural trade liberalization.

A central tenet of the GATT and the WTO multilateral system is the principal of **Most Favored Nation** (MFN) whereby a commitment of a tariff reduction is applied equally to all members. It also requires equal obligations on **non-tariff barriers** (NTBs) (e.g., sanitary and phytosanitary measures and standards set by CODEX).

The CODEX Alimentarius (1963) is a collection of standards, guidelines, and codes of practice established by the Food and Agricultural Organization (FAO) and the World Health Organization (WHO) to protect consumer health, safety, quality, and fairness in the food trade.[55] It addresses animal feed, antimicrobial resistance, biotechnology, contaminants, nutrition and labeling, and pesticides. Historically, if an importing country didn't have its own standards, they would default to CODEX. More recently, however, countries have been developing their own codes of practice and, thus, gradually drifting away from CODEX and the underlying scientific rationale.

International trade in agriculture is also distorted by commodity and income support measures. The GATT and the WTO have attempted to discipline these measures through three codified categories of distortions in the production and trade of agricultural commodities. The first category, the least-distorting support measures—the **green** category—are permitted and include income insurance and safety-net programs, natural disaster relief, a range of structural adjustment assistance programs, certain environmental programs, and certain regional assistance programs. The second category, those which potentially influence production, price, or input use—the **amber** category—are subject to upper limits and must be reported to the WTO. A third category of clearly distorting measures—the **red** category—are totally prohibited and are subject to international trade challenges. Prohibiting any linkage between the amount of support provided to producers and the volume of production, prices, or input use is referred to as **decoupling**. Negotiations to discipline other agricultural trade distortions—for example, state trading and food aid—are continuing.

Both tariffs and NTBs, as well as support measures, have had a strong influence on Canadian agricultural policies and programs, particularly on certain sub-sectors (e.g., dairy and poultry).

5.7 NATIONAL LEVELS OF AGRICULTURAL PROTECTION

Because of socio-political considerations, international trade in agriculture and food products has generally been more constrained than other traded goods. Indeed, the GATT-facilitated liberalization of traded goods, led by the USA, simply exempted many agricultural products. This was followed by the formation of the EEC (later the EC and the EU). The agriculture and food sectors were heavily protected in Europe even prior to the GATT. Other developing countries, including Japan, Switzerland, and the Nordic countries, followed suit with even more restrictive trade barriers. Thus, what we generally have is **managed trade** in agricultural and related food products, and this is still evident in current WTO trade negotiations.

Agricultural support is defined as the annual monetary value of gross transfers to agriculture from consumers and taxpayers arising from governments' policies that support agriculture, regardless of their objectives and their economic impacts. This **producer support estimate** (PSE) can be approximated by calculating a nominal producer protection coefficient (NPPC), defined by the European-based Organization for Economic Cooperation and Development (OECD) as the ratio between the average price received by producers (measured at the farm gate), including the net support per unit of current output, and the border price (again measured at the farm gate).[56] For example, a NPPC of 1.10 suggests that farmers, overall, effectively received prices that were 10% above international market prices. Transfers included in the NPPC are composed of market price support, budgetary payments, and the cost of revenue foregone by the government and other economic agents, but excluding support for general services to the agricultural sector.

Figure 5.11 traces how this NPPC estimate for Canadian agriculture has gradually declined over time. From a high of over 1.40 in 1987 (i.e., an average 40% subsidy), by the mid-1990's it had already dropped to about 1.10 before declining even further in recent years.

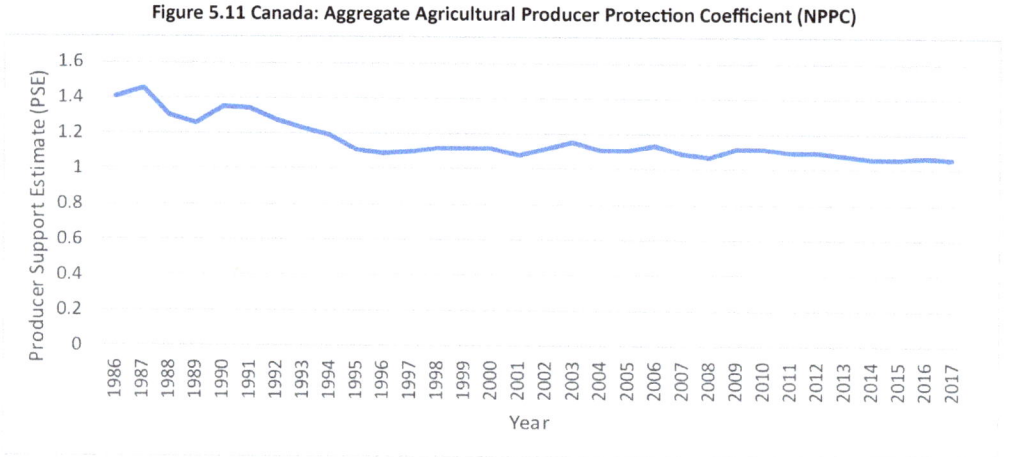

Figure 5.11 Canada: Aggregate Agricultural Producer Protection Coefficient (NPPC)

Source: Basic data from *OECD.Stat*, Monitory and Evaluation, 2018

How, then, does Canada currently rank against our major trading partners in terms of its international competitiveness? Table 5.4 gives Canada a NPPC of 1.06, meaning that our effective farm-gate prices average about 6% more than international prices.

Table 5.4 Nominal Protection Coefficient, Selected Countries, 2017

Country	Coefficient*
S. Korea	2.02
Switzerland	2.00
Japan	1.82
Norway	1.68
China	1.11
Russia	1.09
CANADA	**1.06**
European Union (28 countries)	1.05
United States	1.03
Mexico	1.03
Brazil	1.01
New Zealand	1.01
Australia	1.00

* Farm gate price/border price. All commodities.

Source: OECD Producer & Consumer Support Estimates database, 2017

In short, Canada is **relatively** competitive in the international marketplace. Korea, Switzerland, Japan, and Norway are much more protectionist, while China and Russia are just slightly more protectionist. At the same time, the EU, the USA, and Mexico—all major trading partners—are generally slightly less protectionist, while farmers in Brazil, New Zealand, and Australia have even less nominal protection.

5.8 TRADE AGREEMENTS

Generally consistent with the terms and conditions specified in the **General Agreement on Tariffs and Trade** (GATT), Canada has concluded numerous bilateral and regional trade agreements to gradually expand its commercial footprint abroad while, at the same time, also securing the potential benefits of trade in the domestic market. An estimated 64 percent of Canada's trade in all products is now covered by free-trade agreements.[57]

Within this general framework, Canada now has a number of very important bilateral and regional international agreements that directly impact agriculture.[58]

5.8.1 BILATERAL AGREEMENTS

In the last three decades, numerous bilateral trade agreements have been concluded. These include trade agreements with Israel (1997), Chile (1997), Costa Rica (2002), Peru (2009), Colombia (2011), Jordan (2012), Panama (2013), Honduras (2014), and Korea (2015). Canada is also pursuing free-trade talks with India and the Mercosur trading block (Argentina,

Brazil, Paraguay, Uruguay, and Venezuela).[59] In 2006, Canada only had free-trade agreements with five countries in the entire world. By 2015 (including regional agreements—see following), Canada had concluded negotiations with fifty-one nations.[60] The scope of these agreements varies with respect to the goods traded, services, government procurement, and institutional constraints relative to competitors. Each is unique.

Of the bilateral trade agreements already concluded, the 2015 Canada-South Korea Free Trade Agreement (CKFTA) is probably the most significant. The CKFTA will eliminate the existing 40% tariff on imported Canadian beef within 15 years. At the same time, Canada will eliminate tariffs on South Korean automobiles within two years.

Canada's primary exports to South Korea in 2012 were mineral fuels and oils, ores, slag, ash, wood pulp, and paper products, valued at some $3.7 B. Canada's primary imports from South Korea in 2012 were automobiles, trailers, bicycles, appliances, and electrical machinery and equipment, valued at about $6.3 billion. Our trade balance with South Korea remains decidedly negative.

A Free Trade Agreement (FTA) with China could also be very important to the agriculture and food sector. But attempts to negotiate an FTA with China have not progressed to the extent anticipated. More recently, a diplomatic impasse on some issues has not been helpful.

5.8.2 REGIONAL TRADE AGREEMENTS

The principal regional trade agreements are three:

- **CUSMA – Canada-United States–Mexico Agreement,** 2018. (Also referred to as **NAFTA 2.0, USMCA** in the USA, and **T-MEC** in Mexico.)
- **CPTPP – Comprehensive & Progressive Agreement for Trans-Pacific Partnership** (previously the **TPP, Trans-Pacific Partnership**), 2018.
- **CETA – Comprehensive Economic and Trade Agreement** – between Canada and the twenty-eight EU countries, 2017.

CUSMA – Canada-United States–Mexico Agreement (2018)[61]

CUSMA is a revised **NAFTA**, the North American Free Trade Agreement, which originally came into effect in 1994.

The 1994 NAFTA was an agreement to remove both tariffs and investment barriers between Canada, Mexico, and the United States. NAFTA incorporates the prior agreement between the USA and Canada to remove most tariffs on agricultural trade. Mexico and Canada also had a separate prior agreement on agricultural products that eliminated most tariffs over a fifteen-year period. All the provisions of NAFTA, including the elimination of all tariffs, were implemented by 2008.

Since NAFTA was approved, Canadian agricultural exports to the US and Mexico approximately doubled. In particular, NAFTA helped increase the export of Canadian horticultural crops, oilseed products, pulses (e.g., dried beans), red meats, and processed products. Under NAFTA, Canada tripled canola sales to Mexico between 1994 and 2002, and Mexico is now the second-largest market for Canadian canola. Similarly, beef exports to Mexico have grown from a negligible $4 M in 1997 to over $200 M by 2016.

At the same time, agricultural imports from the United States under NAFTA increased from $4.2 B in 1990 (pre-NAFTA) to $19 B in 2011, making Canada the top export market for US agricultural produce. Canada now accounts for 14 percent of total US agricultural exports. The top import categories are fresh vegetables, fresh fruit, snack foods, processed fruits and vegetables, and red meats. Similarly, with NAFTA and the termination of seasonal fresh fruit and vegetable tariffs of 10–15% on Mexican produce, Canada now imports more tomatoes, guavas, mangos, peppers, and avocados from Mexico.

A recently re-negotiated NAFTA, now called CUSMA or NAFTA 2.0, maintains most of the agriculture-related provisions in the original NAFTA but adds provisions that give US dairy farmers greater access to the Canadian dairy market. Tariff-free access is set at 3.6 percent of the Canadian market, similar to the minimum level of access conceded in the CPTPP (at 3.5%, see immediately following). Canada also agreed to eliminate its Class 6 and Class 7 milk categories and associated pricing schedules for skim milk, skim milk proteins and other components, and ultra-filtered milk within six months after the CUSMA goes into effect. Ultra-filtered milk is mostly used in Canada to make cheese.[62]

CUSMA also includes new rules regarding the grading of wheat. US-grown wheat varieties registered in Canada can now be graded to the same standard as Canada's wheat. This will give Canadian and American farmers the same opportunities to deliver in Canada while still protecting Canada's reputation for quality.[63]

Regarding poultry, the United States gets tariff-free access to Canada for 57,000 tonnes of chicken by year six of the deal, growing one percent per year for an additional ten years. Canada also gave the United States access for ten million dozen eggs and egg equivalent products in year one, growing one percent per year for ten more years.

Two other CUSMA provisions could indirectly have a profound long-term impact on Canadian agriculture, since it requires:

- A country to notify CUSMA members three months in advance if they intend to begin free-trade negotiations with non-market economies (e.g., China).
- Consultation on any domestic monetary policies and exchange rates that might impact another member.

There was a ceremonial signing on November 30, 2018. As of mid-2019, however, Mexico and the US still have to approve the agreement. Mexico has a new president and may not decide to honor it. The US now has a House of Representatives controlled by the Democrats, who may also be reluctant to give President Trump any kind of victory or, at the very least, may demand changes to appease their more protectionist constituencies.

CPTPP – Comprehensive & Progressive Agreement for Trans-Pacific Partnership (2018)

This is a multinational trade agreement involving eleven countries: Canada, Japan, Vietnam, Singapore, Malaysia, Brunei, Australia, New Zealand, Mexico, Peru, and Chile. It is a revised agreement after the recent withdrawal of the USA from the originally proposed TPP. The US withdrawal from the original TPP required numerous changes but most were cosmetic and not substantive changes.

Average annual Canadian agricultural exports to the respective countries in 2014–16 are indicated in Figure 5.3. Agricultural exports to Mexico and Japan dwarf all others.

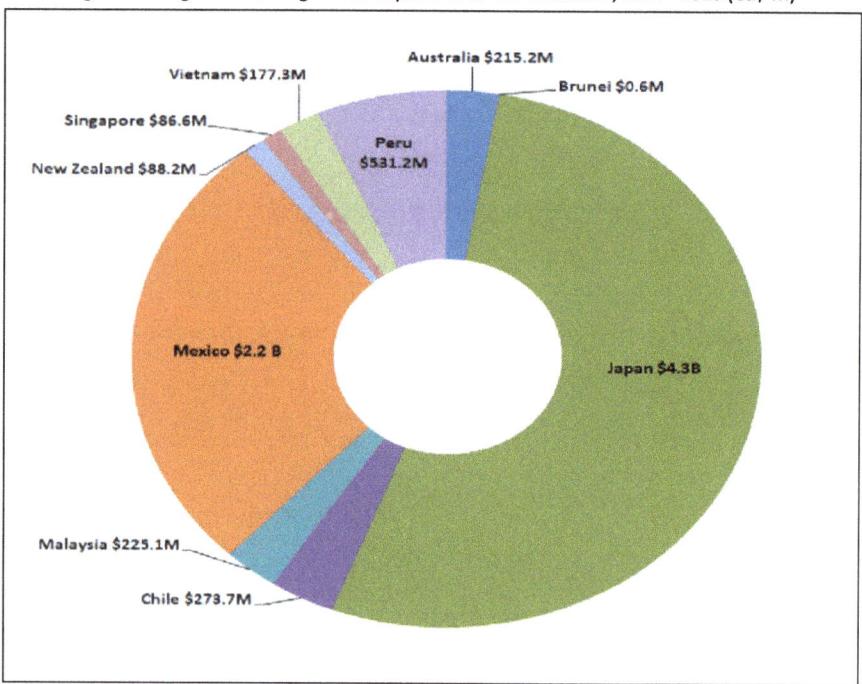

Figure 5.12 Agriculture & Agri-Food Exports to CPTPP Countries, 2014 - 2016 (Cd$ M)

- Australia $215.2M
- Brunei $0.6M
- Vietnam $177.3M
- Singapore $86.6M
- New Zealand $88.2M
- Peru $531.2M
- Mexico $2.2 B
- Japan $4.3B
- Malaysia $225.1M
- Chile $273.7M

Source: GOC

Given the potential for food imports by most of the Asian partners, this agreement opens significant opportunities for Canada. The withdrawal by the US offers a competitive advantage to Canadian interests.

It is expected that the new CPTPP will provide additional market access opportunities for Canadian pork, beef, pulses, fruits and vegetables, malt, grains, cereals, animal feeds, maple syrup, wines and spirits, baked goods, processed grain and pulse products, sugar and chocolate confectionary, and processed foods and beverages.[64]

Importantly, the CPTPP agreement corrects a feed barley tariff differential between Canada and Australia into the Japanese market that had seen Canada at a disadvantage. Australian feed barley was exempted from Japan's state-trading system in 2015, resulting in Australian feed barley imports being duty-free. This advantage led to most of Japanese feed barley imports coming from Australia. Under CPTPP, Canada's feed barley will immediately be treated the same as Australian feed barley. The CPTPP will also immediately provide a Canada-specific annual quota for food-quality wheat destined for Japan and this will ensure that Canadian wheat exports are competitive with wheat exports from other countries, including the USA.[65]

Other specifics regarding the perceived opportunities and benefits of the CPTPP are provided in Table 5.5.

Particularly controversial in Canada was the Canadian commitment to give co-signers access to 3.25 percent of the supply-managed Canadian dairy industry. Under the CPTPP, Canada will provide new market access to supply-managed products by establishing increasingly liberal volume-limited CPTPP wide quotas phased in over five years with additional slow growth until year 13.

The Canadian government ratified the agreement at the end of October 2018. Shortly after, Australia ratified the agreement and the six-member ratification requirement for implementation was met. Thus, the CPTPP came into effect on December 30, 2018, although it is still awaiting ratification by some of the other signatories.

Table 5.5 Opportunities and Benefits of CPTPP for Canada's Agriculture and Agri-Food Exporters

Market Access Highlights

Pork:
- In Japan tariffs of up to 20% on pork products, including sausages, which were subject to the gate price system, will be eliminated within 10 years
- Tariffs of up to 27% in Vietnam on fresh/chilled and frozen pork will be eliminated within nine years

Beef:
- In Japan, tariffs of 38.5% on fresh/chilled and frozen beef, as well as tariffs of 50% on certain offal will be reduced to 9% within 15 years
- In Vietnam, tariffs of up to 31% on fresh/chilled and frozen beef will be eliminated within two years, and tariffs of up to 34% on all other beef products will be eliminated within seven years

Wheat and Barley:
- In Japan, Canada will have access to a Canada-specific quota for food wheat which starts at 40,000 tonnes and grows to 53,000 tonnes within six years. Mark-ups within this country-specific quota will be reduced by 45 or 50 percent
- Canada will have access to a CPTPP-wide quota for food barley which starts at 25,000 tonnes and grows to 65,000 tonnes within six years. Mark-ups applied to the price of food barley by Japan will be reduced by 45% within eight years

Processed Food and Beverages:
- The CPTPP will eliminate or reduce many of the existing tariffs or create tariff rate quotas (TRQ's) on processed foods and non-alcoholic beverages. Including maple syrup, baked goods, processed grain and pulse products, and sugar and chocolate confectionery

Dairy, Poultry and Egg Sectors:
- Canada will provide new market access for supply-managed products. This access will be through volume-limited CPTPP-wide quotas phased in over five years with additional slow growth until Year 13
- There will be an immediate elimination of World Trade Organization (WTO) and CPTPP within-access tariffs for dairy, poultry and egg products
- There will be no reductions to WTO over-access tariffs except for whey powder, margarine, and milk protein substances

Tariff Change Highlights

- **Japan** will eliminate tariffs on close to 32% of tariff lines on agriculture and agri-food products upon entry into force. A further 9% of tariff lines will be provided preferential tariff treatment through permanent quotas and country-specific quotas for Canada. The remaining tariff lines will be provided tariff elimination or reductions over a period of up to 20 years, or reductions of the in-quota or out-of-quota tariff
- **Vietnam** will eliminate tariffs on close to 31% of its tariff lines upon entry into force. A further 67 percent of tariff lines will become duty-free within 15 years, with the remaining being provided preferential treatment through other means (tariff elimination only on in-quota tariff lines)
- **Malaysia** will eliminate tariffs on nearly 92% of its tariff lines upon entry into force. A further 7 percent of tariff lines will become duty-free within 15 years, with the remaining being provided preferential treatment through permanent tariff rate quotas (TRQ's)
- **Australia** will eliminate all of its tariffs on agriculture and agri-food products upon entry into force, except for one tariff line which will be eliminated within four years
- **New Zealand** will eliminate tariffs on almost 99% of its agriculture and agri-food tariff lines upon entry into force, with the remaining being eliminated within five years

Source: GOC, on-line, February 16, 2018

CETA – Canada-EU Trade Agreement (2017)

The European Union, with 500 million people and twenty-seven countries, is the world's largest importer of agriculture and agri-food products—$163 B in 2016 and 16% of all the world's agriculture and agri-food imports.

Under CETA, over 90% of agricultural products will immediately become duty-free, other tariffs are being phased out over seven years, while still other revised tariff-rate quotas (TRQs) are being phased in over the next five years (Table 5.6). CETA also immediately eliminates all tariffs on processed foods, excluding sweet corn and sugar, as well as establishing rules regarding health standards and GMOs.[66]

Table 5.6 Opportunities and Benefits of CETA for Canada's Agriculture and Agri-Food Exporters

Products that become duty-free immediately:	Tariffs phased out over 7 years:
o Pet Food (up to 948 Euros/tonne)	o Durum Wheat (up to 148 Euros/tonne)
o Frozen French Fries (14.4-17.6%)	o Rye and Barley Grain (up to 93 Euros/tonne)
o Sweet Dried Cranberries (17.6%)	o Common Wheat, low-medium quality (up to 95 Euros/tonne)
o Processed Pulses (7.7%)	
o Soybean and Canola Oil (3.2%-9.6%)	o Sweet Corn, frozen (5.1% + 9.40 Euros/100 kg)
o Prepared Vegetables (20%)	o Oats (89 Euros/tonne)
o Fruit Juices (e.g. cranberry, blueberry) (17.6%)	**Tariff Rate Quota (TRQ) established for:**
o Condiments/Sauces (up to 10.2%)	o Beef (50,000 tonnes), 5-year phase-in
o Maple Syrup (8%)	o Pork (80,549 tonnes), 5 year phase-in
o Fresh Cherries (up to 12%, seasonal)	o Bison (3,000 tonnes), immediate
o Fresh Apples (up to 9%, seasonal)	o Sweet Corn, processed (8,000 tonnes, 5 year phase-in
	o Common Wheat (100,000 tonnes), immediate

Source: GOC, on-line, September 19, 2017

Better access to the European beef market is also assured. As a result of CETA, Canada's duty-free access to the EU market will rise from 3,200 tonnes of beef to 37,500 tonnes.[67] However, Europe will continue to enforce its directive against the use of growth promoters (hormones and beta agonists [stimulators]) and require changes in Canadian slaughtering practices.

Support programs are not affected or disciplined under CETA. Barriers provided by support (subsidies) and geographical indications of support programs remain.[68] CETA does, however, give Canada a competitive advantage over the USA and other suppliers to the EU market.

The domestic supply-managed poultry and egg markets were of little interest to European farmers and, therefore, entirely excluded from the agreement, which also met the Canadian objective of maintaining supply management in these sub-sectors.

Canada also reluctantly relinquished 2.25 percent of its dairy quota to imports from the EU. Most threatened are domestic cheeses, the highest valued part of the dairy market. Before CETA, Canada allowed about 21,000 tonnes of foreign cheese into the Canadian market, about 5% of total annual cheese consumption. With CETA, Europe will be allowed to export an additional 18,500 tonnes of cheese into Canada, raising imports to about 9% of total cheese consumption.

To compensate dairy farmers for concessions in CETA, $350 million was made available to help dairy famers modernize operations or expand their product lines.

As of this writing (early 2019), Italy still says that it will not ratify CETA.

5.9 THE GAINS FROM FREER TRADE

The principal of comparative advantage provides the unshakable basis for international trade. The principal states:

> Whether or not one of two regions is absolutely more efficient in the production of every good than is the other, if each specializes in the products in which it has a **comparative advantage** (i.e. greater **relative** efficiency), trade will be mutually profitable to both regions. Real wages of productive factors will rise in both places.
>
> Conversely, a protective tariff, quota or non-tariff barrier (NTB), far from helping the protected factor of production, will instead reduce its real wage by making imports more expensive and by making the whole world less productive through eliminating the efficiency inherent in the best pattern of specialization and division of labor.[69]

Conceptually, this is difficult to understand, so a concrete example is helpful. Assume two regions, Canada and the EU. Both produce wheat and wine and their relative labor requirements are as follows:

Table 5.7 The Principal of Comparative Advantage

Product	In Canada	In Europe
1 unit of wheat	1 day's labour	3 day's labour
1 unit of wine	2 day's labour	4 day's labour

Even though Canada's 1 and 2 days of labour are, respectively, less than Europe's 3 and 4 (implying an **absolute trade advantage**), we have a **comparative advantage** in wheat and Europe has it in wine. Why? Because our 1 divided by its 3 is less than our 2 divided by its 4 (or, equivalently, because our 1:2 is less than its 3:4).

Other motives for international trade include the economies of large-scale production and/or differences in consumer tastes.

There is one major qualification: "in the aggregate" means that there is a net benefit to the country as a whole. But there can be, and usually are, both winners and losers. Much of the push-back to globalization comes from the losers not being adequately compensated (bought off) by the winners, even though this should be possible such that the "winners" still win.

This also underlies the dilemma faced by Canadian trade negotiators who simultaneously try to protect the supply-managed (and, thus, relatively non-competitive) dairy and poultry/egg industries while still trying to secure freer trade for other sectors of the economy. The dairy concessions to market access made in CUSMA, CPTPP, and CETA are, cumulatively, relatively large:

$$CUSMA\ 3.6\% + CPTPP\ 3.25\% + CETA\ 2.25\% = 9.1\%$$

Even with a growing domestic consumer demand, these concessions translate into imports supplying perhaps 18 percent of Canada's dairy products by 2024. The amount of milk displaced would represent the milk currently produced on about 900 Canadian dairy farms (out of 11,000) and dictate corresponding quota reductions (Table 5.8).

Table 5.8 The Impact of CETA, CPTPP & CUSMA on Dairy Quota

Year	M kg Butterfat	% of Quota
2017	-0.2	-0.1%
2018	-1.5	-0.4%
2019	-7.8	-2.0%
2020	-4.9	-1.3%
2021	-4.9	-1.2%
2022	-4.9	-1.2%
2023	-4.1	-1.0%
2024	-2.5	-0.6%

*Assumes 2% dairy quota increase/year.
Source: Dairy Farmers of Canada

The federal government has promised compensation to dairy farmers and has struck one working group to figure out mitigation strategies and another to develop a vision for the Canadian dairy industry.[70]

5.10 FUTURE AGRICULTURAL TRADE PROSPECTS

Canadian farmers are still remarkable producers, now accounting for more than $50 billion worth of exports annually. But in recent years, some have gradually become less productive, less competitive and more financially strapped than their counterparts in various other countries.

Many of the trend lines aren't promising. In the past 15 years, Canada's share of the global wheat market has shrunk to 15 per cent from 23 per cent, and it has dropped from third to ninth in terms of total food exports. Russia recently surpassed Canada as the world's third largest wheat exporter (Figure 5.3), Kazakhstan has doubled its wheat exports, and Brazil has overtaken Canada as the largest agricultural power behind the United States and the European Union. The

Australians, the Americans, the European Union, Brazil, and some of the emerging economies, like China and India, are also becoming increasingly aggressive in the international marketplace.

Canada's present status in a competitive global trading atmosphere gets mixed grades. Our networking and establishment of strong trade ties were at one time exemplary, with Canada known for sound trading practices, quiet effective diplomacy, strong contacts, and assurance of the identity and quality of products. Today, several sectors still maintain this exemplary status, particularly livestock and livestock products, as well as pulses.

Other export sectors have achieved a confusing array of representation, commercial ties and reduced networks for feedback on how to more effectively meet end-user demand. We have traders but very little proactive market development to identify and establish Canada as the preferred source of imports to satisfy and surpass end-user needs abroad. This lack of market development capacity needs to be addressed if agriculture and industry organizations in Canada are going to more successfully compete with other exporting nations, particularly Australia and the United States. The American soybean, wheat, and corn organizations, as well as meat and livestock organizations in Australia, set the standard for market networking and success. Canada must be able to match this USA and Australian markets' development capacity.

The abandonment of institutions like the Canadian Wheat Board, while generally positive, also left a gap in our ability to develop increased market access. The private trade assured us they would take care of it but no other exporting jurisdiction that we compete with lets private traders assume this dominant a role. This will be an area of continuing challenge for Canada as we move forward.

The bottom line is this:

1. We are not the "breadbasket" of the world. We are generally price-takers and, frequently, relatively small players in the international arena. Canadians like to think of their country as an agricultural superpower, where farmers grow vast amounts of food to help feed a hungry planet. But the reality is much different—far from leading the world in agriculture, Canada is falling behind on many fronts.

    a) The export of generic commodities, like wheat, beef, canola seed, petroleum, etc., is threatened by a number of pervasive factors:

    b) The aggressiveness and the relative cost of production of traditional competitors (e.g., with respect to wheat, the USA, Australia, Argentina, and, increasingly, Russia and the Ukraine). Consider the relative price of land in Canada versus, for example, Brazil.

    c) The continued drive for national food self-sufficiency, greatly assisted by "designer" GMO seeds that better adapt to area-specific soil, climate, and technological parameters. Numerous traditional importers in emerging markets are becoming increasingly competitive and less dependent on imports.

    d) The recent decline in the rate of improvement in our own on-going technology-led productivity gains.

e) Our own gradual real cost of production increases embedded in increasing regulation and a gradually changing tax structure (e.g., carbon taxes, labor standards, and safety and health standards not adhered to by many other countries).

f) A periodically disadvantageous exchange rate regime that is closely tied to the US dollar which, in turn, is loosely linked to the strength of our extractive export industries (especially oil and minerals).

g) The current dissatisfaction with many existing trade agreements abroad, as reflected in the recent withdrawal of the USA from the earlier proposed CPTPP.

2. Given the prevailing international climate, the prospects for freer international trade in the immediate future are not very bright. Trade negotiations are made even more difficult because of Canada's dual marketing perspective: protection for the dairy, poultry, and egg industries but freer trade for everything else. This has bedeviled Canadian policy-makers for decades and somehow has to change.

Our best future trade strategy is to continue to aggressively develop more unique, safe, healthy, environmentally friendly, animal-friendly, quality-based products, which have a high income elasticity abroad.[71] These non-GMO niche markets could include more processed livestock-based products (e.g., Kobe beef), various processed grain/oilseed products, as well as various fruit & vegetable products.

A particularly attractive growth area is pulse fractionation to produce and export more plant protein. This is a rapidly growing consumer market, both domestically and internationally, and it has a number of potential auxiliary production advantages—increased crop diversification, adoption of a more regenerative/sustainable cropping regime, and a possible reduction in CO_2 crop emissions.

As per-capita income levels climb in importing countries, the demand for these products will climb even faster. The importance of gradually adding more in-house value to food products through additional grading, cleaning, processing, packaging, and shipping must also be emphasized. An increasingly free and transparent trading regime is also a prerequisite.

At the same time, a parallel strategy can be pursued domestically: additional processing of unique, safe, healthy, environmentally friendly, animal-friendly, quality-based products that have a high domestic income elasticity. Plant protein products are one such example. This also includes facilitating more retail in-store food processing and take-out services. Most grocery stores already have sections that cater to food that can be taken home and served without much or any further processing.

6. Agri-Industry & Value-Added

This chapter provides an abbreviated profile of how agri-food industries are linked to primary producers. These are essentially either **forward linkages** or **backward linkages**. Forward linkages include processors, intermediaries and retailers. Backward linkages are typically suppliers who provide the inputs required by primary agricultural producers.

6.1 AN OVERVIEW

The whole agri-business complex is much larger than the relatively miniscule number of primary agricultural producers (some 193,000) would suggest (Table 3.4). Estimated total employment in the agri-business sector, in total, is slightly over 2 million, about ten times the number of farmers and about 13% of the total payroll employment in Canada. In the Maritimes and the prairies, this estimate is about 16 percent (Table 6.1).

Table 6.1 Agribusiness Employment in Canada, 2015*

Region	Employment*	% Canada
MARITIMES	161,780	16.6%
QUEBEC	428,399	12.5%
ONTARIO	727,019	12.2%
PRAIRIES	502,513	16.4%
B.C./NORTH	234,979	11.3%
CANADA	2,054,690	13.2%

* Agribusiness defined as primary agriculture and fisheries, food & beverage processing, wholesale trade, retail trade & institutional feeding (restaurants, etc.).
Basic data: CANSIM 281-0024 & *Canadian Encyclopaedia*

Focusing solely on food and beverage processing further underlines how integral and relatively large the agri-food industry is in the Canadian economy. Table 6.2 highlights the following:

- The food and beverage processing industry is the second-largest manufacturing industry in Canada in terms of value of production, with shipments worth almost $110 billion in 2016, accounting for about 16% of total manufacturing shipments and about 2% of the national gross domestic product (GDP). It is the largest single manufacturing employer and provides employment for about 250,000 Canadians.[72]

- The largest food and beverage processing industry is meat product manufacturing, which accounts for about 25% of all shipments and sales of $26.7 billion in 2016. Dairy product manufacturing is the second-largest industry with sales of $13.8 billion, followed by grain and oilseed milling with sales of $10.3 billion. Other industries include:
    - Beverage manufacturing ($14.2 billion)
    - Bread and bakery product manufacturing ($6.8 billion)
    - Animal food manufacturing ($7.7 billion)
    - Fruit and vegetable preserving, and specialty food manufacturing ($7.3 billion, 2014)
    - Seafood product preparation and packaging ($4.6 billion, 2014)
    - Sugar and confectionary product manufacturing ($4.1 billion, 2014)
- The food processing industry is the largest manufacturing industry in most provinces. Although food processing is important to the economies of all provinces, Ontario and Quebec account for most of the production with approximately 62% of total sales, The prairies account for about 22%, BC about 8%, and the Maritimes about 7 percent. Meat is the most significant food industry in Quebec, Ontario, Alberta, and British Columbia; grain and oilseed milling are the largest food industries in Manitoba and Saskatchewan; while seafood is most significant in New Brunswick, Nova Scotia, Prince Edward Island, and Newfoundland and Labrador.
- There are now about 6,500 food and beverage processing establishments in Canada. Ninety percent of the establishments have less than 100 employees, 9% have between 100 and 500 employees, while only 1% of establishments have more than 500 employees.

Table 6.2 Food & Beverage Manufacturing, Sales, by Region, 2016, $ Millions

Category	MARITIMES	QUEBEC	ONTARIO	PRAIRIES	B.C./NORTH	CANADA	No. Plants, 1985***
All Manufacturing	25,289.5	148,925.9	297,507.3	94,999.4	48,152.3	620,396.7	
Food & Beverage Manufacturing	6,543.0	28,783.2	39,628.2	22,022.5	9,788.6	108,946.7	4554
% Share of All Manufacturing	25.9%	19.3%	13.3%	23.2%	20.3%	17.6%	
% Regional Share of Canada	6.0%	26.4%	36.4%	20.2%	9.0%	100.0%	
Food Manufacturing	6,313.20	24,364.90	34,121.50	20,589.70	7,800.80	94,777.1	4250
Meat Product Manufacturing (incl. poultry)	x	6,909.4	8,138.7	9,155.0	x	26,677.1	631
Grain & Oilseed Milling	x	1,036.30	3,172.6	5,871.2	x	10,345.9	69
Animal Foods (incl. Livestock Feed)	x	2,152.00	2,737.80	1,605.20	x	7,694.7	554
Dairy Products	x	5,388.20	5,357.60	x	x	13,827.5	394
Bread & Bakery Products	x	2,140.50	2,807.70	775.5	x	6,844.4	1504
Other Food*	x	2,528.00	5,327.20	957.70	x	10,240.1	1098
Beverage Manufacturing**	229.8	4,418.3	5,506.70	1,432.80	1,987.80	14,169.6	304

* Includes manufacturing of coffee, tea, seasonings, dressings, flavourings, specialty foods, sugar and confectionary products, seafood products, cookies, crackers, pasta, and fruit and vegetable preserving.
** Includes tobacco products.   ***Drawn from the Canadian Encyclopaedia, 1985.
x data suppressed to maintain confidentiality.

Related Statistics Canada data also highlights the following:

- The food and beverage processing industry invests about $2 billion annually in capital expenditures; about 80% for machinery and equipment.

- It supplies approximately 75% of all processed food and beverage products available in Canada and is the largest buyer of Canadian agricultural production.

- Canadian processed food and beverage products are exported to about 190 countries with a significant proportion exported to a few countries. In 2016, nearly 90% of the total went to six major markets:
    - United States (70%)
    - China and Japan (>10%)
    - Mexico, Russia and South Korea (>5%)

These manufacturers range from small, traditional, family-run activities that are highly labor-intensive to large, capital-intensive, and highly mechanized industrial processes. Thus, despite the fact that there are over 6,500 food and beverage processing establishments, a few mega-companies greatly affect the operational structure, conduct, and performance of the industry.

Structure, conduct, and performance is also impacted by government taxation policies and energy pricing. This generally handicaps both farmers/ranchers and agricultural processors throughout Canada relative to producers and processors in other countries. (Also see Chapters 5 and 9.)

The top ten food and beverage manufacturers, in terms of 2016 sales, were Kraft-Heinz, Saputo, McCain Foods, La Co-op fédérée, Molson Canada, Agropur Cooperative, Cott Corporation, Maple Leaf Foods, Nestle Canada, and George Weston; all very large, highly diversified, multinational corporations. Their sales range from $2B to $12 B per annum (Figure 6.1). An abbreviated profile of some of these companies is provided in accompanying Table 6.3.

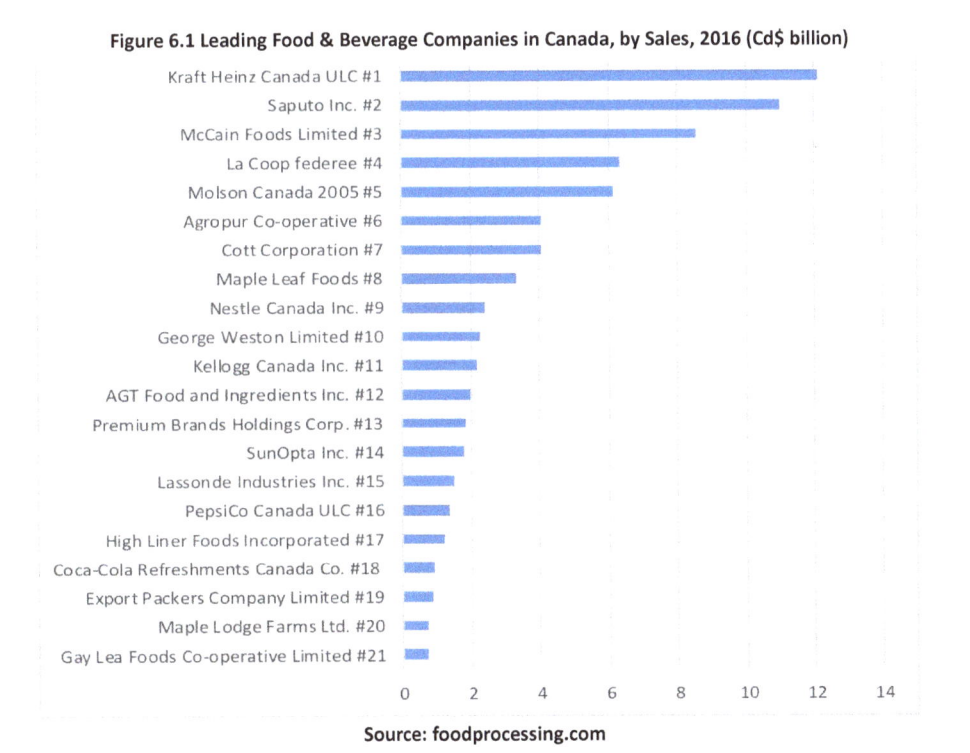

Figure 6.1 Leading Food & Beverage Companies in Canada, by Sales, 2016 (Cd$ billion)

Source: foodprocessing.com

Table 6.3 Profile of Selected Food & Beverage Companies, Canada

Company	Description
Kraft Heinz	Has operations in 40 countries; the third largest food and beverage company in N. America and 5th in the world. Markets in 200 countries. #1 food & beverage company in Canada with over $13 billion in sales. Headquartered in Toronto with facilities in Ingleside, Ontario and Mt. Royal, Quebec. Publically traded; Warren Buffett investor.
Saputo	A Montreal-based dairy company founded in 1954. One of the top 10 dairy processors in the world (incl. USA, Australia, and Argentina) and the largest in Canada. Ranks #19 in Canada/US. Publically traded. #2 in Canada with over $11 B in sales.
McCains	A major Canadian company controlled by the McCain family. Specializes in potato products & frozen foods. Facilities in Grand Falls & Bristal, NB; Anjou, QUE; Etobicoke, ONT; Portage la Prairie & Carberry, MAN; and Coaldale & Chin, AB. Has 53 other plants on six contenents around the world and has sales in excess of Cd$9 B. #3 in Canada ranking. Private.
Molsons	Has five beweries in Canada (Vancouver, Toronto, Montreal, Moncton, & St. John's), 3,000 employees, and $6 B in sales; the 7th largest brewery in the world. #5
Maple Leaf Foods	An amalgamation of Maple Leaf Mills and Canada Packers (1991). Now only produces and sells packaged meats. Owns poultry and hog farms across Canada, as well as Schneider Foods (Kitchener). It's main slaughterhouse is located in Brandon, Manitoba. Public, but controlled by McCain. #8
George Weston	Started in 1882. Evolved from a bread business to a major food processing and distribution company with about $2.5 B in sales. Total of 160 companies, includes Weston Foods, Loblaws, Superstore, and Extra Foods. Public but controlled by the Weston family. #10
Kellogs	American multinational which produces cereals and convenience foods. Manufactures in 18 countries and markets in over 180 countries. Operations in Mississauga (head office), Anjou, QUE; Calgary, AB; and Belleville, ONT. #11

O = mainly livestock-based. See below.

Source: WIKIPEDIA and related

This profile only focuses on food and beverage processing—**forward linkages**. There are also equally important **backward linkages**—the supply of farm inputs—which are examined in Section 6.3, following.

6.2 FORWARD LINKAGES – FOOD & BEVERAGE MANUFACTURING

Forward linkages in both the grain and oilseed sector and the livestock and poultry sector are profiled following.

6.2.1 GRAIN AND OILSEED SECTOR – SELECTED PROFILES

Agri-businesses in the grains and oilseed sector include grain elevators, flour mills, bread and bakery operations, oil crushing plants, barley maltsters, brewers, distilleries, wineries, pulse processors, potato processors, vegetable and fruit processors, sugar processors, and ethanol/biofuel plants.

a. Cereal, Oilseed, and Pulse Elevators

The elevator system in Canada is extensive. It consists of about 33 terminals, 44 process elevators, and 356 primary elevators (Table 6.4).

Table 6.4 Grain & Oilseed Elevators in Canada, 2019

Province	Terminal	Processing*	Malt	Primary	Ethanol	Total Number	Capacity (million tonne)
Nova Scotia	1					1	0.14
Quebec	6					6	1.48
Ontario	17		1			18	2.12
Manitoba	1	8	2	84	1	96	1.84
Saskatchewan		18		186	2	206	4.40
Alberta		9	2	82		93	2.40
B.C.	8		1	4		13	1.23
TOTAL	33	35	6	356	3	433	13.6

*Includes oilseed crushing, milling, distilleries, & pulse processing.

Source: Canada Grain Commission, *Grain Elevators in Canada Licensed by the Canada Grain Commission*, 2019. www.grainscanada.gc.ca

The ocean port terminals are located in Prince Rupert, Surrey, Vancouver (6), Churchill, Thunder Bay (6), Halifax, and along the St. Lawrence (6), There are also smaller terminals at Hamilton (3), Goderich, Owen Sound, Sarnia, Prescott, and Port Colborne.

The relatively new regional primary high-throughput elevators are all located on the prairies and the Fort St. John region of BC. These have replaced the numerous smaller wooden structures (long considered a landmark) previously located in virtually every rail town on the prairies. These facilities generally have 50–140 rail car spots on a main rail line. At least three facilities in Saskatchewan (and one in Manitoba) specialize in pulses. They all grade, buy, and store the grain prior to rail shipment, typically destined for a port terminal. Most also clean and dry grain. The major companies are listed in Table 6.5. Viterra, Richardson, and Cargill have an estimated 75 percent of the total grain market in Western Canada.

Table 6.5 Major Grain Companies in Canada, 2018

Viterra (Glencore Int.)	Anglo-Swiss multinational, also in Australia, with 1/10 of the world grain market. Canada's largest. Viterra is an amalgam of the three Prairie Pools and the UGG, previously Agricore.
Richardson International	"Pioneer" previous trade name. Second largest network on the Prairies. Grain marketing, seed retailing, crop protection, fertilizers, oilseed procssing, and food services packaging. Canadian owned; based in Winnipeg. Private.
Cargill	US-based, operates in 65 countries and employs 142,000 people. Grain marketing, oilseed processing, seed distribution, etc. The largest private agri-food business in the world.
ADM (Archer Daniels Midland)	US-based, operates in 75 countries. Operates 265 processing plants worldwide (incl. ethanol)
Bunge	Founded in the Netherlands in 1818. Now NY-based. Employs 35,000 people in 40 countries, processing oilseeds, wheat, corn, and sugarcane.
Louis Dreyfus	French company founded in 1851, now operates in more than 50 countries. Transports 70m tons of food a year.
Patterson, Parish & Heimbecker	Major players in select Prairie locations.
G3	Emerged from CWB termination. Partial Chinese ownership. Strong growth phase - 3 new inland terminals in Alberta; 3 new inland terminals In Saskatchewan. Port terminal under construction.
Other	Includes Providence Grain Group (7, Canada ownership), Grain Connect (4, Australia-Japanese ownership), and Bunge/COC (Morinville, partial Chinese ownership)

Note: For more details on the traditional big-5, see: Morgan, Dan, *Merchants of Grain,* Penguin, 1979

In addition to inland elevators, all of the larger companies also have their own export terminals. The Port of Vancouver, for example, has Richardson (capacity 5M tonnes), Viterra, Cargill, Parish & Heimbecker (capacity 0.6 M tonnes), Alliance (Patterson-P&H), and G3 (capacity 8M tonnes, 2020).

b. **Flour Mills**[73]

Canada has about fifty-five commercial wheat and oat mills operating in eight Canadian provinces; located about equally between eastern and western Canada. The majority of the wheat milling capacity is in the East, in close proximity to larger urban centers. The majority of the oat milling capacity is situated on the prairies. Canadian mills grind over 3.5 million tonnes of wheat, oats, and barley each year and the major players are: ADM Milling, Ardent Mills, Brant Flour Mills, Grain Millers Inc., Howson & Howson, P&H Milling Group, Pepsi QTG Canada, Prairie Flour Mills, Richardson Milling, and Rogers Foods. Popular brands are Five Roses and Robin Hood.

c. Bread & Bakery Products

The bakery market in Canada it is highly fragmented, with many players and many products. There are approximately 1000 companies and a similar number of new bakery products are launched every year.[74] George Weston, Kellogg, Maple Leaf Foods, and Kraft Foods have about 1/3 of the market (each with an 8% market share), while artisanal bakeries also have about 30% of the market. Bread is increasingly being baked in-house by the major food retailers: Weston (Loblaw, Superstore, Extra Foods), Sobeys-Safeway, and others. Overall, it is a relatively large, but slow-growth $7 B industry[75]

d. Oilseed Crushing Plants

There are now about a dozen crushing plants scattered across the prairies, plus another three in Ontario and Quebec (Figure 6.2). These are generally owned by the major domestic and international grain companies: Viterra, Bunge, ADM, Cargill, Richardson, and Dreyfus.

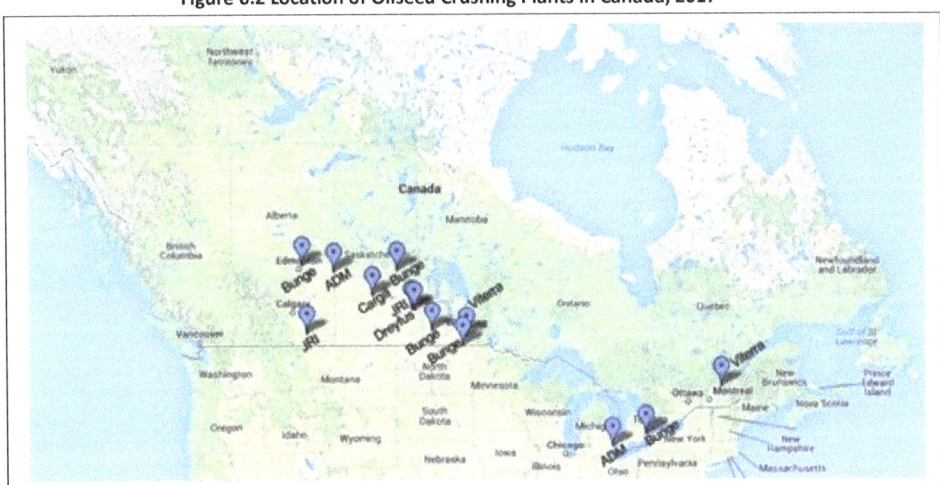

Figure 6.2 Location of Oilseed Crushing Plants in Canada, 2017

Source: copacanada.com

They generally crush canola in the West; soybeans in the East and transform the harvested seed into oil for human consumption/other and meal for livestock feed. The oil is refined to improve its color, flavor, and shelf-life to produce a wide range of consumer food products.

A study released in 2017 estimated that Canadian-grown canola contributes $26.7 billion to the Canadian economy each year, including more than 250,000 Canadian jobs and $11.2 billion in wages (Section 6.4).[76]

e. Barley Malt Plants

The malting barley processing industry produces malt (about 75% of barley) for the beer industry, as well as screenings and sprouts. The major companies are Canada Malting (Thunder Bay and Calgary), Malteurop Canada (Winnipeg), and Rahr (Westcan) Malting (Alix, AB). ConAgra, a major US corporation, is a large shareholder in Canada Malting.

Of all the barley grown by Canadian farmers, less than 25 percent is accepted for malting. About 1.5 M tonnes of Canadian malting barley are sold into the domestic market while export sales average about 500,000 tonnes annually. Domestically, Canadian brewers use about 230,000 tonnes of malt per year to make 1850 million liters of beer.[77]

f. Breweries, Distilleries, and Wineries

Beverage manufacturing in Canada is a $ 14 B industry (Table 6.2). There are two major brewing companies in Canada: Molson-Coors and Labatts. Molson-Coors has six plants (Table 6.3) and Labatts has eight. Molson-Coors also has six regional breweries located in different parts of Canada: Big Rock Brewing and Drummond Brewing (Alberta); Sleeman Brewing and Malting (Ontario and BC); Moosehead Breweries (New Brunswick); Lakeport Brewing, Northern Breweries and Northern Algonquin Breweries (Ontario); and Western Brewing (BC). Labatts is now part of the AB InBev SA/NV conglomerate, the largest brewing company in the world. This is an amalgam of InterBrew (Belgium), AnBev (Brazil), Anheuser-Busch (USA), and SAB Miller (USA). There are, however, also an estimated 540 micro-breweries scattered throughout Canada and these specialty beers are becoming increasingly popular.[78]

The Canadian distilling industry also producers a vast array of spirits (whiskey, rum, vodka, gin, liqueurs, and coolers) but Canadian whiskey is an international standard and is an appellation-protected product (similar to French Champagne). It is distilled from cereal grains, primarily rye and corn. Thirty years ago, Montreal-based Seagram's had 250 brands and was the largest distiller in the world. But by 2002, most of these assets had been sold to the multinationals Diageo, Coca Cola, and Pemod Ricard.[79] The remaining large-scale distillers are Hiram Walker & Sons (Windsor), Highwood Distillers (High River, Alberta), Cirka Distilleries (Montreal), and Pemod Ricard (CC and Wiser's, Montreal) There are also about 33 micro-distillers and craft distillers in Canada.

Canada has also developed a vibrant wine industry. The most recognized areas for producing wines in Canada are southern Ontario (especially the Niagara Peninsula) and the Okanagan Valley in British Columbia. The Niagara Peninsula, with about 11,000 acres of grapes, is the largest viticulture area in Canada. The Okanagan Valley has about 7,500 acres of grapes. There are also small fruit wineries in the Maritimes and the prairies. There are now a total of about 800 licensed wineries in Canada, with almost 500 in Ontario and nearly 300 in BC.[80]

g. Pulse Processing

Canadian pulses are exported to 130 countries around the world and it is the world's largest pea exporter.[81]

Wigmore Farms in Regina is a major pulse processor and exports beans and peas to thirty countries. Alliance Grain Traders Inc., also based in Regina, is another major pulse processor.

Pulse protein fractionation plants are now being developed across the prairies with three already in operation and a fourth (Roquette) anticipated in Portage la Prairie (Figure 6.3). The planned investments will cost in excess of $1 billion.

The global market for specialty food ingredients, such as pea protein, is about US$ 100 billion annually and rapidly growing. In North America, the EU, and even China, the health and nutrition sectors are key markets for ingredients derived from pulses, such as protein and fiber. They can also be used as ingredients to achieve gluten-free status. Plant protein is already used in granola and energy bars, high-protein pasta, baby food, veggie burgers, egg alternatives, beverages, and smoothies. Protein fractions are also used in pet and animal feed.[82]

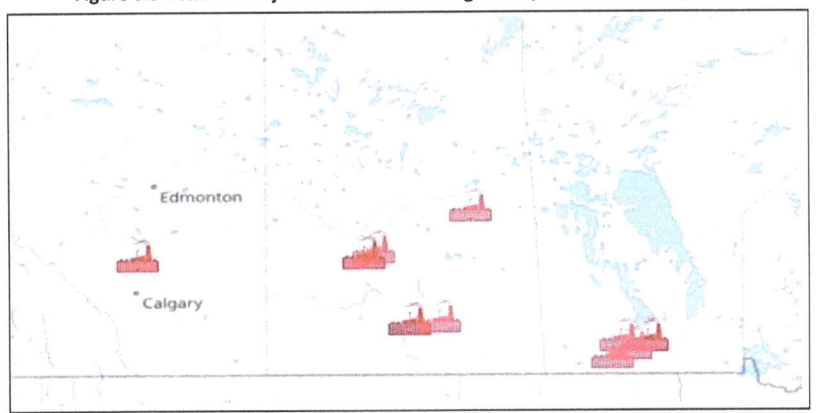

Figure 6.3 Actual & Projected Pulse Processing Plants, Prairie Provinces, 2018

Source: Basic data from the Canada West Foundation, cwf.com

Pulse processing could become a major growth industry because of a number of coalescing factors:

- Rapidly growing national and international market for plant protein (vegans, organic foods, etc.)
- Ready access to vibrant lentil, chickpea, dry pea, and bean production
- Purported health benefits of lentils, beans, and other pulses
- Crop's lower environmental impact, regarding lower CO_2 emissions
- Compatibility with tenets of regenerative/sustainable/diversified cropping practices.

h. Potato Processing Plants

The potato industry is concentrated in New Brunswick-PEI, Ontario, Manitoba, and Alberta.

Headquartered in New Brunswick, McCain Foods (Canada) is an international leader in the frozen food industry and it has two plants in New Brunswick. McCain is the world's largest manufacturer of french fries and frozen potato specialties (e.g., Old Dutch potato chips), as well as making frozen desserts, snacks, appetizers, and authentic Chinese entrees.

McCain also has two facilities in Ontario (Table 6.3). Another major potato processor in New Brunswick is Cavendish Farms, a subsidiary of J. D. Irving Ltd.

In Manitoba, there are three major potato processors: J. R. Simplot, headquartered in Boise, Idaho; Midwest Food Products, and McCain Foods (Canada). McCain has two plants and, together, J.R. Simplot and McCain employ over 1,000 people and contract more than 100,000 acres of potatoes in the province. Simplot represents about 20% of the total Canadian potato market.[83]

In Alberta, the potato processing industry is represented by three major players: Cavendish Farms (a facility previously owned by Maple Leaf Foods), Lamb Weston, and McCain Foods (with two plants). All these plants are located in the Lethbridge-Taber area, drawing their contracted supply from irrigated potato farmers in the area. Edmonton's Little Potato Company has also developed successful niche products.

Frozen potato products are Canada's leading fruit and vegetable preserving and specialty food export, already accounting for 38% of exports in 2014.[84]

i. **Other Vegetable and Fruit Processing**

There are currently over 500 fruit and vegetable preserving, specialty food manufacturing, and frozen food manufacturing establishments in Canada: 210 in Ontario. 130 in Quebec, 70 in BC, and 40 in Alberta.[85] Including potato processing (above), this is a relatively large $7 B industry (Table 6.2).

In the late 1880s, the first fruit and vegetable canning factory was established in Prince Edward County in southeastern Ontario (south of Belleville). By 1900, eight canning plants were operating in the county, one-half of the total in Canada at that time. The major licensed vegetable processors in Ontario today are:

- Bonduelle Ontario Inc.
- Campbell Company of Canada
- Cavendish Appetizers
- Conagra Foods Canada, Inc.
- Countryside Canners Co.
- Highbury Canco Corp.
- Harvest-Pac Products, Inc.
- Lakeview Vegetable Processing, Inc.
- Nationwide Canning Ltd.
- Southcoast IQF
- Sun-Brite Foods, Inc.
- Thomas Canning (Maidstone) Ltd.
- Tomek's Natural Preserves Inc.
- Weil's Food Processing Ltd.

Canadian farmers generally contract to grow certain crops for a specific processor or group of processors and farm-gate prices are typically jointly negotiated. To establish farm-gate prices in Ontario, producers and processors utilize what they call a Final Offer Arbitration (FOA) mechanism.

Most vegetables destined for canning or freezing are now picked by mechanical harvesters, usually owned by the processing companies. Increasingly, this is also true for fruits. For vegetable producers, however, there are on-going challenges. While the overall market for processed tomatoes has been growing, the amount of imports has also been increasing. The market for peas, beans, and sweet corn (at least in Ontario) is in an overall decline.[86]

The processing sector is represented by two Ottawa-based national organizations and their provincial counterparts: the Canadian Food Processor's Association and the Canadian Frozen Food Association.

j. Sugar Processing Plants[87]

Sugar and related confectionary products are also a relatively large $4.1 B industry in Canada (Table 6.2).

There is only one sugar processing facility in Canada and this facility which utilizes sugar beets as a feedstock. Located at Taber, Alberta this facility contracts irrigated sugar beet production from the surrounding area and is operated by Rogers Sugar, a subsidiary of Lantic Inc.

All the other sugar processing facilities in Canada import raw cane sugar: Lantic Sugar in Montreal and Toronto; Redpath in Toronto; and, Rogers in Vancouver.

k. Ethanol & Biodiesel Plants

Since December 2010, the federal government has mandated that gasoline must be at least 5 percent ethanol. In addition, many provinces have equivalent or higher provincial mandates, including a 5% renewable content mandate in Ontario, 7.5% in Saskatchewan, and 8.5% in Manitoba.

There are about twenty-six ethanol and biodiesel plants in Canada, as illustrated in Figure 6.4.[88] There is one in the Maritimes, five in Quebec, twelve in Ontario, seven on the prairies, and one in BC. 2016 bioethanol production in Canada was about 1.75 billion liters and the primary feedstocks remained wheat and corn. 2016 biodiesel production was about 300 million liters and the primary feedstocks are canola oil, soybean oil, cooking oils, animal fat, and recycled oils. To meet legislative mandates, the existing deficit is imported, although wood pellets, a by-product, are exported.

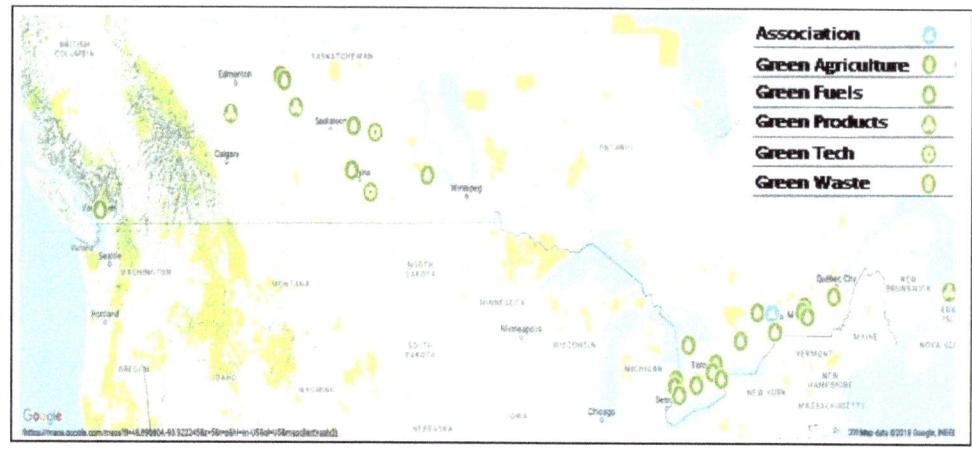

Figure 6.4 Location of Ethanol and Biodiesel Plants in Canada, 2016

Source: Renewable Industries Canada

There is still an on-going debate about the environmental benefits, as well as the economics, of the respective biofuels. According to one report,[89] biodiesel provides 93% more net energy per gallon than is required for its production, whereas ethanol generates only 25% more net energy. Additionally, biodiesel, compared to carbon-based gasoline, is estimated to reduce greenhouse gas emission by 41%, whereas ethanol apparently yields only a 12 percent reduction. Conversely, it is estimated that one acre of corn can produce 420 gallons of ethanol, whereas one acre of soybeans can only produce an estimated 60 gallons of biodiesel.

6.2.2 LIVESTOCK & POULTRY SECTORS – SELECTED PROFILES

Agri-businesses in the livestock and poultry sectors includes beef auction markets, beef feedlots, beef processing plants, pork processing plants, dairy processing plants, and poultry and egg processing plants.

a. **Beef Auction Markets**

Most beef farmers sell their calves through a live (or on-line) auction market where feedlot and other buyers congregate to bid on cattle lots. There are hundreds of such auction markets across Canada but they are concentrated in Alberta where, in 2017, they handled about 2.5 million head. However, even these facilities are gradually consolidating and diminishing in number. In Alberta, for example, the major marts are now Central Livestock (Hrehorets), Nilsson Bros, Balog, and VJV (Vold Jones & Vold).

b. **Beef Feedlots**

Most of the beef feedlots are located in Alberta where, back in 1996, there were still over 400 with sales in excess of $250,000 per annum ($1996). Today, CANFAX lists only 158 feedlots with a minimum capacity of 1000 head. These provide bunk space for 1,327,950 head, 93% of the Western Canadian total. A few major feedlots can accommodate more than 30,000 head and these larger feedlots finish the vast majority of the fat cattle marketed. About one-third of all the Alberta feedlots are located in the irrigated area around Lethbridge. Van Raay and Paskal are particularly large feedlot operators in that area. Western Feedlots (Price) at Strathmore, High River, and Mossleigh is another large 100,000 head feedlot operation.

In Alberta, about one-half of all feed grain (especially barley) is utilized for the backgrounding and fattening of slaughter steers and heifers. Virtually all of this is moved to feedlots by B-train semi-trailers.

c. **Beef Processing Plants**

Meat processing is the largest single food processing sub-sector in Canada, generating about $27 B in sales in 2016 (Table 6.2). This includes beef, pork, poultry, and other meat processing.

In Alberta, the major beef slaughtering plants are now limited to two world-scale plants and two specialty plants:

Table 6.6 Beef Slaughter Plants, Alberta

Company	Location	Daily Kill
Cargill	High River	4125
JBS	Brooks	4000
XL	Calgary	950
Harmony	Balzac	n/a

JBS (a large Brazilian conglomerate) was previously owned by Nilsson Bros. & IBP-Lakeside. Canada Packers, Swift, Burns and other packers have gradually been purchased by other agri-business concerns or simply ceased operations.

Fat cattle usually move by large 6-axle cattle liners, each hauling about 28 tonnes or 45 head from feedlots to slaughter locations.

d. **Pork Processing Plants**

In 2017, there were about 28 federally inspected hog slaughter plants in Canada: Maritimes 0; Quebec 13; Ontario 5; Saskatchewan/Manitoba 5; and Alberta/B.C. 5.[90] The listing for 2016 is provided in Table 6.7.

Table 6.7 Federally-Inspected Hog Processing Plants, 2016

British Colunbia/Alberta:
- Donalds Fine Foods. Langley, BC
- Maple Leaf Foods, Lethbridge, AB
- Lacombe Meat Research Centre, Lacombe, AB
- Sturgeon Valley Park, Sturgeon County, AB
- Olymel, Red Deer, AB
- Trochu Meat Processors, Trochu, AB

Saskatchewan/Manitoba:
- Thunder Creek Pork, Moose Jaw, SK
- Maple Leaf Foods, Brandon, MB
- Winkler Meats, Winkler, MB
- HyLife Foods, Neepawa, MB

Ontario:
- Fearman's Pork, Burlington
- Toronto Abattoirs/Qualiity Meat Packers, TO
- University of Guelph, Guelph
- Conestoga Meat Packers, Breslau
- Great Lakes Specialty Meats, Mitchell

Quebec:
- L. G. Herbert et Fils, Ste-Helene de Bagot
- Agromex, Ange-Gardien
- Les Viandes Du breton, Notre Dame Du Lac
- Atraham Transformation, Yamachiche Cte Maskinonge
- Supraliment S.E.C., St. Esprit
- Olymel. Vallee Jonction
- Olymel, Princeville
- Jacques Forget, St-Louis de Terrebonne
- 9071-3975 Quebec, Yamachiche
- Asia Foods, St-Alexandre
- Viandes Giroux, East Angus
- Xtra Meats, Becancour
- 9181-5688 Quebec (Abattoir Lamarche), Becancour

Atlantic Provinces:
- Antigonish Abattoir, Antigonish, NS

Source: Agriculture & Agri-Food Canada (AAFC)

e. Dairy Processing Plants

Dairy product manufacturing is the second-largest food processing sub-sector in Canada, and generated about $14 B in annual sales in 2016 (Table 6.2). It is now a $17.7 billion a year industry, employs 23,000 Canadians, and accounts for close to 16% of the total shipments in the Canadian food manufacturing industry.[91]

In 2017, there were about 475 dairy processing facilities in Canada, of which about 71% were located in Quebec (41%) and Ontario (30%) (Figure 6.4). This mirrors dairy cow populations. A major processor is Saputo, a Canadian-based company, which is now the ninth largest dairy company in the world with global sales of about US$8.8 B, about one-third of that of Nestle.

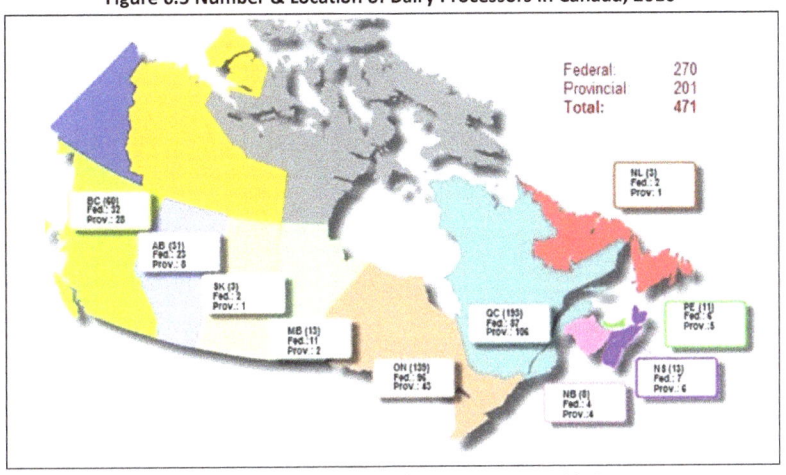

Figure 6.5 Number & Location of Dairy Processors in Canada, 2016

Federal: 270
Provincial: 201
Total: 471

Source: Canadian Dairy Information Centre, 2017

f. Poultry & Egg Processing Plants

Poultry and egg processing facilities include slaughtering plants, hatcheries, egg grading stations, and egg processing plants, and are tabulated in Table 6.8. Their regional distribution generally mirrors provincial broiler, turkey, and egg quotas.

Table 6.8 Number of Poultry & Egg Processing Facilities, Canada, 2016

Facility	Number
Broiler Hatcheries	40
Broiler Hatching Egg Producers	243
Chicken Slaughter Plants	84*
Turkey Hatcheries	11
Turkey Slaughter Plants	50*
Laying Egg Hatcheries	16
Egg Grading Stations	197
Egg Processors	14

*Excludes approx. 60 micro-facilities at Hutterite colonies in Alberta.

Sources: CFC, TFC, EFC, CHEP, FPCC and AAFC

6.3 BACKWARD LINKAGES – FARM SUPPLIERS (FARM INPUTS)

Although easily overlooked, backward linkages with primary agriculture are almost as important to the Canadian economy as forward linkages. Clearly, if farmers are spending, on average, an estimated 82 cents out of every dollar they earn (Table 3.11), the economic viability of these farm sales outlets is directly tied to the farmers' capacity to buy these good and services. And this has an even more pronounced effect on the vitality of rural communities, especially regional rural service centers (sometimes called "growth poles").

Across all farm types, we have estimated (Chapter 3) that chemical fertilizer costs constitute about 22% of total costs, followed closely by livestock and feed purchases (specific to livestock producers) at 15% and 12% respectively. Machinery operating costs (including fuel) and depreciation costs combined make up a relatively large 26% of total farm costs. Seed and pesticide costs are also both relatively large costs for most grain and oilseed producers. We briefly profile some of these major industries immediately following.

6.3.1   GRAIN & OILSEED SECTOR – SELECTED PROFILES

Major input providers in the grain and oilseed sector include the fertilizer companies, the seed and chemical companies, certified seed suppliers, and the machinery companies.

a. Fertilizer Companies

For crop producers, synthetic (chemical) fertilizer costs make up about 15% of total annual cash costs, depending upon the particular type of crop production (Table 3.7).

The major fertilizer companies and their scale of operation is indicated in Figure 6.6. With the recent mega-merger of Saskatchewan-based PotashCorp and Agrium, the combined company easily becomes the largest in the world. Mosaic in Saskatchewan is also a relatively large producer. CF industries (Westco) is also active in Canada, producing gas-heated nitrogen fertilizer near Medicine Hat, AB.

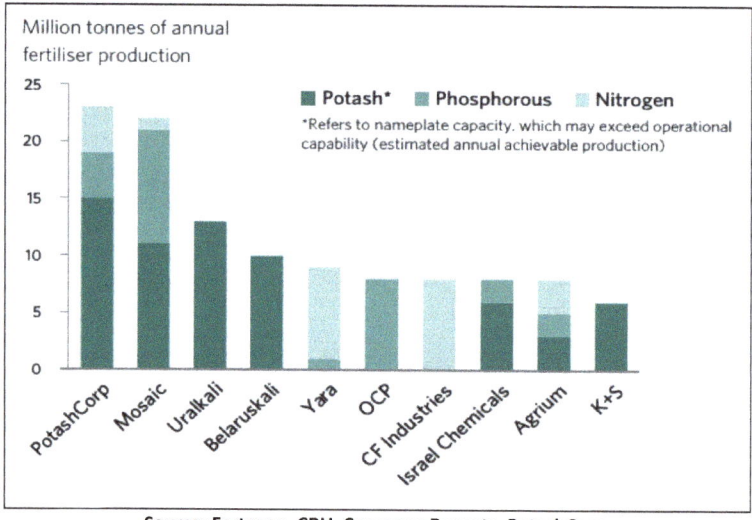

Figure 6.6 World's Ten Largest Fertilizer Companies

Source: Fertecon, CRU, Company Reports, PotashCorp

Canada (with 45% of the world's potash) produces virtually all of its own potash, most of its own nitrogen and sulfur (utilizing natural gas as a heat source), but virtually none of its phosphorus. This generally comes from the southern USA (possibly trans-shipped from Chile or Peru), although 70% of commercial reserves are found in Morocco. In Canada, 95% of all potash is moved by rail and a reported 18% of rail revenue is generated by the fertilizer industry (CFI).

From major storage facilities, fertilizer is usually delivered to local agro-centers in super-Bs and tanker-trucks such that it can subsequently be sold and delivered directly to farmers, most of it during the May–June seeding-spraying period. These numerous agro-centers are increasingly owned and operated by either these same fertilizer companies or the major elevator companies. Agrium (under the Nutrien banner) now has about 550 farm outlets in Western Canada[92]; Richardson Pioneer has about 100.[93] These outlets generally also sell certified seed and crop protection products (herbicides, insecticides, etc.) and some also offer other custom services—spraying, fertilizer applications, advisory services, and so on.

b. **Seed/Chemical Companies**

For crop producers, seed and chemicals (herbicides, insecticides, and fungicides) make up about 20% of total annual cash costs.

Consolidation in the industry continues at a furious pace, as illustrated in Figure 6.7, and as of 2019, this is already outdated with the Syngenta-ChemChina, Dow-Dupont, and Bayer-Monsanto mergers.

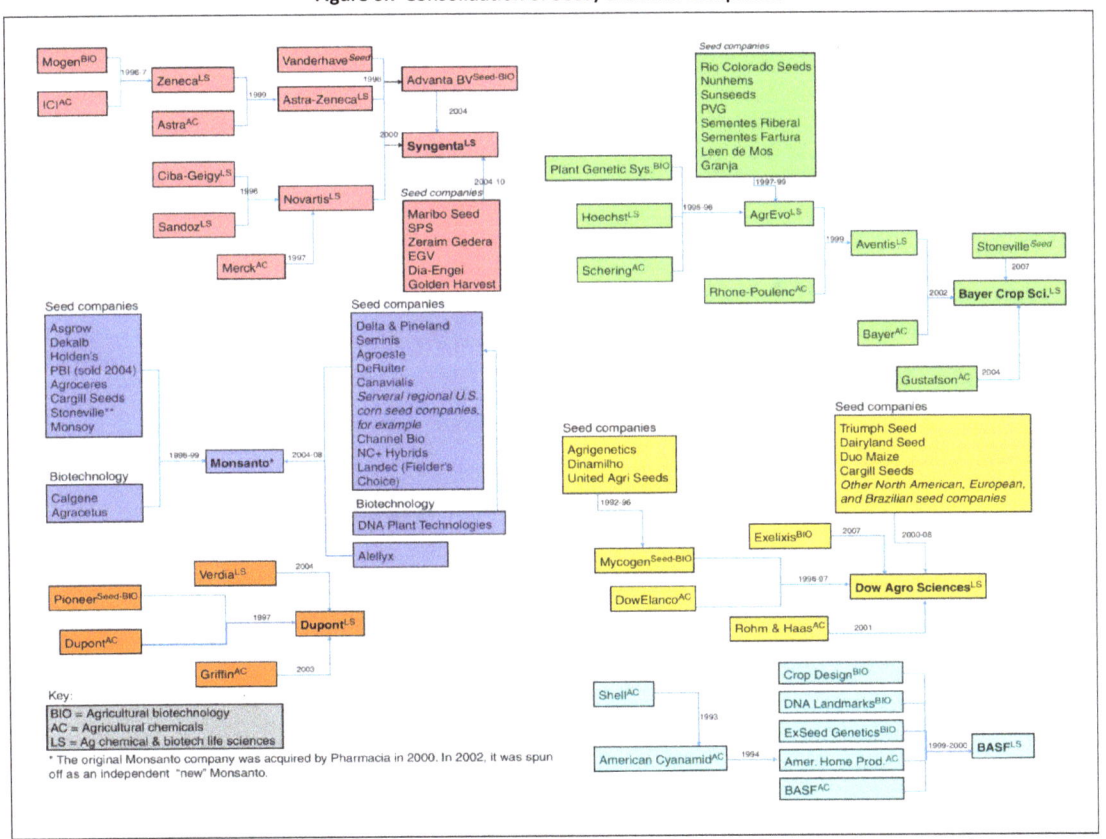

Figure 6.7 Consolidation of Seed/Chemical Companies

Sources: USDA/ERS using Copping (2003), Fernandez-Cornejo (2004), Howard (2009), and company websites

Consolidation and re-branding continues and we are now (2019) down to four very large multinational companies (Figure 6.8):

1. Bayer-Monsanto
2. Syngenta-ChemChina
3. Dow-Dupont (Corteva)
4. BASF

Figure 6.8 Market Share of Seeds and Agrochemicals by the Big Four, 2018

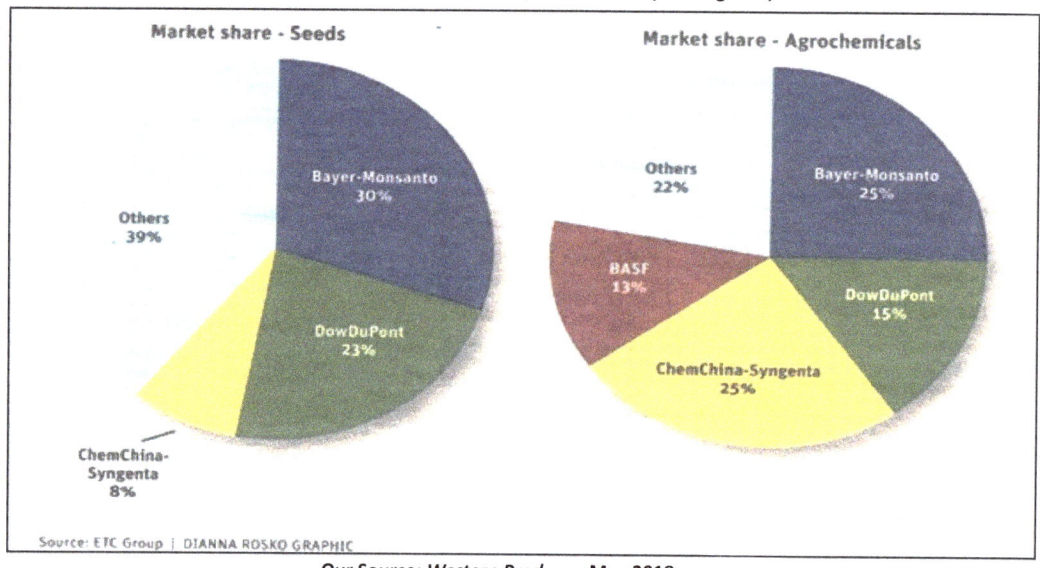

Our Source: *Western Producer,* May 2018

Before the latest round of mergers, their respective seed and agrochemical sales were approximately as indicated in Figure 6.9.

Figure 6.9 The Top 10 Seed and Agrochemical Companies, 2015

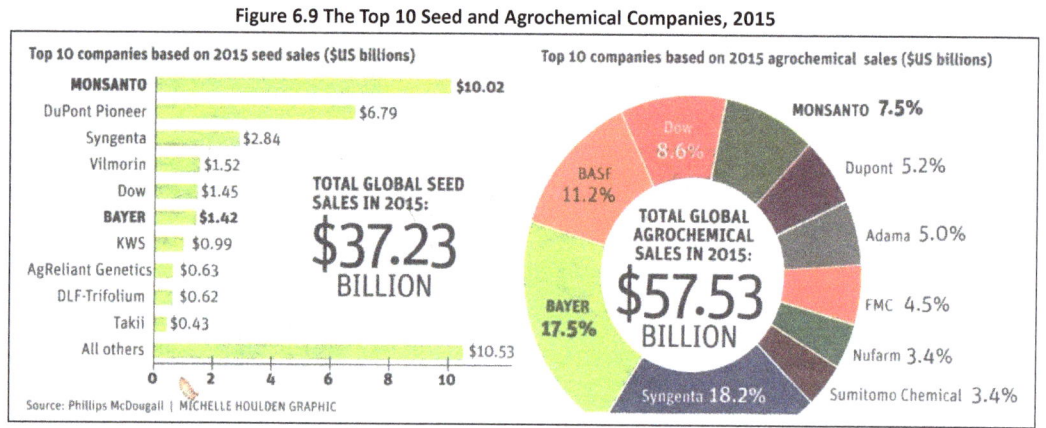

Our Source: *Western Producer*, 2018

In 2017, the big four had agrochemical sales of about $26 B; about 46% of the total market.[94] Bayer and BASF are both German-based, Syngenta is/was Swiss-based, and Dow-Dupont and Monsanto are US-based.

Prior to the latest Bayer-Monsanto mega-merger, Bayer controlled the Liberty Link herbicide-resistant seed and genetics, and Monsanto controlled the Roundup herbicide-resistant seed and genetics. Bayer now owns the Roundup Ready brand while the Liberty Link brand has been sold to BASF. Liberty Link (Invigor) has about 55% of the canola market, Roundup has about 40%, and Clearfield (a non-GMO BASF product) has about 5 percent.[95] Three companies now control about 70 percent of the world's pesticide market. For farmers, this level of market concentration and control is intimidating.

There are numerous smaller seed companies in Canada, including Canterra, Brett Young, and others.[96] Fertilizer-seed retailers are also becoming increasingly active, particularly some large grain and fertilizer companies (e.g., Nutrien).

a. **Certified Seed Suppliers**

Although the major seed and agrochemical companies are the principal source of genetically modified (GM) seed (especially canola, soybean, and corn seed), most other seed is made available through SeCan, the largest supplier of certified seed in Canada. SeCan member companies distribute dominant varieties of cereals (wheat, barley, oats, rye, and triticale), oilseeds (soybeans and flax), pulse crops (pea, lentil, and field bean), and forages.[97] As the principal distributors of varieties developed by government/university researchers, SeCan collects seed royalties to support continued public-sector breeding and research. During the 40 years of its existence and the sale of more than 480 varieties of 27 crops, these cumulative royalty payments now exceed 100 million dollars.

b. **Machinery Companies**

Like the major seed/agrochemical companies, the farm machinery industry has witnessed a similar consolidation process, as illustrated in Figure 6.10:

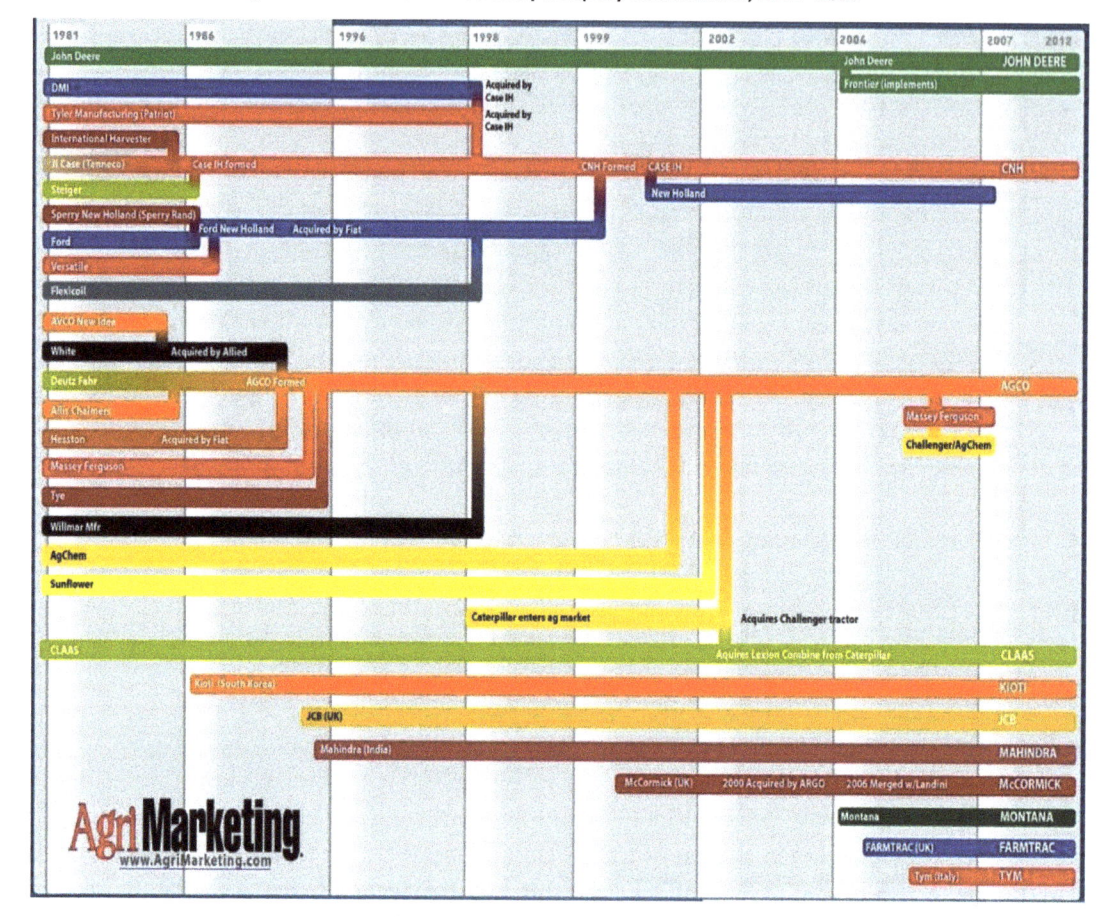

Figure 6.10 Global Farm Machinery Company Consolidation, 1981 - 2012

Our Source: AgriMarketing website

The farm machinery market in Canada is now dominated by just four full-line multinational companies:

1. John Deere
2. CNH Global – Case/IH and Ford/New Holland
3. AGCO – Massey Ferguson, Challenger, etc.
4. Buhler – Versatile, Ezee-On, Farm King, etc.

John Deere and CNH are the two Goliaths (Figure 6.11).

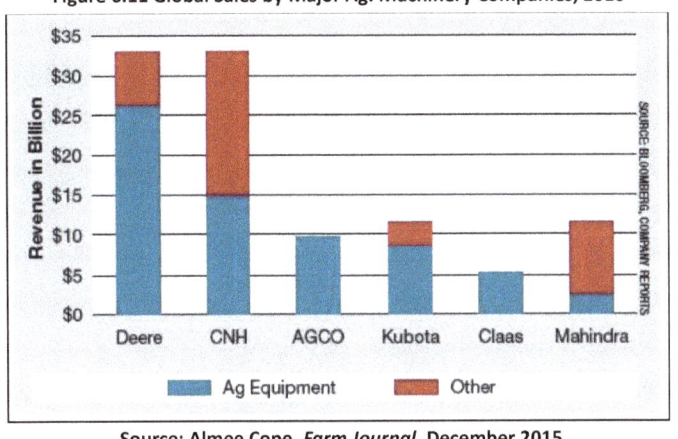

Figure 6.11 Global Sales by Major Ag. Machinery Companies, 2016

Source: Almee Cope, *Farm Journal,* December 2015

However, only Buhler (relatively small, with Russian ownership), headquartered in Winnipeg, manufacturers both large four-wheel tractors and a combine (Versatile) in Canada. Claus/Lexicon is headquartered in Germany and specializes in combines. Both Kubota (Japan) and Mahindra (India) specialize in smaller tractors (<140 hp) and related equipment.

Canadian farmers typically buy about $2–3 B worth of machinery and implements annually (about 3% of the world total), including an average of 19,000 tractors, 3,500 swathers, 4,000 combines, plus thousands of balers, zero-till drills/grain tanks, grain carts, plows, discers, and other harvesting and tillage tools.[98] But the multinationals make almost nothing in Canada, one exception being a zero-till drill manufactured by CNH in Saskatoon following its purchase of Saskatchewan-based Flexi-Coil. John Deere, after purchasing the intellectual property of Conserva Pak™ in 2007, also continued to contract the manufacture of various zero-till drill components at Indian Head, Saskatchewan.

However, a very large percentage of Canada's major farm implements is imported, almost all of them from the USA, Europe (Claus), and Japan (Kubota). Canada is the largest export market for US agricultural equipment, with US exports totaling US$2.6 billion in 2017.[99] The major companies also out-source and re-export (e.g., John Deere lawn mowers from South Korea).

Canadian-made implements include the very popular Bourgault zero-till drill/grain tank, manufactured in Saskatchewan while Manitoba-based MacDon makes almost all of the swathers marketed in Canada.[100] Vermeer also makes balers, etc., in Eastern Canada. Additionally, there are numerous short-line agricultural equipment manufacturers in Canada and that number continues to grow, especially in Saskatchewan. In 1997, there were already 247 manufacturers of agricultural implements in Canada and almost all of them employed less than 100 workers. There are still dozens, if not hundreds, of these smaller short-line agricultural equipment manufacturers. Medium-sized short-line agricultural implement manufacturers include Degelman, Seed Hawk, Meridian/Sakundiak, Brandt, Honey Bee, and numerous others. [101]

6.3.2   LIVESTOCK & POULTRY SECTORS – SELECTED PROFILES

Major agri-businesses providing inputs for the livestock and poultry sectors include feed mills, veterinary clinics, and animal genetic suppliers.

a. Feed Mills

Processed feeds are a very significant on-farm cost for livestock producers, averaging about 27% of their total annual cash costs, depending upon the type of livestock operation in question (Table 3.7).[102]

There are about 500 feed mills scattered across Canada,[103] usually located near most major service centers and typically with a capacity of approximately 50,000 tonnes. There are 6–8 major commercial feed manufacturers and distributors, including Purina Canada, Shur-Gain Feeds, Master Feeds, and Nutrena (Cargill). There are also numerous independents (e.g., cooperatives) but consolidation continues. These mills typically offer a variety of processed or raw products for livestock producers—ground feed, rolled feed, feed additives, and so on. At the same time, larger beef feedlots, hog operations, and poultry operations often have their own feed mill on-site, now estimated to be about ½ of the total potential processed feed market.

b. Veterinary Clinics

In 2018, there were about 3,224 predominately rural veterinary clinics in Canada[104] and many, if not most, of these specialize in large animals; beef and dairy cattle, horses, and so on. The viability of these clinics is heavily dependent upon the demand for these services by farmers and ranchers.

c. Animal Genetics

Dairy, beef, and hog farmers are increasingly turning to artificial insemination to fertilize female animals with elite male genetic material (i.e., semen) to improve the overall health and productivity of their herd. Figure 6.12 identifies the major national and international providers.

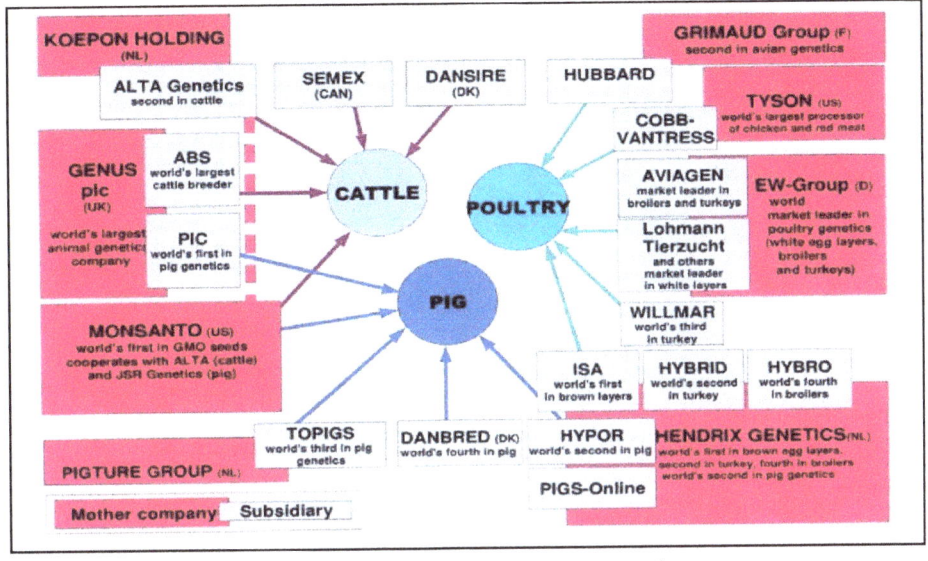

Figure 6.12 Principal Genetics Companies in Global Livestock Breeding

Source: Susanne Gura, *Livestock Genetics Companies*

Semex, in particular, is a major supplier of dairy genetics in Canada, but especially around the world. Alta Genetics is also a major provider. Further details can be found in *Statistics of Canada's Animal Genetics*, 2018 edition.[105]

6.3.3   OTHER INPUT PROVIDERS

There are a whole host of other related businesses in rural and urban areas that directly depend, in whole or in part, on the farming community. These include:

- Fuel & lubricant suppliers – coops, Petro Canada, Esso, Shell, etc. (5–10% of annual farm cash costs; see Table 3.7)
- Part suppliers – Bumper to Bumper, NAPA, etc.
- Auto dealerships – Ford F-150, GMC Silverado, Dodge Ram, etc.
- Large truck dealerships – Kenworth, Freightliner, Peterbilt, Mac, International, Volvo, Western Star, etc.
- Trailers, grain boxes/hoists; grain and livestock trailers – CIM, SWS, Double A, Linden, Doepker, FGI, Lode-King, etc.
- General farm supplies (hardware, grain bins, aeration fans, horse trailers, flat decks, steel fencing/gates, etc.) – e.g., Peavy Mart, Flaman, etc.
- Building suppliers – Home Hardware, Lowes/Rona, Home Depot, etc.
- Specialized electronic providers – grain sensors, seeding monitors, farm surveillance systems, drones for crop monitoring, animal trackers, rumen monitors, etc.
- Building contractors (general, carpenters, electrical, plumbing, roofing, siding, stonework, door installers, concrete, etc.)

- Utility providers – gas, electricity, wifi, satellite, etc.
- Farm auto, truck, and agricultural machinery mechanics and welders
- Seed cleaning plants – co-ops, private, and municipal
- Farm consultants for both crops and livestock, offering crop scouting, ration balancing, breeding, financial management, precision crop production, precision livestock production, and related services
- Custom farm operations – seeding, spraying, combining, corral cleaning, manure hauling, grain hauling, etc. And so on.

Most of these goods and services are directly required by farms and ranches. Additionally, the farm family and related rural residents purchase (like everyone else) food, health items, liquor, banking services, clothes, haircuts, recreation, and so on from local retail outlets. Schools, hospitals, municipal and provincial government offices, farm financial institutions (e.g., FCC), etc. also all have a farm-related clientele. (Also see Chapter 7.)

6.4 THE TOTAL IMPACT [106]

Backward and forward linkages, together with primary agriculture, make up the agri-food complex, a relatively large industry that is integral to the socio-economic well-being of virtually all Canadians:

Agri-Industry Providers ⇔ Producers ⇔ Agri-Industry Processors/Intermediaries/Retailers ⇔ Agri-Industry Consumers

Backward (input) linkages and forward (processing + wholesale + retail + food services) linkages with agricultural producers clearly generate a cumulative total impact on the economy, which far exceeds the value of primary production itself. To illustrate, consider the estimated total impact of the canola industry on the Canadian economy. In Figure 6.13, we see (illustrated in black) that although direct canola crop revenue is only about $9.2 billion (Table 3.6), the total revenue generated by the industry (when including the value of the meal for livestock, oil for food and fuel, port activities, transportation, and elevation) is estimated to be about $26.7 billion. This implies a multiplier of about 26.7/9.2 = 2.89. Figure 6.13 also illustrates how this ripples across the country.

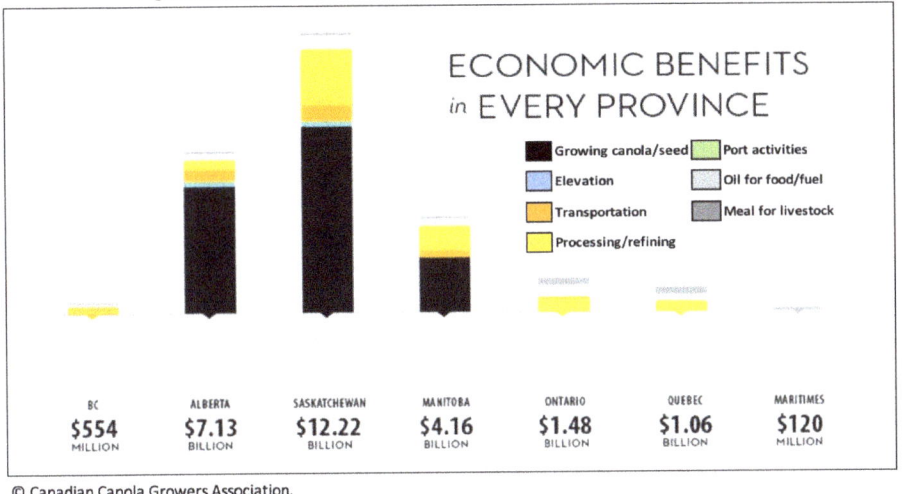

Figure 6.13 Illustration of the Canola Multiplier and Provincial Spin-Offs

ECONOMIC BENEFITS in EVERY PROVINCE

- Growing canola/seed
- Port activities
- Elevation
- Oil for food/fuel
- Transportation
- Meal for livestock
- Processing/refining

BC	ALBERTA	SASKATCHEWAN	MANITOBA	ONTARIO	QUEBEC	MARITIMES
$554 MILLION	$7.13 BILLION	$12.22 BILLION	$4.16 BILLION	$1.48 BILLION	$1.06 BILLION	$120 MILLION

© Canadian Canola Growers Association.

Economic multipliers provide a means to quantitatively measure and assess the impact of changes in an economy. When there is a change in the economy (business opening/closing/expanding/contracting, changes in the demands for a commodity, tax changes, interest rate changes, etc.), it impacts first on those firms, households, and/or governments most directly affected. Then there is a ripple effect as this impact spreads throughout the economy, affecting other firms, households, and governments.

Any change effectively generates three types of effects:

- Initial direct industry impact;
- Indirect ripple effects (sometimes called spin-offs); and,
- An induced income effect as incomes are impacted which, in turn, affects consumption levels.

When an industry increases its output, it must obtain more inputs, which are provided by other industries. The expansion of these other industries means increased demands are placed on their suppliers, and so on through a chain of interdependent industries. This is referred to as the **direct and indirect industry effect.**[107] In addition, as industries increase their production, they increase staff and thus pay more in wages. This increased income in the hands of consumers can generate additional consumption and subsequent increases in industry outputs. This is referred to as the **induced income effect**. The cumulative impact is then commonly referred to as a multiplier with respect to a particular variable. To illustrate, assume we have the following multipliers:

Gross Output 2.692
GDP 0.995
Household Income 0.546
Employment 0.287

Then, for example, a $10,000 expenditure in the agricultural industry should ultimately add about $26,920 to gross output, $9,950 to GDP (value-added), about 0.287 person-years of work, and about $5,480 to household income. And this, of course, varies by sub-sector.

Generally, the larger, more self-sufficient, and more sophisticated an economy (i.e., more integrated), the larger the multiplier. The more Canadian-made inputs a farmer/rancher uses, the greater the value-added in Canada. The greater the Canadian-based processing of agricultural products, the greater the value-added in Canada. **Leakages,** especially imports from outside the designated area or country and the export of unprocessed products, truncate the ripple effect.

Estimates of the agri-food system's total impact on GDP (value-added) and employment in Canada are indicated in Figure 6.14.

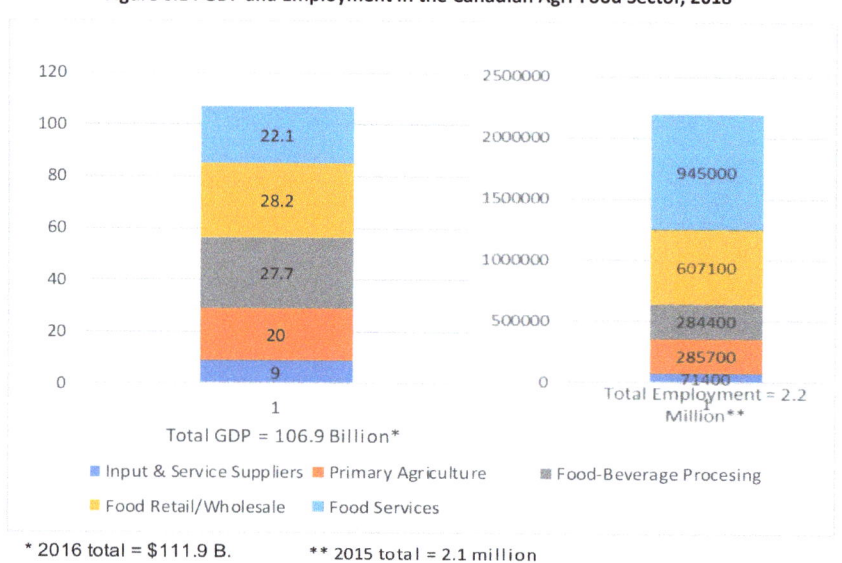

Figure 6.14 GDP and Employment in the Canadian Agri-Food Sector, 2018

Sector	GDP (Billion)	Employment
Input & Service Suppliers	9	71,400
Primary Agriculture	20	285,700
Food-Beverage Processing	27.7	284,400
Food Retail/Wholesale	28.2	607,100
Food Services	22.1	945,000
Total	**106.9 Billion***	**2.2 Million*** *

* 2016 total = $111.9 B. ** 2015 total = 2.1 million

Source: Basic data from AAFC, *An Overview of the Canadian Agriculture and Agri-Food System,* 2015, Ottawa, pp. 23 and 25

In terms of both GDP and employment, primary agriculture only represents about 20% of the entire agri-food complex. In total, the agri-food system contributes about 7% to Canada's GDP and generates employment for about 12% of the entire Canadian labor force.[108]

7. *Rural Socio-Economic Profile*[109]

This chapter describes the economic and demographic characteristics of rural Canada and also identifies issues and opportunities, including resource use conflicts, public services in sparsely populated areas, and the importance of local initiative.

7.1 WHAT IS RURAL?

Rural can be defined in numerous ways and different definitions serve different purposes. There is a longstanding debate over whether "rural" is a geographical concept, a location with boundaries on a map, or whether it is a social representation, a community of interest, a culture, and way of life. As a geographical concept, it is equally ambiguous whether this should be defined in terms of population density, population size, distance from an urban area, or distance to an essential service.[110] In any event, as seen in Figure 7.1, all rural communities share two distinct characteristics: a) a low population density; and/or b) a long distance to density.[111]

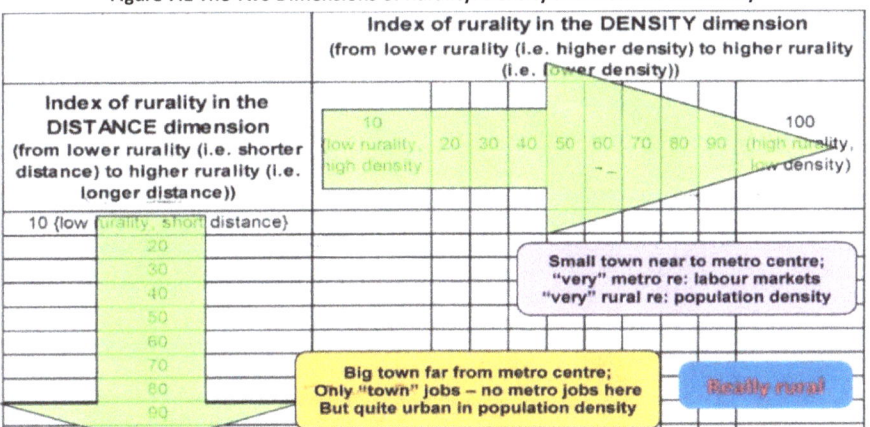

Figure 7.1 The Two Dimensions of Rurality: Density and Distance to Density

Source: Ray Bollman, personal communication

Density indicates the size of economic output or total purchasing power per unit of surface area—say a square kilometer. It is highest in large cities, where economic activity is concentrated, and much lower in rural areas. Density refers to the economic mass per unit of land area, or the geographic compactness of economic activity. It is shorthand for the level of output produced and thus the income generated per unit of land area. It can be measured as the value-added or gross domestic product (GDP) generated per square kilometer of land. Thus, given that a high density requires the geographic concentration of labor and capital, it is highly correlated with both employment and population density. High densities are the defining characteristic of urban settlements and, conversely, low densities are the defining characteristic of rural settlements.

Distance measures the ease of reaching markets and access to opportunity. It refers to the ease or difficulty for goods, services, labor, capital, information and ideas to move from Location A to Location B. It measures how easily capital flows, labor moves, goods are transported, and services are delivered between two locations. For trade in goods and services, distance costs time and money. The placement and quality of transport infrastructure and the availability of transport can dramatically affect the economic distance between any two areas even though the Euclidean (straight-line) distance could be identical. For labor mobility, distance also captures the psychic costs of separation from familiar territory.

There is not a simple urban-rural split. A continuum of densities and distances give rise to a portfolio of interrelated places: major cities, secondary cities, smaller urban centers, towns and villages, and outlying rural areas.[112] But to a large extent, they are symbiotic.

Most of the data presented following is based on decennial Census data tabulated by Statistics Canada, which defines "rural" as individuals living in the countryside outside population centers of 1,000 people or more. This would, therefore, includes all hamlets, villages, or towns with less than 1,000 people, regardless of how they are publicly designated.

7.2 HISTORIC TRENDS

Historic socio-economic trends in rural Canada are generally well-known and well understood. Scholarly articles on the subject are numerous.[113] And the so-called "hollowing-out" of rural Canada is regularly bemoaned by rural commentators and rural-based social activists. And, certainly, looking at rural-urban population trends during the last 175 years is not encouraging, as seen in Figure 7.2.

Figure 7.2 Percent Rural & Urban Population, Canada, 1851 - 2016

Source: GOC, *Population Census*, 1851 - 2016

But, essentially, the growth of rural Canada has simply been slower than urban growth; accentuated by the gradual urbanization of prior rural areas over time. Generally, from a spatial perspective, the greater the distance from a metropolitan area,[114] the lower the historical rate of population growth. (See Figure 7.3.)

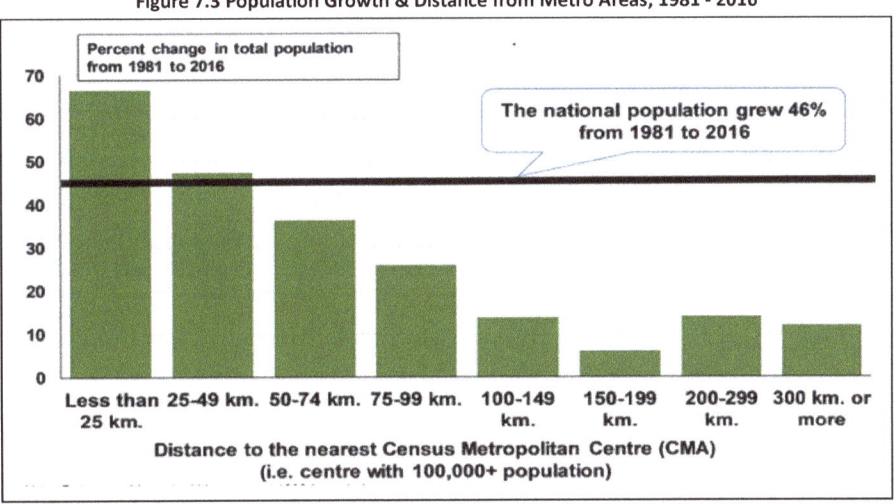

Figure 7.3 Population Growth & Distance from Metro Areas, 1981 - 2016

Note: Data are tabulated with constant 1996 boundaries
Source: Statistics Canada, *Census of Population,* 1981 to 2016
Graphic by Ray Bollman

At the same time, the rural population is still fairly large vis-à-vis the urban population. This is especially true in the Maritimes (44%) and, to a lesser extent, the prairies (21%). (Table 7.1) Rural residents throughout Canada include not only active farmers/ranchers, but also retirees, commuters, government employees (hospitals, schools, counties, etc.), retailers, or employees of a resource-based business. This rural cohort (often with a similar socio-pollical perspective) thus retains a much larger voice in the social-political arena than do farmers/ranchers by themselves.

Table 7.1 Farm, Rural, & Total Population, Canada, 2011

Region	Farm Populaton	Rural Population	TOTAL POPULATION	% FARM	% RURAL
Maritimes	24,155	1,027,600	2,327,638	1%	44%
Quebec	90,735	1,511,525	7,903,001	1%	19%
Ontario	163,435	1,775,670	12,851,821	1%	14%
Prairies	257,015	1,236,765	5,886,906	4%	21%
B.C.	49,840	600,305	4,400,057	1%	14%
CANADA*	585,180	6,151,865	33,369,423	2%	18%

* Excludes the NWT and Yukon.   Also see Table 3.13.

Still, their gradually diminishing role is not very different from the socio-economic evolution of virtually every country in the world.[115] Some communities will ultimately thrive; some will ultimately wither. And this evolution basically has its' own internal dynamic, not unlike the creative destruction inherent to innovation and growth in any free market economy.[116]

In Canada, this dynamic has generally been characterized by the following:

- The relative increase in the value of human time. Thus, we have seen the continuing substitution of capital for labor in primary agriculture. While increasing productivity per person, mechanization has also resulted in farm consolidation and the out-migration of farm families from rural areas.

- The falling price of transporting goods and services, as well as people. Thus, we have seen a change in the shopping patterns of rural dwellers due to both the urbanization of rural tastes and improvements in the transportation system. Rural residents are now prepared to travel farther to obtain additional choice and lower prices. On-line merchants also do a good business in rural communities. Small communities, consequently, lost trade and service outlets while regional shopping centers expanded. Yet, conversely, the falling price of transporting goods does help rural Canada remain competitive in manufacturing.

- An accompanying consolidation of public infrastructure. Smaller schools were closed and students were bussed to larger regional central high schools, which could offer a wider variety of programs. Smaller hospitals with limited facilities were consolidated into regional hospitals (with high-tech and economies of scale) which could provide better health care. Numerous small post offices were closed. Government offices were similarly regionalized. And, most often, new infrastructure requirements (e.g., fiber optics, cable TV), were located in larger centers.

- The price of transferring information continues to fall. Telecommunication prices continue to decline. The internet and other digital linkages are now commonplace. This helps level the playing field and, thus, enhances opportunities in rural areas vis-à-vis more urban areas. Distance has been mitigated by the knowledge economy and the increasing ease of working from anywhere with a combination of power, technology and wifi.

- And, finally, this dynamic is reinforced by rural residents, who frequently want to retire in communities where there is good access to health facilities, recreation facilities, and retail services, while still remaining reasonably close to friends and family. Conversely, professional people (e.g., doctors and government employees) are sometimes reluctant to re-locate to a rural community that does not have similar urban-based amenities. Doctors, in particular, are under-represented.

The rural hollowing-out process is most pronounced with respect to rural youth who face diminished employment opportunities in the rural community. Thus, they often endure the unenviable choice of either remaining in the rural community and being unemployed (Figure 7.4)

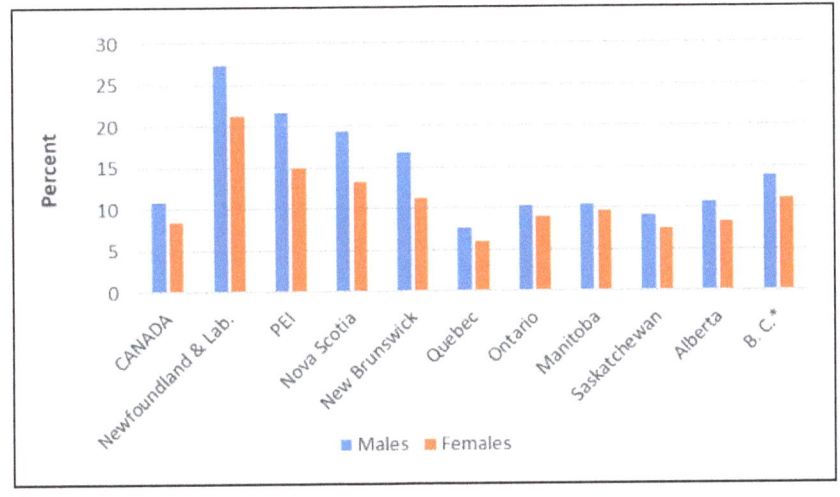

Figure 7.4 Youth Unemployment (15-24 yrs.) in Rural and Small

* 2016.

Source: Statistics Canada. Table 14-10-0106-01

or migrating to an urban center where they can more easily pursue additional education and/or urban-based employment. They also migrate to urban centers because of enhanced social opportunities, more entertainment/recreational opportunities, and more travel opportunities.[117]

Youth unemployment in the Maritimes is particularly high and out-migration throughout rural Canada is especially high for the age 20–30 cohort (Figure 7.5). This brain-drain results in the possible loss of rural Canada's "best and brightest."

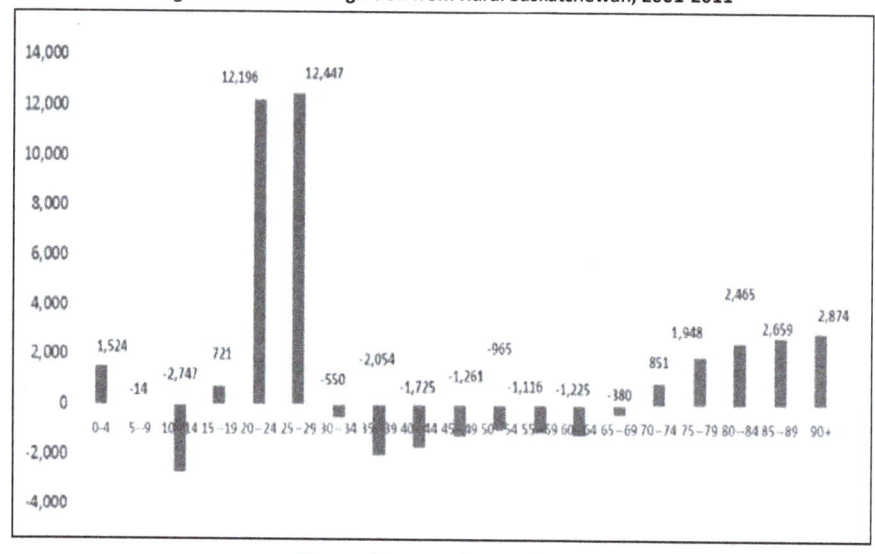

Figure 7.5 Net Out-Migration from Rural Saskatchewan, 2001-2011

Source: Courtesy of Ray Bollman

7.3 HUMAN CAPITAL IN RURAL CANADA

A report by the Canada Council on Learning[118] highlights three on-going educational issues in rural Canada:

- Rural communities generally have higher high school dropout rates (16.4%) relative to urban communities (9.2%);

- Rural communities across the country tend to have lower average levels of education, with urban areas having slightly more than 60% of their population having some post-secondary education while rural communities have slightly less than 50% with some kind of post-secondary education; and,

- Among the 34 (relatively rich) OECD countries, Canada has the largest and hence worst rural-urban gap with respect to levels of education in the workforce.

These findings ae generally consistent with existing data on education levels for farmers/ranchers, as previously discussed in Section 3.5.3.

7.4 RELATED CHALLENGES

Many, if not most, of the related contemporary challenges facing rural Canada have an overarching social dimension. These include issues surrounding: a) labor constraints, b) distinct communities; c) resource use conflicts; d) preserving law and order in remote areas; and, e) climate change adaptation.

7.4.1 LABOR & LEADERSHIP CONSTRAINTS

There are now (post 2008) fewer potential entrants to the non-metro workforce than potential retirees. This has generated thousands of potential rural job vacancies. One way to address this rural labor market shortage would be to attract immigrants, but the vast majority of immigrants to Canada quickly take up residence in major urban areas.[119] Cities generally have more social amenities, afford more anonymity, more easily accommodate immigrant ghettoization, and are generally more culturally diverse.

The federal Temporary Foreign Worker Program (TFWP) is supposed to allow Canadian employers to hire foreign nationals to fill temporary labor and skill shortages when qualified Canadian citizens or permanent residents are not available. But this is not a long-term solution to the increasingly acute labor shortages in rural Canada.

At the same time, leadership in rural communities (like urban communities) has also gradually shifted from numerous service organizations & clubs (Elks, Lions, Rotary, Women's Institute, Agricultural Societies, etc.) to more specific (and very often) local recreational initiatives. Who doesn't know a parent who spends countless (unpaid) hours down at the local hockey rink coaching minor hockey? Part of the former role of service organizations and clubs was to facilitate face-to-face meetings to build community social capital. But, increasingly, it is simply easier and more efficient to mobilize people utilizing electronic media.

With few exceptions, the leadership previously provided by rural churches, both spiritually and socially, has also gradually been eroded. The church is generally no longer the focal point of most rural communities. Before 1971, less than 1 per cent of Canadians ticked the "no religion" box on national surveys, but two generations later nearly a quarter of the population said they were not religious. Now, almost 1/3rd of Canadians assert that they have absolutely no religious affiliation, and only about ¼ of Canadians claim to attend a church at least once a month.[120] The number of rural churches conducting a weekly service or assisting in the provision of community programs (youth drop-in, movie nights, Bible study sessions, etc.) continues to decline.

7.4.2 CULTURALLY DISTINCT CONTIGUOUS COMMUNITIES

Cultural diversity can and should enrich the larger community. But if there is very little cross-pollination, it can also nurture xenophobia and insularity, and lesson the social cohesion and internal growth dynamic of a traditional rural community.

This is most acutely felt with respect to contiguous communities which are each committed to protecting and enriching their own traditional language, culture and social fabric. And this is most evident on the prairies with respect to the on-reserve Aboriginal population and communal Hutterites.

On the prairies, there are over 600,000 Aboriginal residents and they make up about 10%

of the total prairie population (Table 7.2). Now (2018), about 17% of the Saskatchewan-Manitoba population is Aboriginal, and about one-third of Saskatchewan-Manitoba children under 15 years of age are Aboriginal.[121] Nationally, about ½ reside in urban areas while the other ½ reside on approximately 1,264 reserves. There are an estimated 148 reserves/settlements in Alberta, 205 in Saskatchewan, and 234 in Manitoba while BC and Ontario have an additional 316 and 207 reserves, respectively. Yet reserves can sometimes become both a physical and socio-economic ghetto, effectively lessening the long-term socio-economic viability of contiguous non-reserve communities, as well as the long-term socio-economic viability of the reserve itself. Worse, a much higher unemployment rate among Aboriginals and a wide socio-economic gap still persists between the Aboriginal and non-Aboriginal populations, and this gap is similar in both urban and rural areas.

Table 7.2 Aboriginal Population, 2016 (Thousands)

Region	TOTAL ABORIGINAL**	CANADA TOTAL	Percent Aboriginal	No. Reserves***
MARITIMES	129.3	2291.0	6%	67
QUEBEC	182.9	7965.5	2%	58
ONTARIO	374.4	13242.2	3%	207
PRAIRIES	656.9	6289.4	10%	587
B.C.	270.6	4560.2	6%	316
North*	59.7	111.8	53%	29
CANADA	1673.7	34460.1	5%	1264

* Yukon, NWT, & Nunavut.
** North American Indian, Metis, and Inuit.
*** Reserves & settlements > 500 people. Grand total is approx. 3100; 617 First Nations.

Sources: Statistics Canada, 2016 Census; Wikipedia, and related

The establishment of new communal Hutterite colonies can also lesson the long-term viability of traditional rural communities. This largely arises because they provide for their own education and social services and are largely self-sufficient regarding most consumer goods and services. There are about 168 colonies in Alberta, 60 in Saskatchewan, and 107 in Manitoba. At the same time, they are generally very successful farmers/ranchers, expand aggressively, and are representative of almost the only viable and truly communal group in North America today. (Also see Section 3.4, Farm Organization – Legal Structure.)

7.4.3   MULTIPLE RESOURCE USE ON PRIVATE LAND[122]

There are numerous dimensions to the issue of multiple resource use on private land in rural areas. Four of the most common are:

- Invasion of hunters, hikers, skiers, birdwatchers, motorbikes, quads, snowmobiles, etc. In most jurisdictions, this is subject to regulation, but this is still often difficult to enforce (e.g., required signage and the prohibition of motorized vehicles). (Also see Section 10.2.)
- Linear and site-specific disturbances. Sub-surface disturbances (e.g., a pipeline) generally make a one-time payment to the landowner and place a caveat on the property. A linear surface disturbance (e.g., a power line) generally pays an annual rental fee to the

landowner. A site-specific surface encumbrance (e.g., a gas or oil well) also generally pays an annual rental fee to the landowner for the area utilized, including access. At the same time, in some jurisdictions at least, access cannot be denied indefinitely. Similarly, the compensation guidelines for neighbors negatively impacted by a site-specific encumbrance on nearby land are often unclear or non-existent (e.g., a noisy nearby gas compressor station or a noisy nearby wind turbine).

- Abandoned surface and sub-surface facilities on private land. For example, it is estimated that there are now about 155,000 energy wells in Alberta alone that have no economic potential and which will eventually require reclamation.[123] Who pays whom and when?

- Conveyance of pests and disease. Vehicles and construction equipment, in particular, can and do facilitate the spread of pests and soil-borne disease. A good example is the recent spread of clubroot which negatively impacts canola yields (Section 12.1.2). Biosecurity is a growing concern. Timber matting and equipment sterilization is increasingly required.

7.4.4 POLICING LOW-DENSITY POPULATIONS

How to effectively police and maintain law and order in low-density areas is also a persistent source of potential conflict in remote rural areas.[124] Very obviously, when a crime is being committed in a remote rural area, the timely physical presence of a police officer is usually neither practical nor possible. The relevant legislation for self-defense is Section 34(1) of Canada's Criminal Code which says:[125]

> A person is not guilty of an offence if:
>
> a) They believe on reasonable grounds that force is being used against them or another person or that a threat of force is being made against them or another person;
>
> b) The act that constitutes the offence is committed for the purpose of defending or protecting themselves or the other person from that use or threat of force; and
>
> c) The act committed is reasonable in the circumstances.

With respect to the self-defense of property, the courts tend to argue that excessively violent resistance is not "reasonable," whereas with respect to the self-defense of yourself, family, or someone else, the courts tend to have a more lenient interpretation of what is "reasonable."

Gates are easily scaled or removed while video cameras and alarm systems simply record that a crime is in progress. Rural Crime Watch organizations and a Crime Stoppers Tips Line can also help to reduce the prevalence of rural crime.[126] Posting that a rural community has a Crime Watch group, setting up a WhatsApp rural community chat line, and/or establishing neighboring farm/ranch internet linkages (e.g., a Rural Crime Watch smartphone app) might also serve as a deterrent. But to actually stop a rural crime in progress, some adaptation of non-lethal deterrents such as bear spray, paint-ball guns, or stun guns might also be required.

A greater police presence in a sparsely populated area is generally impractical or simply impossible. And addressing the sources of crime is not a short-term solution, although it is still likely the best long-term solution.

A new trespass law in Saskatchewan may serve as a template for similar legislation in other jurisdictions. The default assumption under this law is that people cannot encroach on private property without the permission of landowners. A $5,000 penalty for trespassing without permission will be imposed.[127] Elsewhere, there is a growing demand by rural residents to revise the Criminal Code criteria for "reasonable use of force," enhance the use of electronic monitoring of repeat offenders who are confirmed as a main source of rural crime, and provide federal tax credits to individuals who install crime-prevention measures, among other suggestions.[128]

7.4.5   ADAPTING TO CLIMATE CHANGE[129]

The eventual impact of climate change on Canadian agriculture is expected to be, on average, positive. (See Section 11.5.) But the operative word is "average." Extreme weather events are also expected to become more frequent and these will impact all of rural Canada. Risks associated with extreme weather events could include: a) plant species migration; b) diminished biodiversity; c) destruction of public infrastructure, and d) deteriorating water quality.

The destruction of public infrastructure from more frequent and more extreme weather events could be especially debilitating with respect to transportation networks, telecommunication networks, and electricity and natural gas delivery. The temporary loss of local medical support (hospitals, ambulances) could have especially deadly consequences.

The principal rural challenge is to gradually make this social and physical infrastructure more resistant to weather abnormalities.

7.5   RURAL CONSOLIDATION & GROWTH

Despite persistent socio-economic disruption and displacement, there is still a steady confidence and calm in many, if not most, rural communities today. This can be traced to at least five countervailing and very positive trends:

1. People "dribble" back" to rural Canada as they get older. Again, looking at Figure 7.5, we see that people in the 30 to 50 year cohorts sometimes return to a non-urban area.[130] They may be returning to the family farm or other business, they may be returning to their roots, or they may just prefer to live in a rural community because of its unique amenities and commute to work elsewhere. Young adults often move to rural areas near cities where housing is less expensive, there is more open space for raising children, and one spouse can make the long drive to the city for work. Regardless, what is particularly important is that these people are in the prime of their income-earning years; possibly families with children who attend the local school and people who are more likely to participate in community activities—local sports, service clubs, churches, etc.

2. There is also a socio-economic "blending" process in progress. The physical rural and urban infrastructure is gradually becoming very similar. Today it is increasingly likely that a rural home is heated by natural gas, that it almost certainly has 120V electricity, has a high-quality water and sewage system, has cellular telephone service and satellite TV, and has high-speed wifi to instantly connect to the internet, e-mail server, social media, and dedicated TV. And if that isn't sufficient, it is increasingly likely that they can just jump into their own late-model vehicle, whatever the weather, and quickly access a paved road that will easily deliver them to an urban-based rock concert, NHL hockey game, movie, or upscale dinner in an hour or two. What's not to like?

Figure 7.6 The Rural-Urban Population Interface, 2016

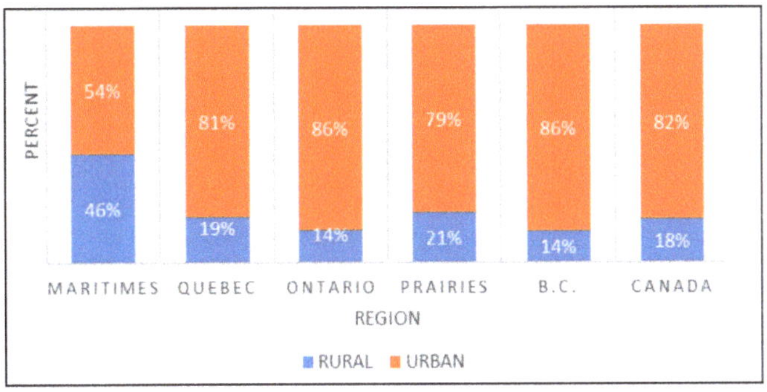

Region	Rural	Urban
Maritimes	46%	54%
Quebec	19%	81%
Ontario	14%	86%
Prairies	21%	79%
B.C.	14%	86%
Canada	18%	82%

Source: Statistics Canada, *Population Census*, 2016

Thus, we see that the rural and urban populations are, both socially and economically, gradually becoming more similar. The country hayseed and the suave urbanite are becoming increasingly indistinguishable from one another. This is especially true in the Maritimes because there are almost as many rural people as urban people. And it is especially true in Ontario and the Ottawa River–St. Lawrence region of Quebec because concentrated rural-urban areas increasingly blend into one another. Higher population densities facilitate socio-economic blending, creating more and more "rubanites."

3. There is an internal vitality built into remaining regional ("growth-pole") communities. Wholesale/retail and related services (i.e., grocery stores, hardware stores, banks, restaurants, accountants, lawyers, hair dressers, etc.) make up about 40% of all rural employment. Three other sectors are also particularly important to rural people: health care/social assistance and educational services (19%), manufacturing (11%), and construction (11%) (See Figure 7.7). Primary producers only make up about 12 percent of national rural employment, although in some regions of the country this would be much higher (e.g., forestry in BC, fisheries in the Maritimes, and oil and gas in Alberta-Saskatchewan).

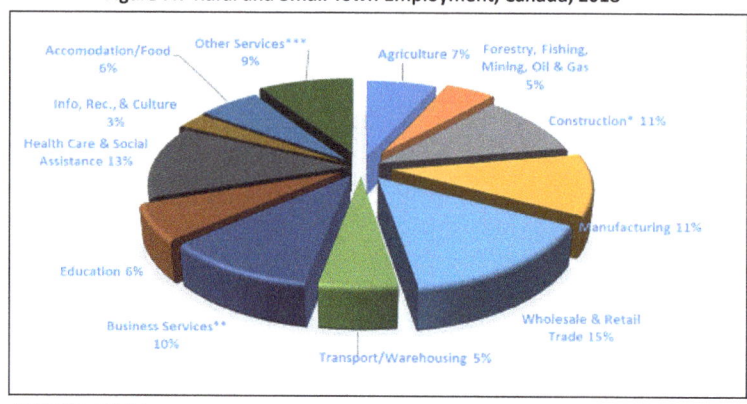

Figure 7.7 Rural and Small Town Employment, Canada, 2018****

- Accomodation/Food 6%
- Other Services*** 9%
- Agriculture 7%
- Forestry, Fishing, Mining, Oil & Gas 5%
- Info, Rec., & Culture 3%
- Health Care & Social Assistance 13%
- Construction* 11%
- Manufacturing 11%
- Education 6%
- Business Services** 10%
- Transport/Warehousing 5%
- Wholesale & Retail Trade 15%

* Includes utilities.
** Finance, real estate, insurance, prof. & technical services, business, bldg., & other support.
*** Includes public administration.
**** Non-census metropoltan area and non-census agglomeration. Self-employed + paid workers.
Total goods-producing = 879,000 (33%); total services = 1,797,000 (67%).

Source: Basic data from Statistics Canada, Table 14-10-0108-01

The trades include local welders, carpenters, mechanics, electricians, plumbers, and numerous others. Transportation includes trucking and school bus operators. Utility-related employees often includes electricity workers (e.g., REA personnel), natural gas workers (e.g., natural gas co-ops), telephone personnel (Telus, Rogers, Bell, etc.), and internet providers. Government includes public school teachers, public health workers, post office workers, public recreation employees, and municipal government employees, although this is not well delineated in Figure 7.7. On the prairies, for example, it is not uncommon for a single county or rural municipality (RM) to employ 30–40 road grader operators just to maintain graveled municipal roads. Construction includes the building trades, road construction, and so on. This rural employment mix mutually reinforces rural self-sufficiency and long-term economic sustainability.

4. In many professions, the laptop computer, e-mail, Skype, search engines, social media, and related applications have radically altered the need for physical proximity in an urban workplace. Anybody who essentially only works on a wifi-connected laptop computer (writers, journalists, bookkeepers, some engineers, architects, economists, etc.) can do this comfortably at home in a rural community. Similarly, physical goods originating in rural communities can now easily be widely marketed on a dedicated website, eliminating both the need to commute to an urban center and the need for physical retail outlets. Increased globalization and new technologies can sometimes also provide rural residents with other new sources of income. A good example is the rental revenue and cheaper power landowners can secure from windfarms. The growing urban demand for healthy recreational opportunities also gives rise to rural corn mazes, bed and breakfast facilities, wild buffalo rides, pick-your-own fruits and vegetables, and so on.

5. The continued importance of manufacturing in rural communities is undeniable. In 2008, there were still almost 400,000 people employed by manufacturers in rural and small-town Canada, about 20% of all manufacturing employment in Canada. (Figure 7.8)

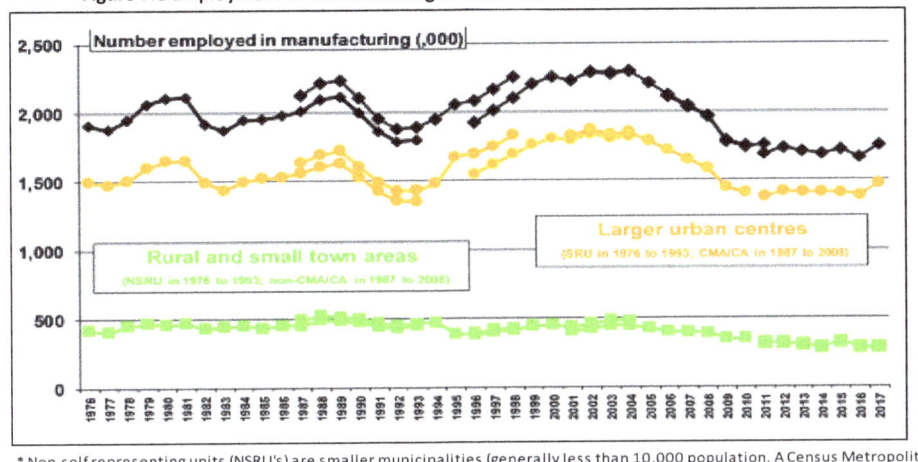

Figure 7.8 Employment in Manufacturing in Rural & Small Town Canada, 1976 - 2017*

* Non-self representing units (NSRU's) are smaller municipalities (generally less than 10,000 population. A Census Metropolitan (CMA) has a core population of 100,000 or more and includes neighbouring municipalities where 50% or more of the workforce comes from the core. A Census Agglomeration (CA) has a core population of 10,000 to 99,999 and includes neighbouring municipalities where 50% or more of the workforce comes from the core.

Source: Statistics Canada, *Labour Force Survey.* Chart by Ray D. Bollman

Sometimes this is resource-based (e.g. oil drilling equipment or meat processing) but often it is not. Everyone knows of at least one company located in an otherwise obscure small town. A typical example is Bourgault Industries, a manufacturer of a very popular airseeder/grain tank. It employs about 600 people near St. Brieux, SK, a town with about 650 residents. Mennonite communities, as well as Hutterite colonies on the prairies also seem very adept at manufacturing. Grain trailers in southern Manitoba, grain boxes in Linden, AB and Humboldt, SK, or trailers, corral panels, and water cannons in Two Hills, AB are indicative. And some Hutterite colonies, for example, make metal sheeting. Rural communities seem particularly adept at fostering smaller, more artisanal manufacturing activities.

Additionally, manufacturing companies sometimes locate outside of the shadow of a major urban center just so that they can more easily secure the personnel required, maintain the physical area required, and/or reduce various costs or production (including local taxes). Other larger manufacturing companies, either through the acquisition of a smaller company or an unique regional advantage, will also sometimes locate in a relatively small town (e.g., Buhler in Vegreville, AB). In any event, manufacturing employment, on average, accounts for maybe 11% of all employment in rural and small-town Canada. (Figure 7.7) This can be the lynch-pin that keeps the wheels from falling off.

Non-metro Canada is still home to 31% of Canada's population, 23% of total employment, and approximately 30% of Canada's GDP.[131]

7.6 GOVERNMENT & COMMUNITY INITIATIVES[132]

A multitude of factors have coalesced to gradually undermine the vitality and long-term sustainability of many rural communities. What can various levels of government do about this?

What can the community itself do about this? Or should everyone just sit back and let it die a "natural" death?

7.6.1   TOP-DOWN

Rural and regional planners (plus urban planners) abound and the policy framework they typically employ usually addresses the central importance of **institutions**, **infrastructure**, and/or **incentives**.[133]

"Institutions" generally refers to policies that are not spatially specific and have effects and outcomes that will vary across locations. These include such policies as the income tax system, intergovernmental fiscal relations, and governance of land and housing markets, as well as education, health care, basic water and sanitation.

"Infrastructure" is shorthand to include all investments that connect places and provide basic business services, such as public transportation and utilities. This includes developing inter-regional highways, pipelines, and airports to promote trade and improve information and communication technologies to increase the flow of information and ideas.

"Incentives" generally refer to spatially targeted measures to stimulate economic growth in lagging areas. These include investment subsidies, tax rebates, location regulators, local infrastructure development, and targeted investment zones.

A Rural Secretariat was established within Agriculture and Agri-Food Canada (AAFC) in 1993 with a broad mandate to carry out research on rural issues and assess implications for rural areas across all departments of the federal government in all decision documents for Cabinet. But this was abandoned in 2013. Regional development agencies such as the Atlantic Canada Opportunities Agency and the Western Economic Diversification Fund marked their 30th anniversary in 2017, but there is a considerable body of research that casts doubt on their basic usefulness.[134] Now, the federal and provincial governments generally design, fund, and execute rural initiatives from a sectoral perspective, not a holistic or cross-sectoral perspective. Piecemeal initiatives typically filter down through the respective Ministries: Health, Education, Economic Development, Environment, Transportation, Municipal Affairs, and so on. The Ministry of Agriculture is generally not very proactive other than providing funding for Agricultural Societies and providing support to 4-H Clubs. (Also see Chapter 9.)

From a national perspective, on-going public policies that pro-actively delay or block the development of resource-based industries effectively amount to an anti-rural development policy. About 500 communities across Canada are heavily dependent upon mining, forestry, and energy for their livelihood.[135]

This is a far cry from the early 1950s, when the Elks collected wheat donations from local farmers, sold the grain, and then contributed endless unpaid hours to building the community hockey arena and accompanying curling rink.

7.6.2   BOTTOM-UP

Over the past few decades control over, and responsibility for, more and more aspects of daily life have slipped from the hands of local rural communities and into the hands of distant decision-makers. "Imported" surrogates, often recipients of very generous salaries and pensions from afar, are frequently assigned the responsibility for implementing Program X, Y, Z. This has often sidelined local leadership, local accountability, and local community engagement, and, more importantly, shifted sources of funding to bureaucratic administrations and discouraging local fund-raising. [136]

There is a growing realization that since every community is unique, most successful social, economic, and environmental initiatives should be crafted and delivered by the community itself. "Government of the people, by the people, and for the people shall not perish from the Earth," as Abraham Lincoln said.

This is intuitively obvious to anyone who has actually lived in a smaller non-metropolitan community. Who puts on the fairs and rodeos? Who puts on the slow-pitch tournaments? Who operates the curling rinks? Who operates the ice skating/hockey rinks? Who operates the swimming pools? Who puts on the breakfasts, dinners, raffles, and bingos? And the list goes on. Very often it's a community-based organization led by a handful of very committed, very hard-working individuals subject to burn-out. While support from local, provincial, and federal governments is generally available for some cost-sharing, volunteer local agencies and personnel are often discouraged by the lengthy and somewhat ambiguous program application formats.

Thousands of community-based organizations in rural Canada remain the principal way many (if not most) rural residents express their altruistic values, build a strong social/business network, build personal friendships and, perhaps most importantly, provide leadership to keep their community strong and healthy and even offer alternatives to increased urbanization. Commitment to a virtuous code of conduct is also very often strongly encouraged. Some principal players:

1. **Agricultural Societies** – nonpartisan, grassroots, volunteer-based organizations whose primary objective is to encourage an awareness of agricultural and to promote improvements in the quality of life of persons living in an agricultural community. Among other things, they: a) hold agricultural exhibitions, b) promote the conservation of natural resources, c) support and provide facilities to encourage activities intended to enrich rural life, and d) conduct fairs and rodeos. There are about 200 in Ontario, 59 in Manitoba, 283 in Alberta, and 58 in Saskatchewan (based on SAASE membership). Agricultural societies are usually supported, financially, by the respective provincial governments (e.g. $11.5 M/year [or, say, $20,000/society) in Alberta).[137]

2. **4-H** – has as its mission engaging youth to reach their fullest potential and focuses on four personal development areas: head, heart, hands, and health. The 4-H motto is "to make the best better" and its slogan is "learn by doing." Their pledge is:

*I pledge my head to clearer thinking
my heart to greater loyalty
my hands to larger service
and my health to better living
for my club, my community, my country and my world.*

Clubs today consist of a wide range of options, each dependent upon local volunteers and each allowing for personal growth and success. Developing public speaking skills and provincial exchanges are encouraged. They also have achievement days, bottle/battery collection days, and roadside clean-up days. 4-H clubs in Canada are usually supported, financially, by their respective provincial governments. Worldwide, there are nearly 6 million active participants in over 50 countries and 25 million alumni.

3. **Women's Institute** – focus, in particular on women and adult education. Its motto is "women inspired to make change count" and it now focuses on food safety, and improving schools and communities. It generally aims to promote an appreciation of rural living and to develop informed citizens through the study of national and international issues. There are about 660 branches across Canada, with about half of its 7,000 members in Ontario.[138] Worldwide, there are more than nine million members in about 70 countries.

4. **Service Clubs** – voluntary non-profit organizations that perform charitable works either by direct hands-on efforts or by raising money for other organizations. A service club is defined firstly by its service mission and secondly by its membership benefits, such as social events, networking, and personal growth opportunities that encourage involvement. Most service clubs consist of community-based groups that share the same name, goals, membership requirements, and meeting structure. This includes the Elks, Rotary, Kiwanis, Lions, Hospital Auxiliary, Royal Canadian Legion, Royal Purple, and Chambers of Commerce:

   Elks – "a quiet network of good deeds" (e.g., raise money for schools and recreational facilities, award college scholarships, help the poor through various charities, sponsor drug awareness and prevention programs). The Elks have more than 1 million members worldwide who "strive to better the communities [they] live in and help those who are in need."

   Rotary – foster the "ideal of service" as a basis of enterprise, encourage high ethical standards in business and the professions, and promote a world fellowship of business and professional people.

   Kiwanis – seeks to provide assistance to the young and elderly, create international understanding and goodwill, help develop community facilities, support agriculture and conservation, support programs to safeguard against crime, and eliminate alcohol and drug abuse. K-Clubs are affiliated non-profit organizations providing support to local associations, services, and community members. Kiwanis has more than 300,000 members in about 8,000 clubs and 70 countries.

> **Lions** – seeks to foster a spirit of "generous consideration" among peoples of the world and to promote good government, good citizenship, and an active interest in civic, social, commercial, and moral welfare. Lions' activities include neighborhood improvement projects, environmental and conservation programs, education and literacy services, aid to the blind and hearing impaired, disaster relief, support for victims of pediatric cancer and their families, and hunger relief.
>
> **Hospital Auxiliary** – committed to accomplishing their goals by enjoying all the benefits of belonging to a charitable organization.
>
> **Royal Canadian Legion** – provides facilities and various support activities for military veterans. Also gives assistance to some public activities.
>
> **Royal Purple** – a national fraternal and charitable women's organization, affiliated with the Elks.

An aging demographic threatens to undermine the vitality of many traditional service organizations. Relying on a dozen or so aging participants cannot continue.

5. **Recreational Activities.** Some recreational activities encourage individualism, narcissism, and instant gratification. But others do, in fact, build comradery, demand team-work, and encourage personal and social responsibility. Is there a community in Canada that doesn't have at least one hockey team that practices or plays 1–2 times per week throughout the winter months? And each team requires numerous dedicated volunteers (coaches, scorekeeper, referees, lunch counter, etc.), as well as access to a community hockey rink. Keeping the required facilities in good working order is also a significant financial challenge.

6. **Churches.** Despite a growing secularism and generally declining church attendance, some churches in some communities are still a very integral part of that community. Aside from providing spiritual guidance, they often provide prayer for the infirmed and carry out other charitable activities. Catholic schools have a church affiliation, as do some charter or alternative schools, and most emphasize community values. Social functions sponsored by churches, such as dinners, raffles, bingos, bake sales, etc., also bring people together.

7. **Boy Scouts** – an association with the stated aim "to help develop well-rounded youth better prepared for success in the world." This includes preparing young people to make ethical and moral choices over their lifetimes by instilling in them values such as trustworthiness, good citizenship, and outdoor skills. Skill-development and activities such as camping, aquatics, and hiking are emphasized. In Canada, in 1965, there were almost 300,000 Scouts and 34,000 volunteers but by 2016, this had dropped to about 62,000 members and 21,000 volunteers, a precipitous decline. The Girl Guides and Brownies are affiliate organizations.

This is bottom-up; organic. It is spontaneously driven by local leadership and hundreds of thousands of committed and dedicated volunteers. This is where mentoring is fostered and honored. This is where life-long bonds are built and an almost Buddha-like affinity for nature and the preservation of the environment is celebrated. These communities intuitively know and understand "Dunbar's number," 150 people, the number that generates the strongest communal bonding; the number we are apparently neurologically limited to when it comes to genuine human relationships.[139]

These are the "somewhere" people; not "anywhere" people.[140] And "somewhere" people cannot easily shift their lives to somewhere else. The socio-economic and environmental health of their local community is their essential lifeline. Their on-going, and sometimes intimidating, challenge is to help unlock their communities' inherent potential to further reinforce their belief in a viable, sustainable rural future for their friends and family.

8. *Agricultural Organizations*

The **Directory of Canadian Agricultural Associations** lists 1,034 national, regional and provincial organizations.[141] Mandates, purpose, governance, and areas of interest and influence range from youth development (4-H, Young Farmers Organization), women in the agriculture and food industry, commodity groups covering every product produced across the country, general organizations along the supply chain, lobby groups with a policy agenda, groups of scientists linked by discipline, regional applied research groups, regulatory bodies, and those focused on various aspects of the environment and organic food production.

Together they represent a multi-dimensional, multi-layered, interlocking Rubiks Cube© where the colors only rarely line up. There are subordinate provincial counterparts, but other than that they are mostly local, regional, and issue-oriented. National agricultural policy is an amalgam of these often-competing interests; sausage-making at its best and worst.

8.1 RURAL COOPERATIVES[142]

Much of the impetus for establishing farm organizations in Canada can be traced to the cooperative movement here and elsewhere. Based on seven core cooperative principles, the most important dictated that members each have one vote regardless of the investment made; anyone could join; surpluses would be distributed to members according to their level of participation; and community well-being was a priority.

The first cooperatives in Canada were formed in the Maritimes and between 1860 and 1900 over 1200 cooperative creameries and cheese factories were established in the Maritimes, Quebec, and Ontario. Before WW1 (1914), many other cooperatives then appeared: Caisse Populaire (a Quebec credit union, 1900); Grain Grower's Grain Company (Prairies, 1906); United Farmers of Alberta (1909)[143]; La Co-op fédérée (Quebec, 1910); Cooperative Elevator Companies (Saskatchewan, 1911 and Alberta, 1913); United Farmers Cooperative (Ontario, 1914); as well as numerous mutual insurance companies and various other fruit, livestock, and tobacco cooperatives. East Coast fishing cooperatives started in the 1920s.

The Alberta and Saskatchewan Wheat Pools were established in 1923; the Manitoba Wheat Pool in 1924. These cooperatives utilized a "cooperative pooling" system whereby members contracted to sell all produce through their cooperative and in return would receive dividends based on the quantity and quality of the produce they supplied. When the Wheat Board was introduced by way of federal legislation (initially in 1920 and then again in 1934 and 1939), this further enabled grain handling cooperatives to strengthen their role in Western Canada. Grain handling infrastructure quickly emerged across the prairies, with each town or village the site of multiple grain elevators, both cooperatives and privately owned.

During the 1940–1990 period, cooperatives generally continued to grow, especially grain and dairy marketing cooperatives, retail cooperatives, and credit unions. And some rural-based cooperatives have continued to flourish in the 21st century:

- Since the mid-1970s, cooperatives have become particularly important in Quebec, both economically and politically.

- Retail cooperatives continue to play a significant role on the prairies and in Atlantic Canada. For example, the Federated Cooperative of Saskatchewan (1944) now has about 365 member cooperatives across the prairies, owns and operates an oil refinery, has retail grocery/hardware/lumber/fuel outlets at most Co-op locations, has numerous bulk fuel outlets, operates various agri-business (fertilizer-chemical) locations, and has annual sales of more than $10 billion—one of the largest cooperatives in Canada. Similarly, the United Farmers of Alberta (UFA) perseveres and now has about 34 farm and ranch supply stores, some 110 fuel outlets, 25 sporting goods stores, and annual sales of more than $2 billion. Rural gas and water cooperatives are yet another success story.

In recent years, however, other rural-based cooperatives have not been so fortunate. Without access to public financing (i.e., no listing on the Toronto stock exchange), their continued growth led to challenging financial times while the ever-increasing breadth and scope of their operations eventually exceeded their management and governance capabilities:

- In 1998, the Alberta and Manitoba Wheat Pools amalgamated to form Agricore which, in turn, merged with the United Grain Growers in 2001 to become Agricore United (a publicly traded company) which, in turn, merged with the Saskatchewan Wheat Pool in 2007 to form Viterra. And then, finally but somewhat predictably, in 2013 Viterra was purchased by Glencore International of Switzerland for $6.1 billion. Glencore is the world's largest and most aggressive commodities trader, worth about $45 billion.[144]

- In the late 1940s, Rural Electrification Associations (REAs) were established across the prairies. These were cooperatives which built and managed their own lines to deliver electricity to rural residents and they did, indeed, prosper for about 50 years. But, gradually, higher operation and maintenance (O&M) costs, increasingly onerous safety regulations, and a membership based in aging rural communities forced many REAs to delegate actual construction and on-going O&M to private electricity providers.

But, flourish or flounder, a collective cooperative impulse and the willingness to sacrifice individual rights for the common good is the foundation upon which many farm organizations in Canadian were built.

8.2 NATIONAL UMBRELLA ORGANIZATIONS[145]

Umbrella organizations, sometimes called general farm organizations, have faced a myriad of challenges ever since their inception, largely related to specific regional interests and often conflicted crop and animal industry interests.

Initially, an organization known as the Canadian Council of Agriculture—made up of provincial organizations from the Maritimes, Ontario, Manitoba, Saskatchewan, and Alberta—was active early in the 20th century. Another movement to establish a national organization to speak for all provinces and all branches of agriculture began in BC in 1935, and was known as the Canadian Chamber of Agriculture. The Canadian Chamber of Agriculture morphed into the Canadian Federation of Agriculture (CFA) in 1935 to attempt to speak for agriculture in the same way that business, labor, and professional groups have their own boards of trade, chambers of commerce, labor unions, and professional associations.

By 1942, the CFA was made up of provincial federations plus the United Grain Growers, Dairy Farmers of Canada, and the Canadian Horticultural Council. There were no individual memberships and the provincial Wheat Pools in Western Canada were included in the provincial federations. The CFA always sought to promote areas of agreement and downplay areas of disagreement.

Today, the CFA, still producer-funded, is composed of all the affiliated provincial federations, some commodity groups, and a half-dozen other agriculture-related organizations (Table 8.1). The strongest provincial affiliates are the Ontario Federation of Agriculture (OFA) and the Quebec-based L'Union des producteurs agricoles (UPA). Central Canada (Ontario and Quebec) has long had strong farmer representation through these two organizations. In Ontario, cooperatives and marketing boards are also very supportive. Membership in the UPA is mandatory.

Table 8.1 Canadian Federation of Agriculture (CFA) - Members, 2018*

1	OFA - Ontario Federation of Agriculture	15	Barley Farmers of Canada
2	UPA - L'Union des producteurs agricoles Union of Agricultural Producers	16	COHA - Canadian Ornamental Horticulture Alliance
		17	Canadian Honey Council
3	Newfoundland/Labrador Federation of Agriculture	18	Mushrooms Canada
4	Nova Scotia Federation of Agriculture	19	CSBPA - Canadian Sugar Beet Producers Assoc.
5	Agricultural Alliance of New Brunswick	20	CFGA - Canadian Forage & Grassland Association
6	KAP-Keystone Agricultural Producers-Manitoba	21	CSGA/ACPS -Canadian Seed Growers Association
7	APAS-Agricultural Producers Assoc. of Saskatchewan		
8	AFA-Alberta Federation of Agriculture	22	Equestrian Canada
9	BCAC -BC Agriculture Council	23	Standardbred Canada
10	Dairy Farmers of Canada	24	Canadian Young Farmers Forum
11	Chicken Farmers of Canada	25	FNA - Farmers of North America
12	Egg Farmers of Canada	26	F.A.R.M.S. - Foreign Agricultural Resource Management Services
13	Turkey Farmers of Canada		
14	Canadian Hatching Egg Producers		

* Subject to errors and/or omissions.

Source: CFA website, 2018

Keystone Agricultural Producers (KAP) is Manitoba's largest general farm organization and it is funded by a check-off at point-of-sale from about 4,000 producers. They have lobbied for better rail movement of grains, removal of the education tax on farmland and buildings, working for improved producer-payment security, and easing restrictions on winter nutrient applications.[146]

The Agricultural Producers of Saskatchewan (APAS) is Saskatchewan's voluntary general farm organization but it differs from other farm organizations in that its 98 members are representatives of all the rural municipalities (RMs) in Saskatchewan and they, in turn, represent about 18,000 agricultural producers in these RMs. There are also eighteen associate members who represent agriculture-related organizations. Almost all of APAS funding comes from the RM membership.[147]

In some of the other provinces, however, especially were membership is voluntary, farm representation is very limited.[148]

The strongest commodity-specific national organizations represent the dairy and poultry industries, both represented by powerful national marketing agencies. Glaringly absent is representation by the beef and pork industries, as well as some of the major grain and oilseed industries, particularly wheat and canola. As such, with this inherent central-Canada bias, the CFA only rarely adequately represents the views of all farmers in Canada, however defined. At the same time, through its high PR-profile and its annual briefs to the federal cabinet, the CFA continues to represent the views of its limited membership and still has a continuing impact on policy development.

The other national farmer organization of historical significance is the National Farmers Union (NFU), wholly funded by membership fees and voluntary donations. Since its inception, the NFU has adhered to an agrarian tradition forged by the Canadian Commonwealth Federation (CCF), the first "socialist" government in North America and principal author of Canada's health system, as well as populist political parties such as the United Farmers of Alberta and Social Credit in both Alberta and BC. The NFU emphasizes the need to "protect" farms from challenges by both commercial interests and government. The NFU also provides a much-needed focus on environmental issues, food safety, the viability of rural communities, improved rural education, and gender equality (Table 8.2). These are all noble objectives, but with voluntary membership, the strength of their national voice is often muted.

Table 8.2 Mission Statement: National Farmers Union - NFU

The only voluntary national farm organization in Canada which is commited to:
- o ensuring family farms are the primary unit of food production
- o promoting environmentally-safe farming practices
- o giving farm women equal voice in shaping farm policy
- o working for fair food prices for both farmers and consumers
- o involving, educating, and empowering rural youth for a better future
- o building healthy, vibrant rural communities
- o ensuring an adequate supply of safe, nutritious food for Canadians
- o solidarity with family farmers

Source: NFU website 2018

The **Canada Grains Council** is another national organization, with broad-based regional representation for grains, oilseeds, pulses, and special crops. Formed in 1969, it tries to coordinate efforts to increase the sale and use of Canadian grain in domestic and world markets. Its membership (Table 8.3) reflects the entire food chain and includes Canadian growers, seed and life science companies, commodity associations, grain companies, and public research institutions.

Table 8.3 Canada Grains Council - Members, 2018

Ag. Producers Assoc. of Saskatchewan (APAS)	Keystone Agricultural Producers (KAP)
*Alberta Barley	Paterson Global Foods
*Barley Council of Canada	Port of Vancouver
Bayer	Producteurs de Grains du Quebec
Cargill	*Pulse Canada
*Canola Council of Canada	Richardson International
*Canadian Canola Growers Assoc.	Saskatchewan Assoc. of Rural Municipalities (SARM)
*Cereals Canada	*Saskatchewan Flax Development Commission
Canadian National Millers Assoc.	Saskatchewan Wheat Development Corp. (SWDC)
Canadian Seed Growers Assoc. (CSGA)	*Soy Canada
Canadian Seed Trade Assoc.	Syngenta
G3 Canada	*Viterra
*Grain Farmers of Ontario	Western Canadian Wheat Growers
*Grain Growers of Canada	Western Grains Research Foundation
Inland Terminals Assoc.	

* Directors

Source: CGC website, 2018

More recently, the **Grain Growers of Canada** is a fourth organization which has emerged with strong broadly - based regional representation. It consists of fourteen grain, oilseed, and pulse groups nominally representing more than 50,000 producers, as well as many private commercial organizations nationally. It is funded primarily through the membership and is essentially a free-market advocacy group focused on influencing federal and provincial government policy. Its single most important achievement to date has been to see the end of the Canadian Wheat Board monopoly. It advocates for more research into seed, value-added processing, and grain by-products.

A fifth umbrella organization is **Cereals Canada.** This is a relatively new organization (2013) with broadly based regional representation that wants to fill in the coordination and market development gaps resulting from the end of the Canadian Wheat Board monopoly. Cereals Canada represents about twenty different producer organizations, crop development and seed companies, grain handlers, exporters and processors.[149] It has a 2018–19 budget of about $1.3 million and will focus on market development, innovation, and industry leadership. It also has representation on the Canada Grains Council. The Team Canada sales teams are examples of the market development work spearheaded by Cereals Canada. To reduce overlap, it is also in talks to merge with the **Canadian International Grains Institute (CIGI, 1972).**[150]

The objectives, membership, and personnel in the **Canada Grains Council**, **Grain Growers of Canada**, and **Cereals Canada** are very similar.

Yet another group of farm-related organizations focus on the Canadian seed industry: the Canadian Seed Growers' Association (CSGA), the Canadian Seed Trade Association (CSTA), the Canadian Seed Institute (CSI), the Commercial Seed Analysts Association of Canada (CSAAC), the Canadian Plant Technology Agency (CPTA) and CropLife Canada.[151] The CSGA, CSTA, CSI, CSAAC, and CPTA are currently (2018) exploring a merger, plus a formal alignment with CropLife Canada, to create a streamlined organization for information management, advocacy, and service to amplify the impact of their various complementary functions. They already cooperate in a Seed Synergy Collaboration Project.[152] The current debate regarding farm-saved seed royalties is being chaired by Agriculture Canada (AAFC) and the Canadian Food Inspection Agency.[153]

Broadly based umbrella organizations, however, are generally not financially viable if they lack a strong and sustainable funding source. The fees from voluntary membership are usually inadequate. Moreover, as valuable and necessary as these organizations were in the past, many people in the agricultural community now concentrate on narrow, commodity-based issues or special interests rather than on the welfare of everyone in the agricultural community. The pendulum has swung away from the philosophy of the collective and the willingness to sacrifice individual rights for the common good.

Additionally, gradually, there were simply more and more organizations competing for the agricultural communities' time and money. Active membership, with increasing affluence, also succumbed to the growing popularity of other recreational and educational pursuits. As a result, farmers and ranchers increasingly let a relatively few organizational activists represent them in multiple organizations. For these financially reimbursed farm representatives, multiple memberships, multiple administrative roles, and overlapping directorships became commonplace.

The commodity organizations that were able to establish government-legislated producer/processor levies (check-offs) and thus be assured a relatively large and predictable annual budget have gradually become more powerful.

8.3 COMMODITY GROUPS

As the umbrella farm organizations have faded, commodity groups have gradually become more powerful. The underlying *raisons d'être* for this multitude of "farm" groups are two:

- Firstly, there is an inherent and seemingly irreconcilable difference in their respective market philosophies. One is essentially protectionist and inward-looking (especially dairy and poultry); the other is essentially expansionist and export-focused (especially beef, pork, pulses, and grains/oilseeds).

- Secondly, there is sometimes an inherent conflict between some sub-sectors (e.g., beef and pork producers want inexpensive feed grains whereas grain producers want high feed grain prices.[154]

National legislation such as the Agricultural Products Marketing Board Act (1947) and the National Farm Products Marketing Act (1972) enabled the establishment of commodity-based organizations and in its second coming, enabled the emergence of supply management for dairy, egg, chicken, turkey, and hatching eggs. Provinces, beginning in 1947 and continuing today, each developed their own Marketing Acts, which enabled the emergence of commodity organizations and the ability to establish check-offs or levies payable by farmers, most often at point-of-sale. The supply management organizational structure quickly emerged after the second edition of the National Farm Products Marketing Act and national organizations fostered by provincial marketing boards were borne in all provinces.

The ability to finance commodity organizations through levies quickly enhanced the position of commodity groups. Generally, levies have been structured as compulsory but over time, except for the supply-managed sectors, have gradually moved to a practice where the levies may be requested to be returned to the farmer/rancher. Since most other organizations must depend on membership fees and voluntary donations, the financial strength of commodity organizations has created a strong position to effectively lobby, influence and develop programs of interest to the benefit of their sector across Canada.

Yet, perhaps unavoidably, their respective missions sometimes overlap, both domestically and internationally. More organizational integration would generate administrative efficiencies and more market development capability, both nationally and internationally. Another anomaly is that the funding structure (i.e., check-offs) over time tends to reinforce the prominence of the principal commodities and, correspondingly, disadvantage the commodities currently of lesser importance. This is exacerbated by the R&D and trade focus of private companies (e.g., seed and chemical companies) as well as the declining public-sector funding for minor crops. Thus, the rich get richer and the poor get poorer. The table is set.

Additionally, there is some friction between some commodity groups and other value-chain members regarding the degree to which farmer-based commodity groups interface with the end consumer, especially in export markets. As more consumers want to know more about the food they buy and eat, the interface between the farmer and the consumer becomes increasingly important. Transparency, consumer confidence, and shared outcomes are of mutual interest.

The principal commodity groups are briefly described in accompanying Table 8.4. And of these, the dairy, poultry, canola (CCC), pulse, and beef (CCA) organizations are particularly powerful. The commonality among virtually all commodity groups is that they have a **value-chain** perspective; they all generally focus on the entire industry, every participant between the field and the fork.

Table 8.4 List of Principal Commodity Groups, 2018*

Commodity	Organization	Description
Beef	CCA - Canadian Cattlemens Association Canada Meat Council	Provides beef produces with a national and international voice. Brings forward key policy issues to the federal government and implements strategies and tactics in pursuit of their vision. Has three Divisions: a) Beef Cattle Research Council, which determines R&D priorities for the beef cattle industry and administers the research funding allocation of the National Check-off; b) CANFAX to provide expert analysis of markets and trends; and c) CANFAX Research Services to provide the beef industry with statistical information and economic analysis. (Ottawa) The Canadian Meat Council is Canada's national trade associaton for the federally inspected meat packers and processors.
Pork	CPC-Canadian Pork Council (Ottawa)	Provides a national voice for hog producers in Canada. A federation of nine provincial pork industry associations representing 7,000 farms. CPC focuses on four strategic priorities: market access, industry competitiveness, consumer confidence, and public policy advocacy. (Ottawa)
Dairy	Dairy Farmers of Canada	Virtually every aspect of the dairy industry is managed by the Dairy Farmers of Canada, with oversight by the Canadian Dairy Commission, a Crown corporation. The legal basis is the federal Agricultural Products Marketing Act (1966). Marketing Boards set production quotas, utilize a cost-plus pricing formula, and restrict imports. Production quotas are largely based on domestic consumption requirements. They also facilitate industry R&D and dairy product promotion. (Ottawa)
Poultry	Chicken Farmers of Canada Turkey Farmers of Canada Egg Farmers of Canada Canadian Hatchng Egg Producers	Virtually every aspect of the poultry industry is managed by the Chicken, Turkey, and Egg Farmers of Canada. The legal basis is the federal Agricultural Products Marketing Act (1966). The respective Marketing Boards set production quotas, utilize a cost-plus pricing formula, and restrict imports. Production quotas are largely based on domestic consumption requirements. The Boards also facilitate industry R&D and poultry product promotion. (Ottawa)
Wheat	Provincial Wheat Commissions Western Canadian Wheat Growers Assoc. Ontario Wheat Growers Marketing Board (OWGMB)	The Alberta Wheat Commission applies a levy of $1.09 per MT on all wheat sold; largely dedicated to wheat research. The Saskatchewan Wheat Development Commission (24,000 growers) collects a $0.52/tonne check-off.The WCWGA is a voluntary advocacy organization which supports market competition in the sale of wheat and barley and, thus, lobbied for the demise of the CWB. Previously called the Palliser Wheat Growers Assoc. (Saskatoon). The OWGMB in Ontario is similar to the now-defunct CWB on the Prairies.
Canola	CCC-Canola Council of Canada CCGA-Canadian Canola Growers Association	The CCC is a federation of provincial associations, canola crushers, and elevators. Focuses on a value-chain approach, R&D, market development, and public policy advocacy. Funded by a $1.00 levy per tonne of canola (elevators and processors). (Winnipeg) The Canadian Canola Growers Association represents the interests of five provincial canola associations (43,000 growers) and is funded through its own business operations, including acting as an administrator of the Advanced Payment Program for 45 field crops and livestock commodities.

.... Continued

Table 8.4 List of Principal Commodity Groups, 2018* (Continued)

Soybeans	Soy Canada (Canadian Soybean Council + Canadian Soybean Exporters Assoc. (CSEA))	A value-chain approach representing seed companies, commodity exporters, and processors. Focuses on expanding markets, soybean quality, and increasing processing capacity. Coordinates industry strategies advocates for funding, and promotes research adoption. (Ottawa)
Flax	The Flax Council of Canada Saskatchewan Flax Development Commission Manitoba Flax Growers Association	A national organization that promoted Canadian flax and flax products for nutritional and industrial uses in domestic and international markets. Recently (2018) hired the Canola Council of Canada to work on its behalf in foreign markets.
Barley	Barley Council of Canada	A mission of the whole barley value-chain to ensure the long-term profitability and sustainable growth of Canada's barley industry. It focuses on research and market development and the provincial organizations have a barley check-off at the point of sale. Members include: Saskatchewan Barley Development Commission, Alberta Barley (11,000 growers), Atlantic Grains Council, Grain Farmers of Ontario, Manitoba Industry Association, Manitoba Wheat and Barley Assoc., National Cattle Feeders Assoc., Producteurs de Grains du Quebec, SECAN, Western Barley Growers Assoc., Brewing & Malting Barley Research Insititute, Canadian Malting Barley Technical Centre, Beer Canada, Cargill, and FP Genetics. (Winnipeg)
Pulses	Pulse Canada	The national industry association that represents growers, processors, and traders of pulse crops in Canada. Direction and funding is provided by the Alberta Pulse Growers Commission, Saskatchewan Pulse Growers, Manitoba Pulse Growers Assoc., the Ontario Bean Growers, and the processors and exporters of of lentils, beans, and chickpeas that are members of the Canadian Special Crops Assoc. (CSCA). The focus is on market access, market growth and innovation, transportation, environment and industry initiatives. Also federal-provincial funding. (Winnipeg)
Corn	Ontario Corn Producers Assoc. (OCPA) Seed Corn Growers of Ontario Manitoba Corn Growers Assoc.	Voluntary producer associations. The Manitoba Corn Growers Association (1200 members) is funded by a refundable producer check-off, 80% of which goes to research and market development. The Seed Corn Growers of Ontario have a Marketing Board. (Chatham)
Vegetables	Canadian Horticultural Council - CHC Canadian Potato Council /Potatoes Canada Vegetable Processors of Canada	The CHC is an Ottawa-based voluntary, not-for-profit national association that represents fruit and vegetable growers across Canada involved in the production of over 120 different types of crops on over 14,000 farms. The Canadian Potato Council/Potatoes Canada represents about 1300 potato producers in eight provinces. The Potato Growers of Alberta (a marketing board) imposes a levy on table potatoes of $25 per harvested acre, $35/harvested acre for seed potatoes, and 7 cents for each 100 pounds sold for processing. For vegetable processors, some provinces also have marketing boards, e.g. Alberta Vegetable Growers (Processing)
Sugar	Alberta Sugar Beet Marketing Board	An organization of sugar producers

* Subject to errors and/or omissions.

Source: Drawn from the respective websites, 2018

8.3.1 NATIONAL MARKETING AGENCIES

Two commodity groups have truly national marketing agencies with comprehensive supply management: the dairy industry and the poultry industry. The respective marketing boards are:

- Milk Producers
- Chicken Producers
- Egg Producers
- Turkey Producers
- Hatching Egg Producers

Milk Producers:

This is a supply-managed industry and details are provided in Chapter 4, Section 4.4 and, briefly, in Table 8.4 above.

All milk producers must pay a non-refundable levy or "check-off" (sometimes called a "service charge") to fund their respective provincial marketing boards. These boards, in turn, fund their own administration, promotional and lobbying activities, as well as some R&D. The Canadian Dairy Commission (CDC) is funded by the federal government (under the Minister of Agriculture and Agri-Food), producers, and the "marketplace."

Figure 8.1 Provincial Dairy Producers: Expenditure Pattern

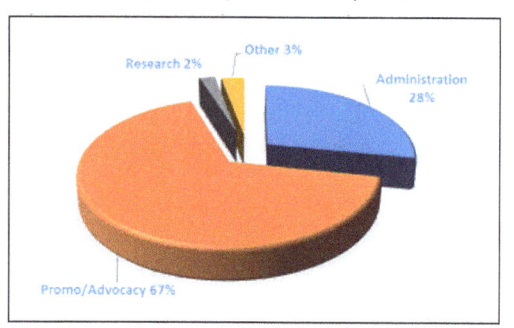

- Administration 28%
- Promo/Advocacy 67%
- Research 2%
- Other 3%

Source: Basic data from commodity website

The Dairy Farmers of Ontario collect at least three levies: 63.5 cents/hl for administration; $1.50/hl for promotion; and another 8 cents/hl for other needs. Thus, applying this $2.21/hl across Canada might generate about $187 million in revenue. The actual figure is not known. The industry reportedly raises an estimated $100 million nationally from levies on milk for marketing purposes, plus membership dues from farmers.[155] Audited financial statements, however, are not readily available to the public.

Expenditure patterns are approximated in Figure 8.1: nearly 1/3rd on administration; nearly 2/3rds on promo/advocacy; and a very small 2% on research.

Chicken, Egg, & Turkey Producers:

This is also a supply-managed industry and details are provided in Chapter 4, Section 4.5 and, briefly, in Table 8.4 above.

Similar to dairy producers, all producers of broilers, eggs, turkeys, and hatching eggs must pay a non-refundable levy to fund their respective provincial marketing boards. And these boards, in turn, fund their own administration, promotional and lobbying activities, as well as some R&D.

Figure 8.2 Poultry Producer Organizations: Expenditure Pattern

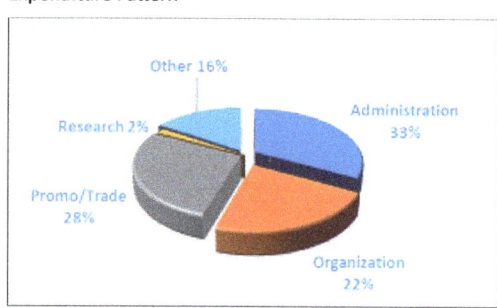

- Administration 33%
- Organization 22%
- Promo/Trade 28%
- Research 2%
- Other 16%

Source: Basic data from commodity website

Table 8.4 indicates that the actual (net) levy revenue collected by the Chicken Farmers of Canada, Egg Farmers of Canada, and the Canadian Hatching Egg Producers was $25–$30 million. This is net of service fees applied to producers to pay for product sales. The expenditure pattern for these three (turkey excluded) groups is approximately 1/3rd on administration; 1/3rd on promotion/advocacy; about 1/5th on organization, and a relatively small 2% on research (Figure 8.2). Audited financial statements are not readily available to the public.

The seven persistent criticisms of national marketing boards and supply management are well-known and are also addressed in Chapter 4:[156]

- Higher consumer prices than if these industries were exposed to the international marketplace
- Production inefficiencies within the industry (largely due to cost+ price-setting) and resource mis-allocations between provinces and regions (because of their institutional rigidity)
- Discrimination against new entrants to the industry because of increasing quota values (See Section 4.4)
- Complications and constraints in international trade negotiations; inherently limiting the benefits which could be secured from fewer trade restrictions on other sectors of the Canadian economy, regarding CETA, CPTPP, CUSMA, etc. (Chapter 5)
- Consolidation, which has greatly reduced the number of dairy and poultry producers
- Capital mis-allocation with quota values now representing 70–90 percent of the capitalization of dairy and poultry farms
- Capital disparities among farmers, regionally and nationally, as highly profitable dairy and poultry farm businesses have a capital base not available to most other farm businesses.

8.3.2   PROVINCIAL MARKETING BOARDS & COMMISSIONS[157]

The National Farm Products Marketing Act requires provincial commodity groups to establish national producer organizations to enable check-off levies. The net result is a multitude of provincial organizations with duplicating administrative, program, and operational expenses.[158] Inside and outside of Canada, this often looks like an opaque, overly complex and challenging environment in which to establish commercial linkages. Several smaller commodity groups by-pass this issue by collecting provincial levies but sharing administrative and program expenditures; a thriftiness dictated by smaller budgets (e.g., Prairie Oat Growers Association).

For the supply-managed industries (dairy and poultry), levies are mandatory and applicable to total production. For the remaining commodity groups, a major impetus for the strengthening of these groups has come from federal-provincial legislation that has allowed them to establish provincial producer-processor levies—check-offs at the point of sale. These service charges imposed on producers can be either refundable or non-refundable. This has given the respective non-supply-managed commodity groups a stable and relatively large annual budget which they can then devote to their various objectives—public relations, domestic and/or international product development, political advocacy, research, and so on. Table 8.5 is indicative of how much muscle commodity organizations can potentially exert. And for the rapidly growing sub-sectors, such as canola, pulses, and soybeans, their budgets grow proportionately.

Table 8.5 Producer Checkoffs & Potential Revenue, 2018

Commodity	Ontario Levy - License Fee	Alberta Service Fee - Levy*	Est. National Quantity**	Potential Revenue $M***	
Dairy	$2.21/hl		84.7 B	187.2	
Broilers*****			1.1 B	7.2	2014
Eggs*****			701.5 M	18.7	2015
Hatch Eggs*****			102.9 M	2.2	2016
Beef		$2/hd. refundable; $1hd. CCA/transaction	2.7 M head X 1.75 turnover	14.2	
Hogs		$1.00/hd.****	21 M	21.0	
Canola		$1.00/tonne	19.6 M	19.6	
Wheat	$0.89/tonne	$1.09/tonne	32.1 M	35.0	
Barley	$1.29/tonne	$1.20/tonne	8.8 M	10.6	
Oats	$1.49/tonne		3.7	5.5	
Pulses		1% of sales	$3,754.5 M	37.5	
Soybeans	$1.52/tonne		6.6 M	10.0	
Corn	$0.44/tonne		13.9 M	6.1	

* Currently still generally refundable. Alberta Beef now refunds 86%; Alberta Canola < 7%.
** Quantities ignore production-marketing distinction.
*** Based on Alberta levies except for oats, soybeans and corn. Assumed national
**** Plus 25 cents on hogs <50 kg. sold outside Canada.   ***** Est. actual.

Sources: Grain Farmers of Ontario (website), Stats Can Table 001-0017,
Canadian Beef Grading Agency; *Annual Reports* CFC, EFC, and CHEP;
and Dairy Farmers of Ontario, *Annual Report*, 2017

a. **Beef:**

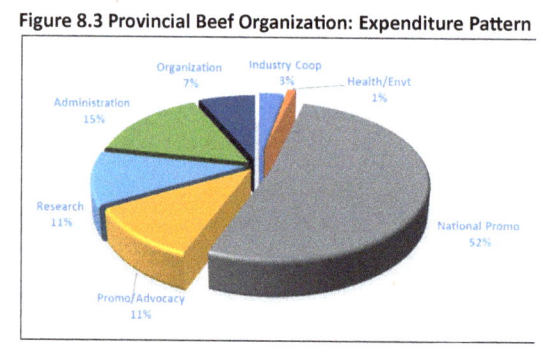

Figure 8.3 Provincial Beef Organization: Expenditure Pattern

(Pie chart: National Promo 52%, Promo/Advocacy 11%, Research 11%, Administration 15%, Organization 7%, Industry Coop 3%, Health/Envt 1%)

Source: Basic data from commodity website

The Canadian Cattlemen Association (CCA), and its provincial counterpart organizations provide beef producers with strong representation in Ottawa and elsewhere. (See Table 8.5) How these provincial counterparts spend their check-off revenue is estimated in Figure 8.3. Aside from revenue generated by a $1/head/sale and forwarded directly to Ottawa, the provincial organization also supplements this by contributing about ½ of the provincial check-off to the national Canadian Beef Cattle Market Development and Promotion Agency. Expenditures on the other major items are: administration 15%, research 11%, provincial advocacy & promotion 11%, and provincial organization 7 percent.

Ancillary to this is the Cattle Feeders Association. Their purpose is to represent the unique interests of the feedlot sector which, at times, differ from the cow-calf sector on provincial and national issues.

b. Hogs:

The annual budget of the Canadian Pork Council (CPC) is not readily available.

c. Canola:

The Canola Council of Canada and its provincial counterpart organizations also have strong representation in Ottawa and elsewhere. (Also see Table 8.5.) On the basis of a $1/tonne check-off, their provincial expenditure patterns are approximately as indicated in Figure 8.4. Five items—research 24%, national advocacy 22%, provincial advocacy 19%, administration 18%, and the provincial organization 17 percent—reflect their expenditure pattern.

Figure 8.4 Provincial Oilseed Organization: Expenditure Pattern

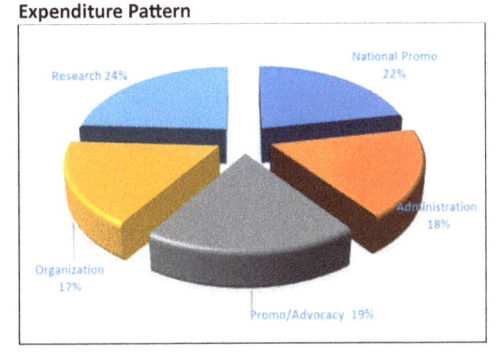

Source: Basic data from commodity website

The Canola Council of Canada has been value-chain oriented. Recent producer challenges to this industry approach threaten to diminish the collaborative nature of this alliance and future struggles to maintain unity appear likely.

The 2018 withdrawal of Richardson International from the CCC has prompted a realignment of CCC priorities. The Council's core budget in 2019 will be $5.2 million, down from $8.7 million in 2017. Its role in Canola Performance Trials (CPT) and provincial disease and pest surveys are being scaled back.[159] It is also getting out of canola promotion activities in established markets.[160]

d. Wheat, Barley and Oats:

The annual budgets of the various wheat and oat organizations are not easily accessed. There is no umbrella group that indisputably represents all wheat growers in Western Canada, although the Grain Farmers of Ontario provides effective representation for wheat, barley and oat producers in Ontario. Levies applied to wheat and barley producers in both Ontario and Alberta are similar: about $1/tonne for wheat and $1.25/tonne for barley. Oat producers in Ontario are assessed $1.49/tonne. The emergence of Grain Growers of Canada to create a collaborative framework not only for cereal grains but inclusive of agri-business and the livestock industry, a first in national agricultural organizations, is encouraging.

The ability of farmers to request a return of their paid levies after the fact is increasingly challenging many of the grain commodity groups. With the increased size of grain farms, especially in Western Canada, the large levies paid by individual farmers or incorporated entities is resulting in much higher requests for refunds. The barley and wheat organizations have recently integrated their operations in Alberta.

e. **Soybeans and Corn:**

Soybean and corn producers in Ontario are assessed a levy of $1.52/tonne and $0.44/tonne, respectively. They are well represented by the Grain Farmers of Ontario, an umbrella group.[161] This jointly administered levy was challenging in its initial stages but has proven to be very successful since corn and soybeans are generally grown on the same farms.

8.3.3 COMMODITY-GROUP RESEARCH

The principal criticism regarding non-supply-managed commodity groups funded by a mandatory check-off is that they often lack financial transparency and direct member accountability.[162] Their respective websites typically only provide participating producers with a mission statement, a list of nominal members, and considerably less financial information than good governance would require. Many producers who actually finance these organizations cannot easily determine what some commodity groups either spend or accomplish.

With respect to a wide array of prairie crops, much of their producer-generated research expenditure in Western Canada is funneled through the well-represented Western Grains Research Foundation (WGRF). (Table 8.6)

Table 8.6 Western Grains Research Foundation (WGRF)

MISSION:
Primarily a research organization which coordinates and helps set grain rresearch priorities with member organizations.
MEMBERS:
Agriculture & Agri-Food Canada (AAFC) — Prairie Oat Growers Association
Agricultural Producers Association of Saskatchewan — Saskatchewan Barley Development Commission
Alberta Grain Producers Assoc., Canadian Seed — Saskatchwan Flax Development Commission
Canadian Canola Growers Association — Western Barley Growers Association
Canadian Seed Growers Association — Western Canadian Wheat Growers Association
Keystone Agricultural Producers — Western Pulse Growers
Manitoba Wheat & Barley Growers Association — Western Winter Cereal Producers
National Farmers Union

Source: WGRF website

The WGRF coordinates and helps prioritize cereal, oilseed, and pulse R&D funding generated by commodity levies (check-offs) generated in Western Canada and then, in turn, leverages this funding by securing matching funds from either the private sector and/or various governments, as well as a substantial endowment and the excess revenues generated by the Canadian National (CN) & CP railways in moving grain. The WGRF budget indicates a research budget of about $20 million annually, some 80% of which goes to varietal development and agronomy (Table 8.7). The WGRF essentially brings the major crop organizations and public research organizations together to establish research priorities and then conduct that research.

This research funding generally goes to university researchers on the prairies and the University of Guelph (Ontario), institutions that have always at least partially relied on outside funding.

The University of Saskatchewan (U of S) is a key player, especially in varietal development. The Crop Development Centre at the U of S, and its faculty, are nationally recognized for their contributions. While some private research has been funded, research funding generally goes to university researchers.

But almost 2/3rds (58%) is going back to the federal Agriculture and Agri-Food Canada ministry; the very ministry that has historically funded most of its own research directly through the federal budget (Figure 8.5). Thus, there is a very real concern in the agricultural community that funds for research generated by producer levies are substituting rather than supplementing prior federally funded research. In recent years, direct government funding for agricultural research has, indeed, gradually eroded to levels well below those of our international competitors (Section 9.8).

Table 8.7 Western Grains Research Foundation, Budget, 2016, $,000

Item	Endowment	Wheat Fund	Barley Fund
REVENUE			
Endowment Fund	5,863.50		
Producer Levies		6,626.1	844.0
Royalties		1,734.5	141.0
Other Income	5,013.20	1,529.9	190.1
	10,876.70	9,890.5	1,175.1
RESEARCH EXPENDITURES			
Variety Development	3,125.5		
Agronomy	2,296.9		
Minor Crop Development	402.1		
Crop Risk Management	205.3		
Crop Utilzation	115.4		
	6,722.0	10,794.90	1,192.80
OPERATING EXPENDITURES			
Office	603.6	352.1	50.3
Professional Fees/per Diems	81.2	40.7	7
Travel/Commun.	185.7	227.2	21.9
Scholarships	112.0	7.0	1
	982.5	627.0	80.2
Percent of Research Expenses	15%	6%	7%

Source: WGRF *Annual Report*, 2016

Figure 8.5 WGRF Research Fund Destinations, 2016*

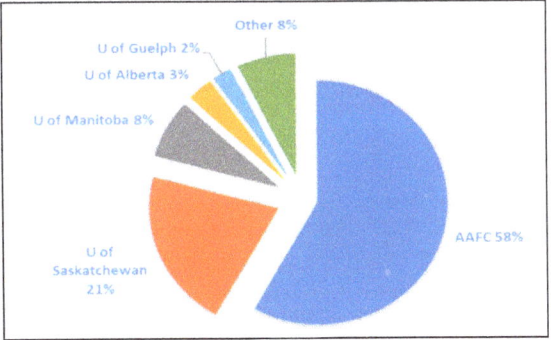

- AAFC 58%
- U of Saskatchewan 21%
- U of Manitoba 8%
- U of Alberta 3%
- U of Guelph 2%
- Other 8%

* Wheat and barley funds.

Basic data from WGRF website

8.4 SUPPLIER & PROCESSOR ORGANIZATIONS

A host of national and regional organizations representing sectors beyond basic agricultural production include the Canadian Meat Council, Dairy Processors, Poultry Processors, provincial food processing organizations, the Western Grain Elevator Association, and the Canadian Seed Trade Association, which has recently moved to integrate with the Canadian Seed Growers Association. (Also see Chapter 6.) The primary purpose of most is simply to retain the capacity to deal with issues and challenges arising from trade, the regulatory framework, and common commercial interests. Maintaining good relations up and down the value chain can result in collaborative efforts through trade alliances and regulatory initiatives.

8.5 FINAL COMMENTS

Agricultural organizations in Canada have evolved, devolved, and re-appeared over our 150-year history as a nation. The agricultural industry, like our country, has moved from building a foundation of people and infrastructure to now striving to embrace global and technological realities to create yet another new era of prosperity in the agricultural community.

The role of farm organizations in this process is mixed, acknowledging that at times strong leadership has provided direction and moved sectors in directions beneficial and lasting. Clearly, there has often been a struggle arising from the independent nature of farmers and the need for collaborative, collective, cooperative initiatives. Compared to many other countries, Canada's non-governmental organizations in agriculture are still in a developmental mode; still seeking more effective ways to provide a better understanding of their mission to an increasingly disconnected public audience.

Thus, the plethora of agricultural organizations continues to grow, largely fed by a membership with a strong commitment to a particular issue. Even as farm numbers have declined, the number of organizations has continued to grow, the net result being a focus on narrow interests and an inability to focus on national policy issues driven by an industry with a shared vision. At the same time, one striking feature of the 21st century, at least to 2020, has been the gradual decline in farmer participation, interest, and commitment to organizations that supposedly represent their interests. Not unlike the lower percentage of voters participating in our democracy, agricultural organizations struggle to sustain even the most basic requirements of member participation to ensure good governance. Only rarely can the various organizations successfully confront this challenge; the UPA in Quebec and the supply-managed organizations being the exceptions—most likely because of their significant roles, responsibilities, and accountability to their memberships.

9. Role Of Government[163]

Faced with a varied climate and geography, accompanied by historically different settlement patterns, the agricultural sector in each region of the country is very unique, each with its own set of challenges. But the principal unifying factor throughout Canada's relatively brief history has been the facilitating role of the omnipresent government, particularly the federal government. From the colonial era to the present, agricultural development has been state-supported and sometimes subordinated to other interests.[164] Prairie agriculture, in particular, has long been treated as a problem child because of its distance from markets and its relatively volatile temperate climate.

9.1 HISTORICAL OVERVIEW, 1870–2019

The Constitution Act (originally the BNA Act), 1867, initially created a federal dominion between four provinces, Ontario, Quebec, New Brunswick, and Nova Scotia, which was similar in principle to that of the United Kingdom. It specifies the federal parliamentary structure, the House of Commons, the Senate, the justice system (common law excluding Quebec), and the taxation system. More specifically:

- The federal and provincial governments share power over agriculture and immigration (Section 95). Either level government can make laws in this area but in the case of conflict, federal law applies.
- The provinces have power over civil and contract law (Section 92).[165]
- Up to the farm gate, the federal and provincial governments have equal powers, but beyond the farm gate federal and provincial trade and commerce legislation and provincial civil and contract law prevails. The federal government is generally responsible for interprovincial commerce (transport, communications, etc.) and provincial tariffs are prohibited.
- The provinces have power over education, subject to religious minority protection (Section 93).
- The BNA Act does not expressly grant jurisdiction over health care or related social programs (Sections 91 and 94). In practice, this is shared according to the fiscal capacity of the respective governments.
- The Natural Resources Act (1930, incorporated into 1982 BNA changes) transferred

control over Crown lands and natural resources in the prairie provinces and BC to the respective provincial governments.

9.1.1   THE DECADES 1870–1950

Particularly notable during this period were five federal initiatives:

- Freight rate protection for farmers with the establishment of the artificially low **Crow's Nest Pass Agreement, 1897**, for railroads to facilitate grain exports from the Prairies

- The establishment of the **Canadian Wheat Board (CWB), 1935**, for the pooled pricing of prairie wheat and (after 1941), managing the sale of all wheat, oats, and barley from western Canada, as well as the sole control (by license) of imports and exports of these grains and products throughout Canada[166]

- Periodic *ad hoc* responses to a perceived crisis in the agricultural sector, (e.g., coping with the Dirty Thirties, which resulted in the re-possession and re-grassing of much of the Palliser Triangle in Saskatchewan and Alberta and the establishment of the **Prairie Farm Rehabilitation Administration (PFRA) in 1935**).

- The **Feed Freight Assistance Act, 1941**, subsidizing feed grain rail transport from the prairies, east, west, and north (NWT)

- The establishment of the **Agricultural Products Marketing Board Act, 1947**, which empowered provincial marketing agencies to sell agricultural products to other provinces and countries. (By 1970, there were about seventy provincial marketing boards and this is still the basis for marketing arrangements in Canada.)

These initiatives were accompanied by the widespread formation of farm cooperatives throughout Canada, especially on the prairies during the 1910–1930 period.

Table 9.1 Federal Legislative Timeline, Canada, 1860 - 1950

YEAR	POLITICAL	AGRICULTURE	COOPERATIVES	BOARDS, COMM. & CORPS	TRANSPORT
1860			1860-1900: East 1200 coops + mutual insurance		
1861			First store, NS		
1867	Confederation: Quebec,				
1870	Manitoba joined, #5. NWT formed				
1871	BC joined, #6				
1873	PEI joined, #7				
1876	Keewatin NWT, created				
1887		Experimental Farm Stations Act			
1896		General Inspection Act (grain)			
1897					Crow's Nest Pass Agreement
1898	Yukon Territory created				
1900		Manitoba Grain Act	Caisse Popular		
1905	Alta & Sask. joined, #8 & #9				
1906		Federal Meat Inspection Act	Grain Growers Grain Co.		
1909			Co-operative Union of Canada		
1910			Co-operative Federee (Que)		
1911			Sask. Coop Elevator Co.		
1912		Canada Grain Act; Board of Grain Commissioners	Pre-WWI: numerous purchasing & marketing coops		
1913			Alta. Farmers Coop Co.		
1914			United Farmers Cooperative (Ont.)		
1917			United Grain Growers		
1919		1st Canadian Wheat Board			
1923			23-24: 3 Prairie Grain Pools		
1926				Produce Marketing Act (BC)	
1931	Statute of Westminster. Independence & resources to provinces		1930's: Canadian Cooperator, UFA, Western Producer, Wheat Pool, Credit Union (NS), other		
1935		Prairie Farm Rehabilitation Admin. (PFRA)			
1939		Prairie Farm Adjustment Act			
1941					Feed Freight Assistance Act
1945	National Emergency Powers Act				
1947				Agricultural Products Marketing Board Act	
1948			Late 40's: Trust Companies (Que), Ont. & Sask. wholesalers		

9.1.2   THE DECADES 1950–1980

In the post-WWII period and throughout most of the 1950s, conditions in the agricultural community were generally good, or at least relatively benign. But gradually the storm clouds formed: growing stockpiles of grain, weaker and incessantly volatile product prices, and a persistent and pernicious on-farm cost-price squeeze. In this increasingly hostile environment, the particularly notable federal legislative initiatives during this period were the following (Table 9.2):

- The **Agricultural Stabilization Act (ASA)**, 1958, which initially guaranteed minimum prices (i.e., 80 % of a 10-year average) in every province for nine commodities (as amended in the 1970s): slaughter cattle, hogs, industrial milk and cream, lambs and wool, corn and soybeans, as well as wheat, oats, and barley outside the CWB area. (In the 1970s, this was modified to 90% of a 5-year average). This was the first Act of its kind in Canada that utilized a set formula to calculate the direct support payments to farmers.[167] Its intended replacement was the Farm Income Protection Act of 1990. (See Section 9.3.)

- The **Crop Insurance Act**, 1959, which, for the first time, potentially made available to farmers, in every province, all-risk crop insurance. This established a formula for the conditional funding of provincial programs; the first agricultural program to introduce a cost-sharing template between the federal and provincial governments.

- **The Farm Credit Act**, 1959, the legal basis for the emergence of the federal Farm Credit Corporation (FCC), still a major source of credit for farmers. It was initially established because private banks were historically reluctant to lend to agriculture without the capacity to seize land in the event of a bankruptcy.

- The **Agricultural and Rural Development Act (ARDA),** 1962, which was a joint federal/provincial initiative to facilitate the development and conservation of rural land and water. It included conducting a major land inventory and direct payments to farmers for withdrawing cropland from production.

- The **Canadian Dairy Commission Act,** 1967, established the Canadian Dairy Commission (CDC), which chairs the Canadian Milk Supply Management Committee (CMSMC), made up of provincial Dairy Boards and provincial officials. The CDC has authority over industrial milk (Classes 2 through 5). The CMSMC coordinates pricing and production quotas so as not to exceed Canadian fluid milk/cream requirements. Fluid milk prices and volumes (Class 1) are set by the individual provinces within the "western pool" and the "P5" (eastern provinces) with coordination between the two pool areas. All milk is measured as butterfat, not milk volumes or weights. The CDC is a Crown Corporation, which reports to the Minister of Agriculture and Agri-Food.

- The **National Farm Products Marketing Act**, 1972, which set out the mechanisms under which tobacco, poultry, and egg products could be marketed. Similar to the Canadian Dairy Commission Act (see above), this Act allowed for the establishment of a (federal) coordinating Board empowered to approve provincial levies and quotas. It is also a Crown Corporation, responsible to the Minister of Agriculture and Agri-Food. The provincial boards were established under provincial legislation.

- The **Western Grain Stabilization Act**, 1976, was established to stabilize producers' net proceeds from the production and sale of wheat, oats, barley, rye, flaxseed, canola, and mustard seed produced in the prairie provinces and the Peace River region of BC. In 1988, seven additional crops were covered: triticale, mixed grains, sunflower, safflower, buckwheat, peas, lentils, faba beans, and canary seed. It was replaced by the Agricultural Marketing Programs Act, 1997 (Section 9.3).

Table 9.2 Federal Legislative Timeline, Canada, 1951 - 1980

YEAR	POLITICAL	AGRICULTURE	COOPERATIVES	BOARDS, COMM. & CORPS	TRANSPORT
1958		Agricultural Stabilization Act (ASA)		Agricultural Stabilization Board	
1959		Prairie Grain Advance Payments (PGAP) Act; Crop Insurance Act (now AgriInsurance); Farm Credit Corporation Act			
1962		The Agricultural and Rural Development Act (ARDA)			
1967		The Canadian Dairy Commission Act		Canadian Dairy Commission	
1970		Lower Inventories for Tomorrow Program (LIFT); Animal Diseses & Protection Act			
1972		National Farm Products Marketing Act			
1975		ASA and PGAP Amendments			
1976		Western Grain Stabilization Act (WGSA)			
1977		Advance Payments for Crops Act (APCA) -East			

9.1.3　THE DECADES 1980–2000

North America generally witnessed deteriorating macroeconomic conditions from the mid-1970s to the late 1980s. Rampant inflation and skyrocketing interest rates (reaching 20%+/annum) only exacerbated the continuing on-farm cost-price squeeze. Farm bankruptcies were once again commonplace and, not surprisingly, the family farm and rural Canada then felt increasingly threatened by this intimidating social and macro-economic dynamic. As rural Canada was seemingly being sapped of both its strength and its vitality, the voice of the agricultural sector as well as other sectors became increasingly shrill. And once again the federal government responded, gradually turning away from crisis-inspired *ad hoc* programs to more on-going formula-based funding. The principal farm-related initiatives during this 1980–2000 period were(Table 9.3):

- The **Western Grains Transportation Act**, 1983, succeeding the Crow's Nest Pass Agreement (1897) by paying a Crow Benefit directly to the railways and putting a ceiling on rail grain transport rates on the prairies; an initial ceiling not to exceed 10 percent of the world price of grain. A continuing (lesser) subsidy was also to be paid directly to the railways.[168]

- The **Agricultural Products Cooperative Marketing Act**, 1985, was an Act to assist and encourage cooperative marketing of agricultural products. This laid out the terms and conditions, as well as the mechanics of advance payments, to improve on-farm cash flows. Repealed in 1997.

- **Farm Income Protection Act (FIPA)**, 1991, which consolidated the Agricultural Stabilization Act (1958) and the Crop Insurance Act (1959), as well as repealing the Western Grain Stabilization Act (1976). This was the umbrella legislation for GRIP and NISA.

- **Gross Revenue Insurance Plan (GRIP),** 1991, was established under the Farm Income Protection Act (FIPA, 1991). GRIP built on the existing all-risk Crop Insurance Program by providing producers with revenue protection, offering both price and yield protection. The insured level of gross revenue was derived by multiplying the producers' long-term average yields for each crop planted by the target prices for each crop. Government payments were then made when a producer's market revenue fell below this predetermined revenue target. All grain, oilseed, and specialty crop producers were eligible to participate in GRIP.[169] It was both intrusive and bureaucratic since it required government officials to verify production levels on-site. The program was terminated in 1996 although some provinces (e.g., Ontario) continued to fund this program as it was established under GRIP.

- **Net Income Stabilization Account (NISA),** 1991, was a companion program to GRIP (above), which was also established under the Farm Income Protection Act (FIPA). The purpose of NISA was to encourage producers to save a portion of their income for use during periods of reduced income. Producers could deposit 3% of their eligible net sales annually into their NISA account and receive matching government contributions. All NISA (bank) deposits earned a 3% interest rate bonus in addition to the regular rates offered by the financial institution where the account was held. Excepting supply-managed products (poultry and dairy), most other agricultural commodities qualified. The program was terminated in 2002 although money could be withdrawn until 2009. It was re-started as AgriInvest (Section 9.4).

- **Western Grains Transition Program**, 1995, a one-time payout to western grain farmers for the termination of all rail subsidies embedded in the WGTA (1983). Consistent with this initiative, the **Feed Freight Assistance Program** (1941) was also terminated in 1996 with a one-time payout.

- The **Pest Control Products Act**, 1995, which consolidated all responsibility for federal pesticide regulation with Health Canada. The products regulated includes all pesticides and pest management agents used in every industry and household. The provinces can, however, also regulate chemicals within the federal parameters, i.e. they can be stricter. e.g. neonic use in Ontario.

- The **Agricultural Marketing Programs Act**, 1997, consolidated the following Acts: the Advance Payments for Crops Act; Prairie Grain Advance Payments Act (1959); Agricultural Products Marketing Board Act (1947); and the Agricultural Products Cooperative Marketing Act (1985). This basically unified the terms and conditions associated with advance payments (prior to physical delivery) and made the respective commodity organizations responsible for program administration. The objective was improved cash flow for all farmers and this still continues today. The CWB's responsibility for wheat and barley grain advances was transferred to canola producer organizations after the CWB ceased operations in 2011.

- **Agricultural Income Disaster Assistance (AIDA)**, 1998, and the **Canadian Farm Income Program (CFIP)**, 2000. These were two similar programs. Each was designed to provide financial assistance to farmers (based on anyone who submitted a farm income tax return) facing dramatic income declines for reasons beyond their control, such as flooding, disease, price collapse, or rapidly rising input costs.[170] AIDA was in place in 1998 and 1999; CFIP covered the claim years 2000, 2001, and 2002 under the existing FIPA legislation (see above).

Table 9.3 Federal Legislative Timeline, Canada, 1981 - 2000

YEAR	POLITICAL	AGRICULTURE	COOPERATIVES	BOARDS, COMM. & CORPS	TRANSPORT
1983					Western Grains Transportation Act
1985		Canadian Agricultural Products Act; Special Canadian Grains Program (SCGP I)	Agricultural Products Cooperative Marketing Act (APCMA)		
1989		Farm Support & Adjustment Measures (FSAM)			
1991		Farm Income Protection Act (FIPA) - GRIP & NISA; consolidated ASA & repealed WGSA			
1995		Pest Control Products Act; Western Grains Transition Program			Western Grain Transportation Act, amended
1996					Feed Freight Assistance terminated
1997		Agricultultural Marketing Programs Act (APCA+PGAP+APB+APCMA)			
1998		Agricultural Income Disaster Assistance (AIDA)			
1999	Nunavut NWT, created				
2000		Canadian Farm Income Program (CFIP)			

9.1.4   THE 21ST CENTURY[171]

Canada's national Agricultural Policy Framework in the 21st century for the free-market component (i.e., excluding dairy and poultry) has basically tried to redesign the earlier GRIP, NISA, and crop insurance programs, still prioritizing risk management, and farm income enhancement, but now more rigorously adhering to the terms of the 1994 WTO Agricultural Agreement. Each requires a contribution from producers, the provincial government, and the federal government.

The first of these was the **Canadian Agricultural Income Stabilization (CAIS) Program,** 2003, which was designed to replace NISA and integrate income stabilization and disaster protection. It was available to all farmers, included a government contribution, and allowed participants to select a protection level for their farms and then make the necessary deposits to provide that insurance. CAIS payments were then made when the farmer suffered a loss. This program was terminated in 2007.

Thereafter, four basic programs were designed to replace all of the prior programs that focused on income stabilization, disaster protection, and income enhancement. Called safety nets, all of these came into effect in 2006–2008:

- **AgriStability** – a program to protect overall farm income. It is based on a farmer's financial information and stabilizes the farmer's income in the event of a decline in his/her production margin. AgriStability is based on margins, the difference between revenue and eligible costs. The current margin is compared to an historical five-year average reference margin and administrators verify this with information collected through farm income tax filings from the Canada Revenue Agency (CRA). Participating farmers can receive an AgriStability payment when their current year production falls below 70% of their reference margin. This is essentially a GRIP-CAIS upgrade. AgriStability is delivered in Nova Scotia, New Brunswick, NL, Manitoba, and the Yukon by the federal government. In BC, Alberta, Saskatchewan, Ontario, Quebec, and PEI, it is delivered by the respective provinces. With the associated Agri-Quebec Plus program, there is additional financial assistance available.

- **AgriInvest** – a self-managed producer government account designed to help producers manage small income declines, as well as making investments to manage risk and generate income. A farmer can annually deposit up to 100% of his/her Allowable Net Sales for the year and receive a matching government contribution of 1% of Allowable Net Sales. At a participating financial institution, this account has two components: Fund 1 holds farmer deposits; Fund 2 holds the matching government contributions and interest. Withdrawals must first come from Fund 2 and these are taxable as investment income (i.e., not as farm income). Subsequently, withdrawals from Fund 1 are not taxable. This is similar to the earlier NISA program, but now farmer deposits are not limited to 3% of net sales and regardless of how much the farmer contributes, the government will only contribute 1% of net sales (instead of 3%). This is also audited utilizing CRA data.

- **All-Risk Insurance** (now called **AgriInsurance**) – a very old (1959) and venerable program that has continually expanded to include ever-more crops (e.g., pasture) and livestock production. It has also expanded to cover more insurance levels and, in some provinces, to also now include commodity-specific price insurance. This is administered by the province, with a 36% and 24% financial contribution from the federal and provincial government, respectively. Thus, since farm premiums typically cover about 40 percent of the total cost of this insurance, this program (like those in the USA, EU, and elsewhere) also provides a major indirect subsidy to farmers.

- **AgriRecovery** – introduced in 2006 to deal with regional or national disasters –flooding, droughts, disease, and so on. This is essentially a catch-all program that is operationalized when the AgriStability, AgriInvest, and AgriInsurance programs cannot respond adequately to the disaster at hand. It is specifically designed to respond to asset losses; not income losses. (Earlier AIDA and CFIP disaster programs focused on income losses.)

Table 9.4 Federal Legislative Timeline, Canada, 2001 - 2018

YEAR	POLITICAL	AGRICULTURE	COOPERATIVES	BOARDS, COMM. & CORPS	TRANSPORT
2002		Net Income Stabilization Account (NISA) terminated			
2003		Canadian Farm Income Program (CAIS)			
2006		AgriRecovery			
2007		AgriStability			
2008		AgriInvest			
2011		Marketing Freedom for Grain Producers Act (CWB)			

Accompanying this suite of programs, the **Farm Credit Corporation (FCC**, 1959), a Crown Corporation (see Figure 9.1), also played a vital role in providing different kinds of credit (but especially long-term credit) to farmers at very competitive (fixed and variable) terms and interest rates. Although once exclusively a farm lender, FCC is now also providing funding to finance on-farm diversification projects and value-added agricultural operations beyond the farm gate. They have offices in almost every agricultural trading center throughout Canada, and have about 150,000 loans with more than $25 billion outstanding.

AAFC also has an increasing focus on **food safety** (Section 9.4) and **environmental sustainability** (Chapter 10), accompanied by an on-going commitment to aggressive **international marketing and trade** (Chapter 5). As of April 2018, the Canadian Agricultural Partnership (CAP) replaced the prior Growing Forward 2 programs. This consists of fifteen cost-shared funding initiatives for 2018–2022, including an Environmental Stewardship and Climate Change program.

9.2 ORGANIZATION OF AGRICULTURE AND AGRI-FOOD CANADA (AAFC)

The organizational structure of AAFC often changes. Figure 9.1 nevertheless provides us with an indication of its general structure. Figure 9.1 is especially useful because it highlights how the various Boards, Commissions, and Secretariats are directly responsible to the Minister:

- **Canadian Dairy Commission (CDC, 1966)** – oversees supply management for the dairy industry. Mandated to provide efficient producers of milk and cream with the opportunity to obtain a fair return to their labor and investment. Also chairs the Canadian Milk Supply Management Committee (CMSMC), which coordinates the management of industrial milk supplies in Canada and integration of the fluid milk market between provinces. Responsible for determining the price that farmers receive for raw milk. (Also see Section 4.4.)

- **Farm Products Council of Canada (FPCC)** – oversees the supply management of eggs, poultry, and chicken. It supervises the operations of the Egg Farmers of Canada, Turkey Farmers of Canada, Chicken Farmers of Canada, and Canadian Hatching Egg Producers. (Also see Section 4.5.)

- **Canadian Food Inspection Agency (CFIA)** – enforces all health and safety standards under the food and drug regulations. Now reports directly to the Minister of Health (Health Canada). (Also see Section 9.3 following.)

- **Canadian Grain Commission (CGC, Winnipeg)** – responsible for regulation in the grain industry. It covers all country and terminal/inland elevators in the former CWB area, but only terminal elevators elsewhere in Canada. (Also see Section 9.3 following.)

- **Farm Credit Corporation** – provides specialized and personalized financial services to farming operations. Canada's largest agricultural credit provider. (Section 9.4.)

- **Canadian Para-Mutual Agency** – regulates and supervises para-mutual betting on horse racing at race tracks across the country. (Also see Section 9.3 following.)

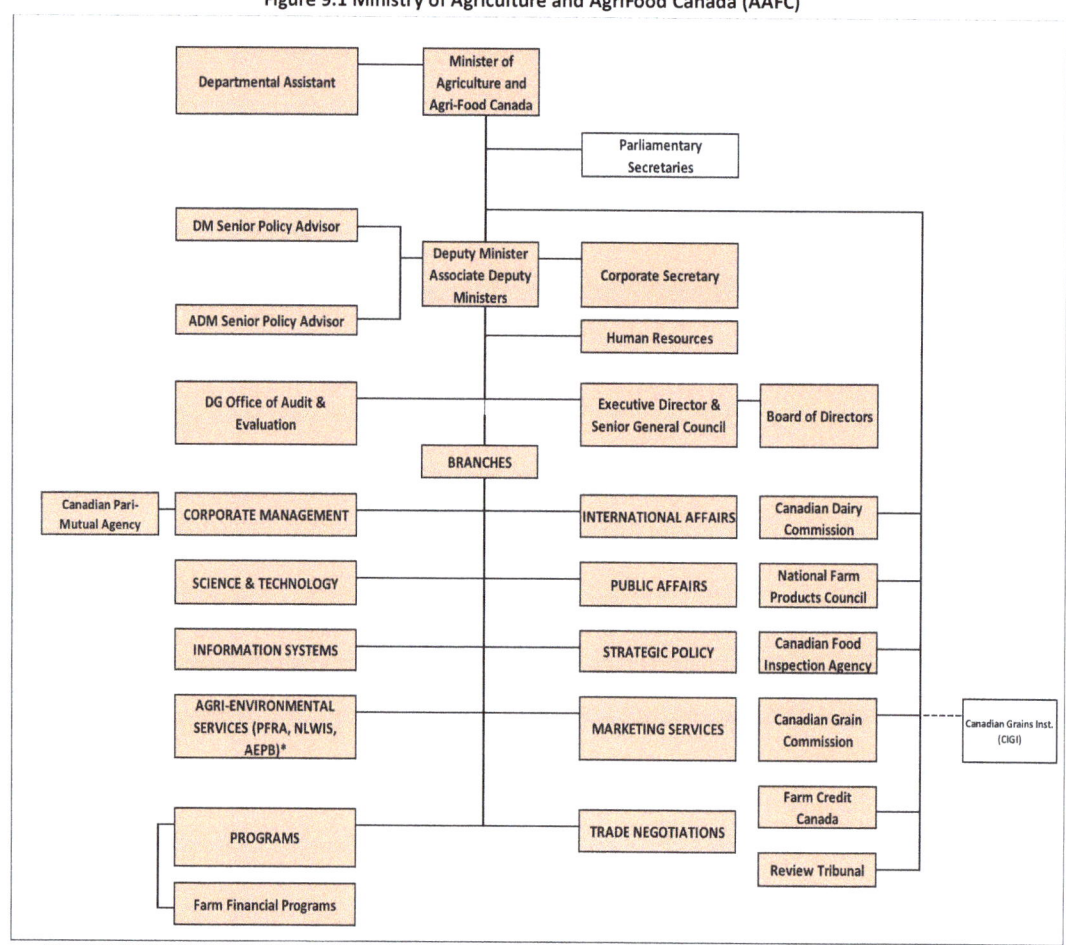

Figure 9.1 Ministry of Agriculture and AgriFood Canada (AAFC)

*PFRA = Prairie Farm Rehabilitation Administration; NLWIS = National Land and Water Information Service; AEPB Agri-Environmental Policy Bureau.

Basic Source: *Agriculture and Agri-Food Canada Ogranization Chart,* February 1, 2018. Website.
Also see GEDS and *Agriculture and Agri-Food Canada Portfolio,* website, 2016.

9.3 THE REGULATORY ENVIRONMENT

The vast array of federal regulations address, in particular, public health and safety, environmental standards, and agriculture-related commerce. The principal federal regulatory agencies are the following:

- **Canadian Food Inspection Agency (CFIA)** – enforces all health and safety standards under the food and drug regulations and is also responsible for the administration of non-health and safety regulations concerning food advertising, labeling, and advertising. Designated GMOs, functional foods, organic foods, etc., as well as the nutritional composition of foods, all come under the CFIA. Specifically, with respect to livestock and poultry, the CFIA administers the Livestock and Poultry Carcass Grading Regulations under the authority of the Canada Agricultural Products Act (1992). (See AAFC Organization Chart, Figure 9.1 above.)

- **Public Health Agency of Canada** – provides specialized scientific advice to detect and prevent risks to Canada's food supply and public health. This includes GMO and related issues.

- **Canadian Grain Commission (CGC, Winnipeg)** – responsible for regulation in the grain industry. It provides a full range of inspection, weighing, analytical, and entomology services. (See AAFC Organization Chart, Figure 9.1 above.)

- **Canadian Para-Mutual Agency** – regulates and supervises para-mutual betting on horse racing at race tracks across the country. It is funded by a levy of 0.8% on each bet placed on horse races in Canada. (See AAFC Organization Chart, Figure 9.1 above.)

- **Pest Management Regulatory Agency (PMRA)** of Health Canada is responsible for administering the Pest Control Products Act and Regulations (PCPA), registering pest control products, re-evaluating registered products, and setting maximum residue limits under the Food and Drug Act. Potential providers must provide all the scientific studies necessary for determining that a product is acceptable in terms of safety, merit, and value. Pesticides imported into Canada are regulated nationally under the same PCPA.

- **The Canada Agricultural Review Tribunal (CART, 1983)** is an independent, quasi-judicial regulatory appellate tribunal that is mandated to balance the rights of Canadians with the protection of health and well-being of Canadian consumers and the economic vibrancy of Canadian agriculture and agri-food industries. The tribunal currently reviews notices of violation issued by the Canada Border Services Agency, the Canadian Food Inspection Agency, and the Pest Management Regulatory Agency. It also reviews certain decisions of the Minister of Health and the Minister of Agriculture and Agri-Food. (Wikipedia)

- **Transport Canada** establishes emission control regulations for all motorized vehicles including those engaged in agricultural operations, if applicable. The umbrella legislation is the Canada Environmental Protection Act (1999).

- The **Agreement on Internal Trade (AIT, 1995)**, between all Canadian jurisdictions is aimed at reducing and eliminating barriers to free movement of persons, goods, services, and investment within Canada. The provinces of Alberta and British Columbia signed a Trade, Investment, and Labour Mobility Agreement (TILMA) in 2006, with a goal of creating a seamless economic region between the two provinces by eliminating many of the still-existing interprovincial barriers. But barriers persist, especially regarding the provision of services. With the supply-managed industries (dairy and poultry), there is virtually no interprovincial trade.

9.4 THE RAILWAYS[172]

Probably nothing influenced settlement patterns and the development of the grain export trade on the prairies more than the railways. As the railways were being built across the prairies, day-by-day, a wave of homesteaders arrived and agricultural development followed almost immediately. Towns and villages and the iconic prairie grain elevator quickly sprang up at almost every siding, usually 10–12 miles apart. Grain was then shipped by rail in relatively small, inefficient box cars.

The evolution of the rail transportation system and the international grain trade in western Canada is largely the result of a succession of interventions by the federal government:

- 1885. Completion of the transcontinental (Canadian Pacific – CP) railroad through the southern part of the prairies. This was a condition of BC entering Confederation. Government ceded sixteen of every thirty-six sections of homestead land to CP (Section 2.4).

- 1897. Statutory commitment to the Crow's Nest Pass freight rates (**"Crow Rate"**). Among other things, CP agreed to provide reduced rail rates for farmers' grain shipped east to the Great Lakes and for farm machinery shipped west from central Canada "forever." This was later made applicable to all railroads and extended to include the Churchill and Pacific ports.

- 1918. Canadian National (CN) was formed as a government-owned corporation (1918) through the amalgamation of several financially troubled, privately owned more northerly railways, most notably the Grand Trunk Pacific and the Canadian Northern. By 1923, CN had become the largest railway in the world, operating more than 35,000 km (21,700 miles) of track and employing more than 100,000 people.

- 1941. The **Feed Grain Freight Assistance Program** was established to subsidize the transportation and storage of feed grains from the prairie provinces to eastern Canada and BC.

- 1971–1982. A period of discretionary federal payments to the railways to compensate them for maintaining the Crow Rate to transport grain. Branch lines subsidies grew from $37 M in 1971 to $322 M in 1982.[173] This was a decade when the federal government, the Saskatchewan and Alberta provincial governments, and the CWB also purchased hopper cars for the railways to accelerate replacement of the old, inefficient box cars (8,000, 2,000, and 2,000, respectively).

- 1977. The Hall Commission's "Grain and Rail in Western Canada," which recommended the abandonment during 1977–81 of 2,165 miles of grain-related prairie branch lines and the retention of other branch lines until 2000. This initiated the rationalization of branch lines to improve railway efficiencies but increased farm trucking costs.

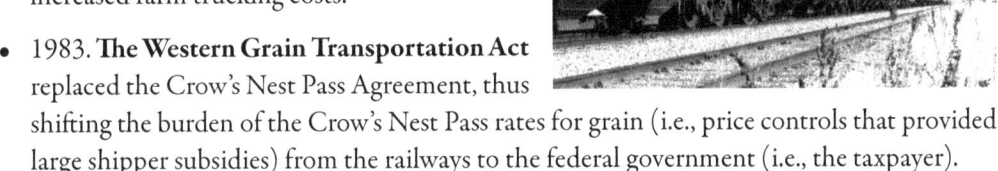

- 1983. **The Western Grain Transportation Act** replaced the Crow's Nest Pass Agreement, thus shifting the burden of the Crow's Nest Pass rates for grain (i.e., price controls that provided large shipper subsidies) from the railways to the federal government (i.e., the taxpayer).

- 1987. **National Transportation Act (NTA)** allowed shippers located on only one of the railways' lines greater access to the other's line. As well, it allowed the confidential negotiation of rates and encouraged a reliance on market forces and arbitration rather than on regulation of most non-grain rates. It also allowed both railways somewhat greater freedom to abandon uneconomic branch lines.

- 1995. The **Crow's Nest Pass Rates** were abolished and the **Western Grain Transition Payment Program** provided one-time payments to farmers to assist them in making the transition away from subsidized shipping.[174]

- 1995. The government-owned Canadian National Railway (CNR) was privatized and listed on the Toronto Stock Exchange.

- 1996. The **Canada Transport Act** extended the 1987 NTA provisions and allowed both railways to eliminate low-density lines, resulting in either total abandonment of the line or the sale to a short-line railway. The viability of rail-abandoned communities was jeopardized and the grain-elevator system was re-configured to only include high through-put inland terminals along still-existing main lines.

- 1999. **Bill C-34** replaced the regulation of maximum rates with a regulation of maximum revenues (i.e., a revenue cap) that CP and CN could earn from the movement of grain. (See: Transportation Appeal Tribunal of Canada Act (2001).)

- 2011. Bill C-18, the **Marketing Freedom for Grain Farmers Act** ended the Canadian Wheat Board's monopoly over the export of wheat and barley on the prairies. The CWB had also participated in the allocation of CN/CP grain cars to elevator points.

- 2013. The **Fair Rail Freight Service Act** established more performance-based criteria for export grain shipments, including regulations under which fines for unacceptable CP/CN performance could be imposed.

The federal government has constitutional authority over interprovincial transport and, as such, the regulation of railways is largely the responsibility of Transport Canada (Ministry of Transport). The principal implementing agencies are the Canadian Transportation Agency and the Transportation Safety Board of Canada. The railways are represented by the Railway Association of Canada.

Today, of all the railway networks in the world, Canada's is the third largest and transports the fourth largest volume of goods. Every year, Canadian railways move 70% of the country's surface goods (including 40% of its exports) and carry about 70 million people. Combined, they have about 72,093 km (44,797 miles) of functioning railway track. Both CN and CP now extend deep into the United States and both are now highly profitable private companies.[175]

Both CN and CP are now supporting the construction and operation of inland grain terminals by providing circular 125 hopper car "spots" and dedicated locomotives to quickly and efficiently load and move grain to export positions and then re-cycle the hopper cars back to the prairies for re-loading. The round-trip to and from the Pacific coast typically takes 8–9 days. Increased farm trucking costs and increased road maintenance costs are not incorporated into this business model.

The tempestuous relationship between the prairie grain industry and the rail transportation industry is due to the grain industry's almost total dependence on the satisfactory performance of, essentially, only two rail companies, CN and CP. CP dominates the south, CN the north. As grain competes with oil, coal, iron ore, fertilizer, and other goods shipped by rail, the timeliness of railway service and service rates are often an issue. Prairie grain farmers argue that since the majority of Canadian grain is shipped to market by rail, the reliability of the Canadian grain handling system must be assured if Canada's competitiveness in domestic and global markets is to be maintained or enhanced. The love-hate relationship continues.

9.5 AGRICULTURAL RESEARCH[176]

Agricultural research and development (R&D) are the life-blood of a growing and sustainable agricultural sector in Canada and elsewhere. There are numerous actors, funders, and partners in this R&D process, but particularly the federal and provincial governments, the private sector, industry groups, post-secondary educational institutions, non-governmental organizations (NGOs), and international organizations.

Figure 9.2 Federal Publis Agriculture R&D Spending, 20185-2016

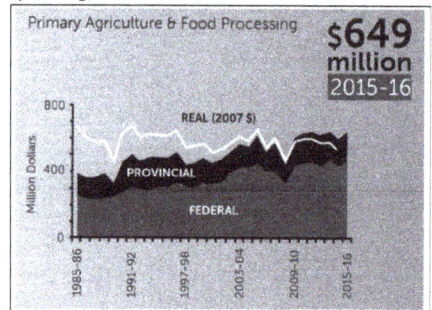

Source: AAFC, *An Overview of the Canadian Agriculture and Agri-Food System*, Ottawa, 2016

9.5.1 FEDERAL GOVERNMENT

The federal government has been a major contributor to agricultural R&D in Canada. Federal agricultural research stations are numerous and extensive funding is allocated there and elsewhere.

In 2015–16, this amounted to some $649 million; about 1% of gross farm receipts. But with the new proprietary legislation in Canada and the proprietary research of the mega seed and chemical companies (Chapter 6), this public funding is very gradually being withdrawn and the number of senior professional staff is gradually declining. In recent years, the available resources from government has tightened and, in real dollar terms, actually declined. (See the white line, Figure 9.2.)[177]

9.5.2 PROVINCIAL GOVERNMENTS

Provincial support for agricultural research and related activities (education and extension) in 2015–16 amounted to about $440 million, with their contribution generally reflecting the relative importance of agriculture in their respective provinces (Table 9.5).

Table 9.5 Provincial Agriculture R&D Expenditure, 2015 - 2016

PROVINCE	$ MILLION
PEI	6.7
NS	40.9
NB	2.0
NL	6.4
QUEBEC	52.0
ONTARIO	104.5
MANITOBA	58.4
SASKATCHEWAN	38.8
ALBERTA	108.9
BC	20.7
TOTAL	439.3

Source: AIC, *An Overview of the Canadian Agricultural Innovation System,* 2016

This is about 2/3rds of what the federal government contributed in this same year but it has also been gradually declining (in un-inflated $) since 2012.

9.5.3 PRIVATE SECTOR R&D

Private-sector agricultural R&D conducted in Canada is estimated to have dropped to about $73 **million** in 2015; a relatively modest investment (Figure 9.3).[178] But the operative words are "conducted in." The multinational Bayer Crop Sciences, for example, had a R&D budget of about Cd$6 **billion** in 2015–16. This is almost 10 times Canadian-based public agriculture R&D expenditures (see above) and it is just one among many such companies. Obviously, much of this private R&D research is ultimately adopted by numerous countries, Canada included, and, indirectly, paid for by the consumer.

Figure 9.3 Canada Private Agriculture R&D Spending, 1981-2015*

*Exclude forestry, fishing, and hunting.
Source: Statistics Canada

9.5.4 INDUSTRY GROUPS

Commodity-specific and industry associations also coordinate and direct funds toward research and act as intermediaries by supporting related extension activities. They are often fairly effective at forging public-private-producer partnerships to leverage available R&D funding and, as such, are becoming increasingly important R&D players. (See especially, Tables 8.5 and 8.7.)

9.5.5 UNIVERSITIES AND COLLEGES

Universities and colleges are the key providers of cutting-edge technical knowledge in agriculture. Colleges and technical schools are particularly adept at conducting applied R&D and extending it to farmers and agri-business. And they increasingly seek private-sector partnerships to supplement on-going financial constraints. Canada-based research is also supplemented by Canadian studies abroad; a knowledge-base that is often imported back to Canada.

9.5.6 NGOS AND INTERNATIONAL LINKAGES

The principal public-sector international R&D linkages are CGIAR and the IDRC. CGIAR (Consultative Group on International Agricultural Research) is a highly regarded international agricultural research organization with research facilities in numerous countries—Mexico, Peru, the Philippines, Lebanon, Nigeria, India, Colombia, Sri Lanka, Ethiopia, Kenya, etc. It is well-known for its effective coordination of research to improve the genetics (e.g., dwarf wheat), agronomy, and farming systems most appropriate to the countries and regions in question. The Canada-based IDRC (International Development Research Centre) also funds agricultural research abroad, usually in conjunction with other international research initiatives.

9.5.7 INTERNATIONAL COMPARISONS

Does Canada spend enough on agricultural R&D? This currently amounts to about 4.6 percent of GDP and this places us, internationally, in about the middle of the pack. Numerous critics bemoan this:

> Public sector research funding is a pitiful fraction of what some of our competitors like China and South Korea invest, and well behind what similar nations like the US lay out.[179]

According to Figure 9.4, the best performer is New Zealand (at 0.11%) and the worst is the USA. But it is still difficult to anticipate the long-term consequences of our seemingly mediocre R&D performance. Increasingly, global companies are taking their applied R&D to all corners of their worldwide operations. GMO soybeans (developed in the USA) can be found in Tajikistan, Canada, Brazil, and dozens of other countries. Private proprietary research is increasingly being transmitted everywhere, quickly and seamlessly, with local producers and consumers eventually absorbing the cost of this research in their product purchases.

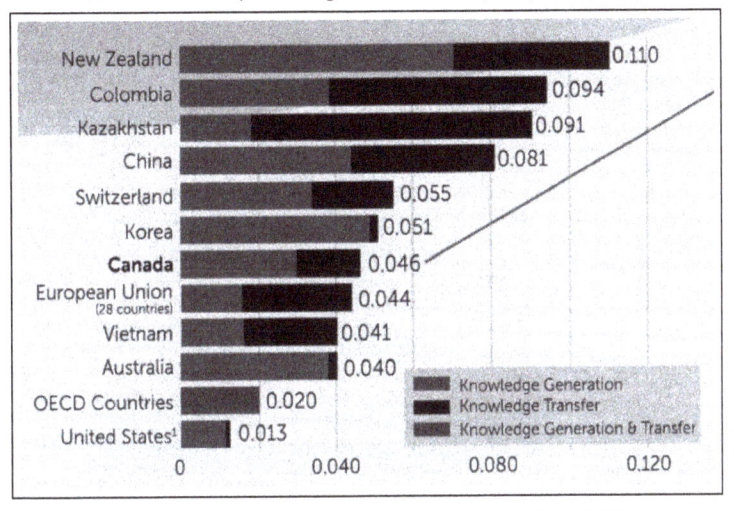

Figure 9.4 Public Expenditure in Agriculture R&D and Extension, Seleted Countries, Percentage of Total Country GDP, 2015 (X100)

New Zealand	0.110
Colombia	0.094
Kazakhstan	0.091
China	0.081
Switzerland	0.055
Korea	0.051
Canada	0.046
European Union (28 countries)	0.044
Vietnam	0.041
Australia	0.040
OECD Countries	0.020
United States[1]	0.013

Legend: Knowledge Generation, Knowledge Transfer, Knowledge Generation & Transfer

Source: OECD, Producer & Consumer Support database, 2016.
Our Source: AIC, *An Overview of the Canadian Agricultureal Innovation System*, Ottawa, 2016.

9.6 PROVINCIAL GOVERNMENTS

Although agriculture is, constitutionally, largely a provincial responsibility, we have seen in the foregoing that the federal government has, especially in recent years, generally taken the lead. Money is power. Yet the federal government abhors funding a mishmash of regional programs and has therefore moved over the past 25 years to assure that industry participants and provincial governments are treated the same across Canada. Thus, more and more on-going programs are national in scope and have a similar ***modus operandi*** in all participating provinces. And cost-sharing is now well-established and on-going. Aside from producer contributions, the mutually-agreed-upon cost-sharing formula is 60% federal and 40% provincial.

With respect to farm income maintenance, prior to the 1970s, the provinces generally relied on the establishment of provincial marketing boards with varying degrees of power in the marketplace. Agricultural extension services (a mainstay), agricultural technical schools, colleges of agriculture, and provincial agricultural research stations are also the exclusive purview of the province.

Purely provincial government programs in recent years basically fall into three categories: (a) ***ad hoc***, and (b) resource management, and (c) research and extension.

9.6.1 AD HOC PROGRAMS

These are difficult to categorize and rigorously tabulate, since there are, literally, hundreds. But they have generally been in response to disasters, specific crisis, or, more generally, income supplements and/or funding directed at improving production and marketing efficiencies. (Table 9.6)

Table 9.6 Tabulation of *Ad Hoc Provincial* Government Programs, Post 1990

PROGRAM TYPE	REGION					ALL REGIONS 3/
	Maritimes	Quebec	Ontario	Prairies	B.C.	
Disaster 1/	6	9	2	35	5	2
Special 2/	1	2	2	8	3	5
Efficiency	32	3	15	14	11	22
TOTAL	39	14	19	57	19	29

1/ Droughts, floods, winterkill, predators, excess rain.   148
2/ BSE, potato cyst nematode, Aleutian Disease, TB, Golden Nematode, Duponchellia (pest) bee losses, PVYN, codling moth, PMWS (post-weaning wasting).
3/ Independent of region-specific initiatives. Total region = region+all.

Basic Source: Statistics Canada, *Direct Payments to Agricultural Producers.* (website)

Many of these income supplements or initiatives to improve production or marketing efficiencies were considered transitional payments, or program adjustments. The prairies easily lead in the disaster category, with Alberta being the most generous. The Maritimes leads in the income supplement category, with numerous income supplement programs also being made available to all farmers, Canada-wide.

9.6.2   RESOURCE-BASED PROGRAMS

Because provinces have jurisdiction over the natural resources within their borders (including land), provincial Ministries of Agriculture have tended to focus on land and water conservation programs that would further enhance agricultural productivity. With financial and engineering assistance from the government, this resulted in extensive irrigation development in Alberta-Saskatchewan and widespread land drainage in Ontario and Quebec. (See Chapter 10 for details.)

There are also now fourteen federal-provincial environmental programs created under the 2008–09 federal-provincial Agricultural Policy Framework (APF) to help primary agriculture and the agri-food sector achieve environmental sustainability with respect to soil, air, water, and biodiversity. These include Environmental Farm Planning (EFP), the National Farm Stewardship Program (NFSP), Greencover Canada, and the National Water Supply Expansion Program (NWSEP). The EFP is a voluntary process used by individuals and managers to systematically identify environmental risks and benefits from their own farming operation and to develop an action plan to mitigate these risks. The NFSP, Greencover Program, and the NWSEP provide technical and financial assistance to land managers/owners related to targeted areas of environmental sustainability. (Also see Chapter 10.)

The most proactive province in terms of resource-based programs is Alberta, most of it being directed to irrigation development and rehabilitation in southern Alberta. This also includes extensive funding from Alberta Environment. Until recently, funding for new irrigation infrastructure was shared 60%–40% by the province and farmers, respectively. Periodically, extensive funding by the province is still provided for irrigation rehabilitation; principally primary and secondary canals. The Ontario and Quebec governments have also committed significant funding to agricultural land drainage. (Also see Chapter 10.)

9.6.3 RESEARCH – EXTENSION

Most provinces continue to have extensive applied research facilities. But their role has gradually diminished in recent years. Funding (Section 9.4.2) has slowly been diverted elsewhere and the number of professional staff and world-class researchers has diminished.

Other than in Saskatchewan and Quebec, publicly funded extension services have also gradually declined in importance. In the remaining provinces, the hands-on rural agricultural extension agent and the hands-on home-economist are no longer posted in a rural district office. Now the few remaining extension personnel may be located in a regional office, have a toll-free phone-in service, or simply maintain various web sites. Increasingly, proprietary and site-specific farm advisory services are provided by private-sector input suppliers (fertilizer, chemical, seed, feed, equipment, machinery, etc.), buyers (elevators, processors, etc.), commodity groups (Section 9.4.3), and/or private farm management consultants.

Many of the **federal** research stations also informally support provincial extension activities: field demonstrations, tours, research reports, public presentations, and so on.

9.6.4 CREDIT

Paralleling the credit programs available through Farm Credit Canada (see Figure 9.1), many provinces also operate a **Farm Loan Program**, which typically provides long-term loans for the purchase or improvement of land, equipment, machinery, breeding livestock, production quotas, shares in a farm company, and so on. A maximum loan is perhaps $5 million and the terms and conditions are often 20 years at a fixed interest rate. Another feature is sometimes a Beginning Farmer Incentive, with an interest rate reduction (say 1.5% below market rates) applicable to beginning farmers.

9.7 SUPPORT MEASURES FOR AGRICULTURE

The urban perception is that farmers are always checking their mail box, impatiently waiting for yet another check to arrive from the provincial or federal government. This is largely a myth. But they are still receiving numerous indirect (and well-disguised) program subsidies.

9.7.1 TOTAL DIRECT AND INDIRECT SUPPORT

The OECD defines agricultural support as the annual monetary value of gross transfers to agriculture from both consumers and taxpayers arising from government policies that support agriculture, regardless of their objectives and their economic impacts. Transfers include market price support, budgetary payments, and the cost of revenue foregone by the government and other economic agents. This includes indirect consumer transfers to dairy producers (about $1/2 billion/annum[180]), direct budgetary payments, and public R&D/extension support.

During the last 32 years, this has ranged from a low of about $4 billion/year in 1997 to a high of almost $9 billion/year a decade earlier. In 2016 and 2017, this leveled off at about $6 billion/year; approximately 10% of total farm revenue.

Figure 9.5 Direct & Indirect Producer Support, 1986 - 2017

Source: OECD, *Producer and Consumer Support Estimates* database, 2018

9.7.2   DIRECT BUDGETARY PAYMENTS

Direct government payments to farmers/ranchers have fluctuated wildly during the last five decades, climbing from almost nothing in the early 1970s to about $5 billion in the early 2000s, before gradually declining again to $2–$2.5 billion/year (Figure 9.6).

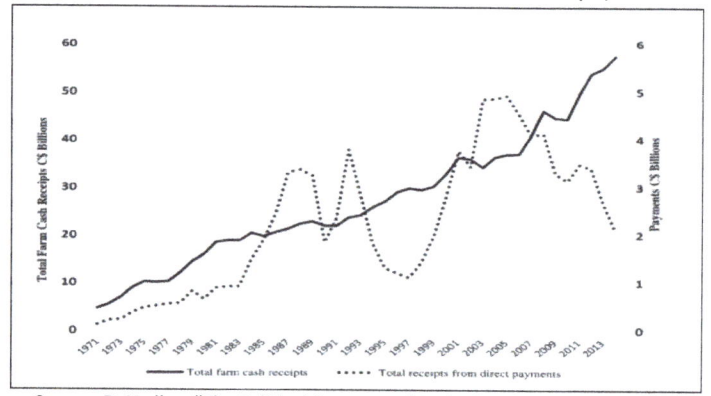

Figure 9.6 Direct Government Payments and Total Farm Cash Receipts, 1971 -2014

Source: D. Hedley, "The Political Economy of Agricultural Policy in Canada," in: *Handbook of International Food & Agricultural Policies,* Chapter 6, forthcoming.

Almost 80% per cent of all direct government spending on agriculture (including the bureaucracy) now goes toward so-called "business risk management" initiatives, which are, in part, income support payments. These are: subsidized crop/livestock insurance (45%, almost all of it crop insurance), AgriStability (19%), and subsidized AgriInvest (15 percent). During 2012–2016, direct budgetary payments to farmers/ranchers averaged about $2.4 billion per year. (Table 9.7 and Figure 9.7) Details regarding each of these programs can be found in Section 9.4, preceding.

Table 9.7 Direct Payments to Agricultural Producers, Average/Yr. 2012 - 2016

Item	$, 000	Percent
Crop/Livestock Insurance	1,056,538	45%
AgriStability	456,627	19%
AgriInvest	351,618	15%
Agri-Quebec	82,006	3%
Provincial Stabilization	299,094	13%
Other	127,716	5%
TOTAL	2,373,599	100%

Source: Statistics Canada, #002-0076

Figure 9.7 Direct Payments to Agriculture, 2012 - 2016

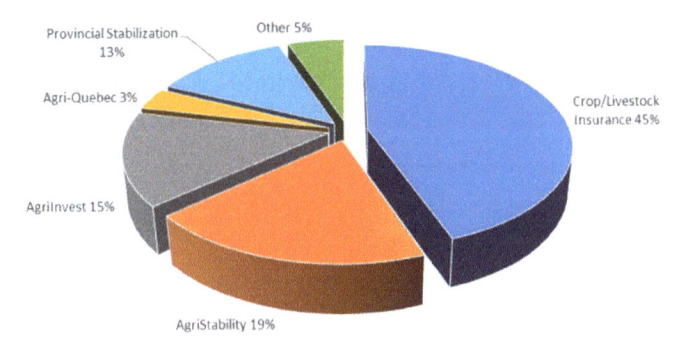

Source: Table 9.7

A total average annual direct payment of $2.4 billion only represents about 4 percent of gross farm revenue (Table 3.6), but in terms of **net** farm income (Table 3.11), this climbs to, maybe, 15–20 percent. This is a very substantial public contribution to net farm income.

9.8 THE TRANSITION

Until after WWII, agricultural development was largely government-driven. We were nation building. Then rapid technological and socio-economic change allowed the agricultural community itself (and particularly producers) to largely become the voice of rural Canada whereby government simply tried to respond, often reluctantly, belatedly, and in an *ad hoc* manner. But ever so gradually, a government policy framework evolved (early in the 21st century) that could essentially be put on auto-pilot. The table was set and just in time. The political power of the agricultural community by the turn of the century was almost exhausted.

Now the domestic consumer and the international marketplace is largely setting the future agenda for Canadian agriculture. There is a growing emphasis on the importance of nutrition, food safety, food substitutes, pollution-free production practices, humane livestock management, environmental sustainability, and a raft of other related consumer-driven concerns and controversies. There are new spokespersons in the private sector and new spokespersons in the public sector. A new era and a new agenda. Bring it on.

10. *Resource Managment In Agriculture*

10.1 OVERVIEW

"Resource management" most often refers to the responsibility of government to ensure that natural resources under their jurisdiction are managed in a sustainable manner. This is also generally in reference to Crown land owned by the federal or provincial governments since, in Canada, 41% of all the land is federal Crown land and 48% is provincial Crown land.[181]

Most federal Crown land is in the Canadian territories (NWT, Nunavut, Yukon) while about 4% (17 M ha.) is located in the provinces. This is mostly national parks, Aboriginal lands, and military bases. Most provincial Crown land, on the other hand, is generally either provincial parks, boreal forest, or wilderness, with over 90% of the boreal forest in Canada located on provincial Crown lands. Provincial lands account for about 60% of the area of Alberta, 94% of the land in BC, 95% of Newfoundland/Labrador, and 48% of New Brunswick.

Less than 11% of Canada's land, mostly within 800 km (500 miles) of the USA border, is in private hands, but this private land generally has much higher population densities and much more intensive resource utilization. About two-thirds of this is agricultural land (64.2 M ha; 158.7 M acres) and about one-third (34.2 M ha; 84.5 M acres) is in other uses, primarily urbanization and fragmentation (Figure 10.1).

Figure 10.1 Land Ownership in Canada

- Private Farm Land 7%
- Other Private Land 4%
- Federal Crown Land 41%
- Provincial Crown Land 48%

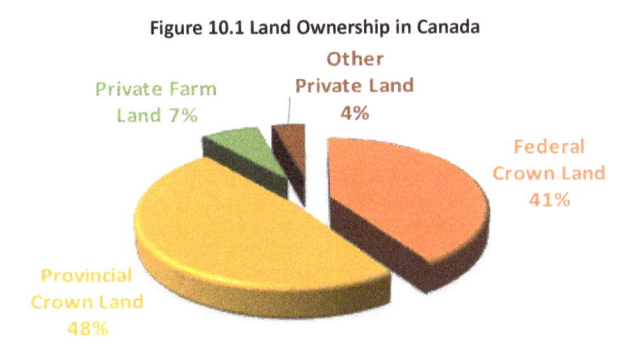

Source: Basic data from Statistics Canada

This chapter extends the definition of "resource management" to include private land and, specifically, privately owned agricultural land. Farmers and ranchers are, indeed, guardians, stewards, and custodians of the land. They nurture it, caress it, and sustain it.

10.2 AGRICULTURAL RESOURCE PROFILE

Land use in agriculture is illustrated in Figure 10.2 and highlights the following:

- Crop production and fallow only utilizes 60% of the total agricultural land
- Thirty percent of all agricultural land is native or tame grassland (pasture); land that is not seeded to an annual crop
- Seven percent is woodlots and wetlands

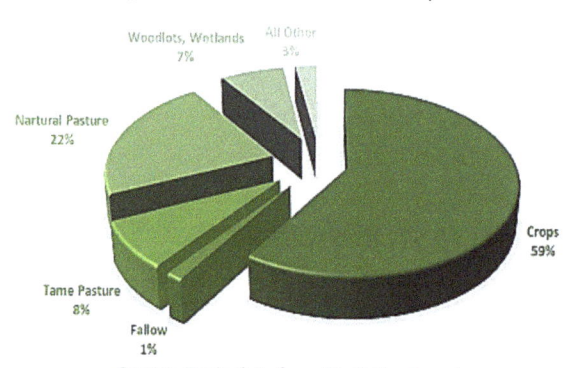

Figure 10.2 Use of Private Farm Land, 2016

- Woodlots, Wetlands 7%
- All Other 3%
- Nartural Pasture 22%
- Tame Pasture 8%
- Fallow 1%
- Crops 59%

Source: Basic data from Statistics Canada

10.2.1 CROP LAND

Sustainable resource management, with respect to annual crop production, has historically focused on the control of soil erosion (wind and water) and maintaining the tilth, organic matter, and nutritional health of the soil. Early monoculture systems wasted land and ruined soil health by depleting the ground of nutrients that plants need to grow. Early methods of crop protection often involved either excessive tillage—plowing or working the soil too much—or using inorganic chemicals like sulfur, mercury, and arsenic compounds to fight pests and diseases. Many of these older chemicals are no longer used because of their toxicity or inability to break down in the environment. Additionally, on the Canadian prairies, until the 1980s, many farmers sometimes burned their stubble fields (i.e., what remained of the grain after it was harvested) to facilitate clearing the fields for planting in the next crop year. Now, a no-till or zero tillage method is widely employed. Left over material from the harvested crop is left on the field and the seeds are planted directly into the soil. This technique increases the amount of water and organic matter (nutrients) in the soil and lessens soil erosion. Reduced tillage and safe synthetic fertilizer additives are now generally accepted practices.

Two other important land management issues are **land clearing** and **wetland drainage**. Brushing is still commonplace (at least on the prairies) and is not generally subject to provincial regulation (Section 10.6.3). Hence, trees along fence lines, shelterbelts, bush patches, etc. continue to disappear to better accommodate increasingly large commercial farming operations. For the same reason, from a farmer's perspective, the drainage of temporary wetlands (sloughs) on crop land is often highly desirable. This, however, is generally subject to increasingly strict provincial legislation (Section 10.2.3).

Prairie Pothole Region

A related resource management issue regarding private crop land is how to also accommodate other potential users of this fragile ecological landscape. Hunters, hikers, recreation vehicles (motorbikes, quads, skidoos), and bird watchers who enter private lands without permission can cause damage and conflict.[182]

10.2.2 PASTURE LAND

Pasture land management has greatly improved over the years. Rotational grazing, prohibiting livestock from polluting water sources, and otherwise working to maintain a healthy, diverse landscape is now commonplace.[183]

Large tracts of Crown land are also leased to (generally) eco-sensitive ranchers in the foothills of the Rocky Mountains (Alberta). They actively support a volunteer Cows and Fish Program, which is striving to foster a better understanding of how improvements in grazing management of riparian areas can enhance landscape health and productivity for the benefit of both cattle producers and others who use and value riparian areas.

A related and growing concern, once again, however, is how to also accommodate other desired users of private lands—hunters, hikers, recreation vehicles (motorbikes, quads, skidoos), bird watchers, and so on.

10.2.3 WOODLOTS AND WETLANDS[184]

Woodlots and wetlands make up about 7 percent of all agricultural land in Canada, some 4.5 M ha (11 million acres), and yet, historically, this area has received very little attention. Over 162,000 ha (400,000 acres) are endangered wetlands on the Canadian prairies.

The enhancement and maintenance of private woodlots, in particular, has received very little attention. Indeed, the destruction of existing trees on agricultural land is still commonplace on the prairies, usually "brushed" to more easily accommodate extensive farming operations.

With respect to wetlands, it is important to first distinguish between permanent, semi-permanent, and temporary wetlands (sometimes called "potholes" –as per accompanying illustration). The bed and shore of all permanent and naturally occurring water bodies (e.g., lakes, streams, rivers) are owned by the province. This extends to the bank of a permanent water body, a line along the upper limit of the bed and shore.

On the other hand, temporary and semi-permanent wetlands are owned by the farmer/rancher and it is these water bodies that continue to be converted to agriculture across the prairies (and the northern US). From a farmer's perspective, it is very often profitable to do so.[185] At least 40% of these wetlands have already been lost to Canadian agriculture in the last 50 years and, despite increasingly strict provincial regulations to limit this land conversion, the losses continue.[186]

Yet temporary wetlands serve a multitude of critical ecological functions. First, they act as kidneys; filters that reduce the quantity of nutrients and other pollutants potentially entering streams, rivers, lakes, and groundwater. And, secondly, they are required habitat for 50% of all North American ducks (12 species), as well as an estimated 600 species of plants and animals, fifty of which are at risk. According to one estimate, a Canadian wetland is worth almost $6,000/ha. ($2,400 per acre) per year but just $2,000/ha. ($800/acre) per year if converted to farmland.[187] (Also see Table 10.4.)

Until very recently (see Section 10.9.3), the only major player to try to stop, or at least slow down, this ecological destruction was Ducks Unlimited, a large American NGO strongly supported by US duck-hunters.[188] They are saving wetlands and associated habitat by both land purchases and land owner agreements.

10.3 LAND DEGRADATION[189]

There are numerous concerns and interests regarding soil degradation and soil conservation, including the esthetics of certain land forms, the productive value of the soil, and the impact of soil changes on other sites and future choices. There are, in essence, four major land degradation issues:[190]

- Soil salinization
- Soil erosion by wind and water
- Soil acidification
- Organic matter loss/nutrient decreases

10.3.1 SOIL SALINIZATION

Saline soils are generally classified as being either "primary" or "secondary." **Primary** saline soils are geological saline soils that were saline before agriculture began and have never had any productive capability. They are attributed to the many sand and gravel aquifers arising from all of the water released in the warm interglacial periods.[191] **Secondary** or man-induced saline soils may be attributable to agricultural practices or to changing surface water runoff patterns. These soils are the result of the addition, redistribution, or concentration of soluble salt by ground water or surface water. This includes saline seeps and expansion of saline or alkali sloughs. As the salt content increases, so does the EC (electrical conductivity). A high EC coefficient (say >4dS/m) indicates the presence of a saline soil.

Saline soils form when the water table is close enough to the soil surface to allow capillary action to raise the groundwater to the soil surface (Figure 10.3).[192] In general, water must be within 2 m (6 ft) of the soil surface for this to occur, but the critical depth (i.e., the depth beyond which water cannot wick to the soil surface) varies with soil texture.

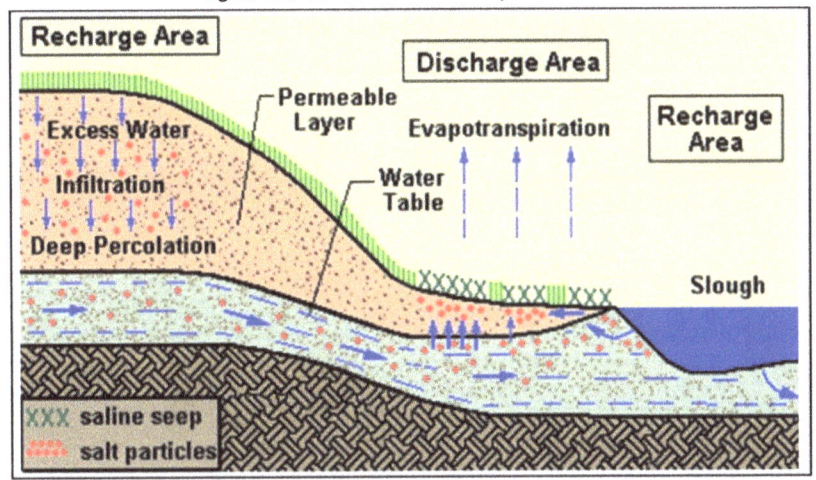

Figure 10.3 Generalized Saline Seep Mechanism

Source: L. Leskiw

Saline soils have a high enough concentration of soluble salts to impact crop yields and crop choice (Table 10.1). Sodic soils have high levels of sodium (expressed as a high sodium adsorption ratio, SAR), low levels of soluble salts, and poor soil structure and this too has a negative effect on crop growth and yield. Soils with high levels of both soluble salts and sodium are referred to as saline-sodic soils (Figure 10.4).

A white "bathtub ring" is a visual saline area around the edges of saline sloughs throughout the prairies. Both natural and man-induced, there are approximately 2.2 million hectares (5.4 M acres) of saline land on the prairies and this is increasing at a rate of about 1% per year.[193] This represents about 3.4 % of the agricultural land base in Canada (Table 3.2).

Table 10.1 The Effect of Salinity on Crop Yields

E.C.*	Degree of Salinity	Hazard for Crop Growth	Plant Response	Relative Tolerance of Crops**	Relative Yield *** Wheat	Relative Yield *** Barley	Relative Yield *** Canola
0 to 2	Non-saline	Very low	Negligable				
2 to 4	Slightly saline	Low	Restricted yield of sensitive crops	Beans, peas, corn, soybean, sunflowers, clovers, and timothy	66%	81%	62%
4 to 8	Medium saline	Medium	Restricted yield of many crops	Cereals, canola, flax, bromegrass, alfalfa, sweet clover, and trefoil	37%-52%	59%-70%	30%-47%
8 to 16	Severely saline	High	Only a few crops yield satisfactorily	Wheatgrass and wild rye	22%	49%	14%
>16	Very severely saline	Very high	Only a few salt-tolerant grasses grow satisfactorily				

* E. C. = Electrical Conductivity.   ** Still expected to yield at least 50% of normal.
*** For E.C. values of 4, 6, 8, and 10 and assuming no compensating inputs.

Sources: Manitoba; and Holm, H. M., Soil Salinity, A Study of Crop Tolerances and Cropping Practices, Saskatchewan Ag. Report No. 25M, Regina, 1983.

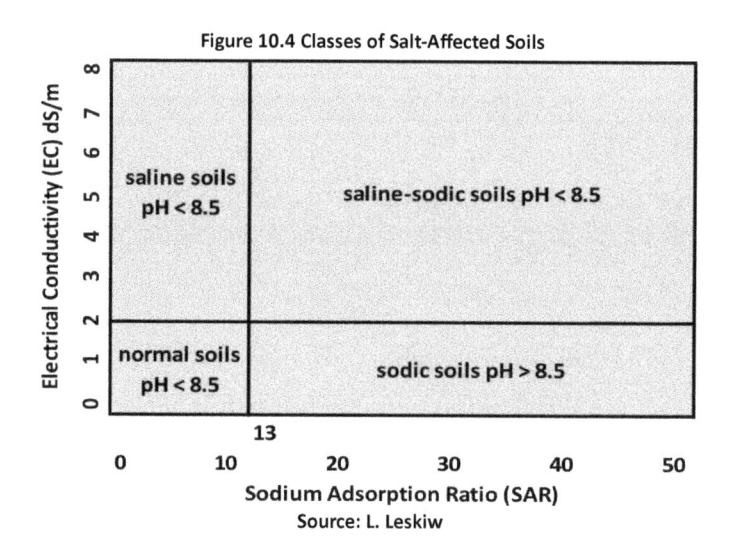

Figure 10.4 Classes of Salt-Affected Soils

Source: L. Leskiw

10.3.2 SOIL EROSION

Erosion removes part or all of the topsoil, and often the upper part of the subsoil, resulting in a soil with low organic matter, poor tilth, low water holding capacity, poor nutrient supplying power, and lower capability for production. This results in poorer growth and a greater risk of more erosion.

Erosion by water is perhaps the most recognized form of soil degradation and is considered by most people to be the issue of greatest concern in soil conservation across Canada, even in the semi-arid parts of Western Canada. Water erosion occurs sporadically but is often a recurrent problem in the same location, resulting in large total losses. Water erosion on glacial landscapes with poorly integrated drainage, which characterizes much of the prairies, often results in the redistribution of topsoil in a small area and often within the same field. This can obscure its severity.

Wind erosion has been a major concern in the "Palliser Triangle" (SW Saskatchewan and SE Alberta) since the 1920s and it continues to be a recurring problem there and elsewhere. Many soil landscapes throughout the prairies have experienced total topsoil losses of between 30 to 50 percent.

Annual erosion losses depend upon the complex interaction of soil, climate, and growth. Numerous hydrologic, weather, soil temperature, nutrient level, tillage, slope, texture, and plant growth characteristics must be considered:

$$A = R*K*L*S*C*P$$

where

A = erosion = tonnes of topsoil/acre/year

R = rainfall factor

K = soil erodibility factor

L = slope length factor

S = slope gradient factor

C = cropping and management, or cover factor

P = erosion control practice factor

The land management factors in this Universal Soil Loss Equation, C and P, emphasize the central role that management plays in this whole process: the type of ground cover maintained and exactly how the crop is grown are equally important. Generally, when a soil is increasingly disrupted (e.g., tillage) and/or increasingly little organic matter (or other nutrient sources) is recycled, wind and water erosion increases. The annual erosion losses due to water are highly dependent upon the ground cover maintained (approximately):

No Ground Cover (fallow)	27 tonnes/ha/year
Cropland	5.4 tonnes/ha/year
Forest	1.1 tonnes/ha/year

At the same time, there is constant soil regeneration and re-deposition. What is tolerable, therefore, depends upon the balance between on-going soil erosion and on-going soil regeneration. Tolerable soil losses on cropland are defined as the maximum rate of annual soil erosion that will permit a high level of crop productivity to be sustained economically and indefinitely. The US Department of Agriculture estimates this to be between 1.2 and 12.1 tonnes/ha/year, depending upon the thickness of the topsoil and the existence of a favorable and sufficient rooting depth. Ten tonnes per acre per year represents the loss of about 2.5 mm (1/10 inch) of topsoil.

For the grain-growing region of western Canada, it is estimated that the area that has suffered a total topsoil loss is about 5.3 million hectares (13 M acres), about 8% of the total **cultivated** land area in Canada. The areas of loss are primarily upper slopes in hummocky, rolling, and undulating landscapes, but even fairly level soils have also lost topsoil due to wind and water erosion.

The impact of wind and water erosion on crop yields is different for different soil types, as indicated in Table 10.2.

Table 10.2 The Effect of Erosion on Crop Yields (Relative Productivity)

Soil Type	Crop	100% loss	50% loss	Not eroded
Brown Chernozemic	Cereals	50%	80%	100%
Dark Brown -Black Chernozemic	Cereals	70%	90%	100%
	Canola	50%	80%	100%
Gray -Dark Gray Luvisol	Cereals	40%	70%	100%
	Canola	20%	50%	100%
Solonetzic	Cereals	10%	40%	100%
	Canola	0%	30%	100%

Source: Knapik, L., et. al., *Agricultural Degradation in Western Canada*, Agriculture Canada, 1984

10.3.3 SOIL ACIDIFICATION

Most of the acid soils that are farmed were acidy before farming due to the natural soil processes of leaching and microbial action. The addition of acid-forming materials to the soils can, however, increase acidity, lowering the pH of near-acid soils into the acid range. The inputs of major concern for causing acidification of soils are nitrogen (N) and sulfur (S) fertilizers and airborne N and S pollutants.

On the prairies, there are about 2.4 M ha (6 M acres) of acid soils in the grain-growing area, about 4% of Canada's improved land area. And this is gradually increasing, although the impact of the added N and S through increasingly heavy chemical fertilizer applications may not be too deleterious because many prairie soils are inherently slightly alkaline. Acidic soils (i.e., a lower pH) reduce yields approximately as indicated in Table 10.3 and lime can be applied to sometimes gradually raise the pH of acidic soils.

Table 10.3 Effects on Crops of Liming Acid Soils

Rating	Soil pH	Direct Effects on Crops	Indirect Effects on Crops
Slightly acid	6.1 to 6.5	No direct effect of liming on most crops.	Liming may improve the physical properties of some medium and fine textured soils (particularly Gray and Dark Gray Luvisolic soils).
		Fields with an average pH just above 6.0 may have areas where the pH is below 6.0. Alfalfa and sweet-clover yields will be increased on the more acid areas.	Improved soil structure and reduced crusting will be particularly beneficial for small-seeded crops such as canola.
Moderately acid	5.6 to 6.0	Improved survival and growth of rhizobium bacteria which fix nitrogen in association with alfalfa and sweet clover.	Liming may improve the physical properties of some medium- and fine-textured soils as indicated above.
		Yields of alfalfa and sweet clover are increased.	Plant availability of phosphorous fertilizers is improved.
			Increased microbial activity and release of plant nutrients.
		Small increases in yield of barley occur in the first two to three years following lime applications with larger increases (25-30 per cent) occurring in subsequent years. Yields of wheat and canola will be increased less than barley. Yields of more acid tolerant crops may be increased as a result of indirect effects of lime as outlined above.	
Strongly acid	5.1 to 5.5	Increased nitrogen fixation and yield of legumes.	Indirect effects as outlined above for moderately acid soils.
		Soluble aluminium and manganese are reduced to nontoxic levels.	
		Yields of most crops are increased as a result of reduced levels of aluminum and manganese and improved availability of phosphorus and other nutrients.	
Very strongly acid	Less than 5.1	Direct effects as outlined above for strongly acid soils.	Indirect effects as outlined above for moderately acid soils.
		Yields of most crops are severely reduced unless the soil is limed. Very strongly acid soils are very infertile. Acid tolerant crops (oats and some grasses) moderately well if adequately fertilized.	

Source: L. Leskiw

10.3.4 ORGANIC MATTER LOSS

Loss of organic matter (OM) from topsoil horizons results in the loss of nutrients (especially nitrogen), loss of nutrient supplying power, and deterioration of physical properties of the soil that are important for crop growth. Lower OM levels correlate with surface crusting, higher bulk densities, slower water infiltration, lower available water capacity, and higher erosion risks. Deterioration of these soil quality factors results in lower crop yields. At the same time, the balance between active and inactive fractions of the OM are even more critical than the total amount of OM in determining how much crop nitrogen (N) and immobilized N are available each year. The important goal in OM and nutrient management is sustainable recycling.

Organic matter is lost by erosion and by microbial action and is diluted by conventional cultivation. The rate of loss is highest in the initial years of cultivation, then gradually decreases before stabilizing after about 80 years.

On the prairies, the Brown Soil Zone had low initial OM levels and this was exacerbated by initial cultivation and summer fallowing. In the Dark Brown and Black Soil Zones, initial OM levels were relatively high but also declined with cultivation and then subsequent summer fallowing. Gray Luvisol (Gray Wooded) Soils have essentially no organic matter in the topsoil (A horizon) in their natural state and generally benefit from cropping by adding organic matter.

Amelioration of OM loss involves good soil management and cropping practices which replace more OM than is being removed. This includes additions of OM by manuring, straw addition, fertilizing to increase growth, the inclusion of forage crops in rotations (especially on Gray Wooded soils), continuous cropping (i.e., less summer fallow), and minimum/zero tillage. The widespread adoption of most of these practices during the last 30–50 years has now generally reversed the earlier OM decline on the prairies. A higher steady-state OM level is now typical on many, if not most, prairie farms.

10.4 URBANIZATION OF AGRICULTURAL LAND

Does the on-going urbanization of agricultural land threaten Canada's long-term capacity to produce food and fiber for domestic and international consumers? This concerns both rural and urban Canadians.

10.4.1 AN OVERVIEW

By the turn of the century, urban uses in Canada had already consumed more than 1.2 M ha (2.9 M acres), one-half of which was on Canada Land Inventory Classes 1,2,3 land, the most productive and reliable agricultural land. This is approximately the size of all of Prince Edward Island. Each year we lose an additional 20,000 to 25,000 ha (50,000–60,000 acres) of prime farmland to urban expansion.[194] For every million people we add to Canada's population, we lose another 530 km² (53,000 ha; 131,000 acres). In Ontario, about 20 percent of Class 1 farmland is now being used for urban purposes; in Alberta the comparable estimate is about 8 percent.[195] This is especially disconcerting because only about 4.5% of Canada's total land area is actually Classes 1,2,3 farmland (Table 10.4).

Table 10.4 Area of Class 1-3 Agricultural Land, Canada

Region	Classes 1-2-3 (km2)*	% Total Land
Maritimes	28,658	5.3%
Quebec	22,039	1.4%
Ontario	72,833	6.8%
Prairies	321,616	16.4%
B.C.	9,486	1.0%
CANADA	454,632	4.5%

* 1 km2 = 100 ha. One hectare = 2.47 acres.
Basic Source: Statistics Canada

The disproportionate urbanization of Classes 1,2,3 farmland is not an accident. Many if not most urban settlements initially chose fertile land locations contiguous to navigable water bodies on which to settle.

The two main factors impacting urban land use are a growing population and the use of more land per urban dwelling. Canada's population growth is predominately through immigration and immigrants overwhelmingly choose urban areas in which to reside. With the increasing use of automobiles, the preferred dwelling is a single detached home, which consumes relatively more land per urban dwelling than apartments and townhouses. The result is urban sprawl onto land that was previously highly productive farmland. This sprawl not only involves housing, industry, and retail outlets but also accompanying physical infrastructure—streets, highways, utility corridors, golf courses, and perhaps even a local airport. Worse, even with legally protected agricultural areas, much of the land is already held by private developers. As a consequence, farmers who lease the land back have even less incentive to make investments to maintain a sustainable farming operation. They also migrate to more distant farmland, inadvertently inflating farmland prices in those areas.

10.4.2 A REGIONAL PERSPECTIVE

How are various provinces addressing the issue of urban sprawl and the need for Canada to preserve its long-term capacity to produce food and fiber for domestic and international consumers? We briefly look at the initiatives in Quebec, Ontario, Alberta, and BC.

Quebec: The *Commission de protection du territoire agricole du Québec* has the responsibility of keeping agricultural zones that are favorable to the practice and development of agricultural activities. In so doing, the Commission safeguards these agricultural zones and helps makes their protection a local priority. It is unique in that the Commission still allows farms and other development to co-exist around Montreal and other major urban centers. The agricultural zones cover an area of 63,000 km^2 (6.3 M ha; 15.6 M acres) in 952 local municipalities.

Ontario: Ontario has 52 percent of Canada's Class 1 land but also some of Canada's highest population densities. Its cities continue to swallow up huge tracts of farmland for their suburbs and urban sprawl. Every year, Toronto develops another 2,500 ha (6,200 acres) of some of Canada's most productive farmland.[196] In the Niagara Fruit Belt, lying between Niagara-on-the Lake, St. Catherine, Lincoln, and Grimsley, the on-going loss of farmland is most acute. Farmers inadvertently contribute to the problem by selling off sections of their farms to housing developers for large sums of money.

The *1981 Regional Niagara Policy Plan* only partially slowed the gradual urbanization process. Prime fruit and grape growing lands were eventually surrendered to accommodate more housing, schools, golf courses, churches, warehouses, industrial parks, and retail outlets, along with the additional land required to accommodate supporting infrastructure (e.g., a better highway from Hamilton to Niagara Falls).

There are now two major greenbelts in Ontario. One surrounds Ottawa and the other is the Golden Horseshoe Greenbelt, which is a 7,300 km² (730,000 ha; 1.8 M acres) band of land that encompasses the rural and agricultural land surrounding the Greater Toronto area, the Niagara Peninsula, and parts of the Bruce Peninsula. This is also larger than Prince Edward Island. Most of the land consists of the Oak Ridge Moraine, an environmentally sensitive land that is a major aquifer for the region, and the Niagara Escarpment, a UNESCO Biosphere Reserve. The Ontario government created the Greenbelt Act in February 2005 to protect this greenspace from all future development, with the exception of limited agricultural use.

One problem with greenbelts in Ontario (as elsewhere), is that they spur the growth of areas much farther away from the urban core, thereby actually increasing urban sprawl onto more distant agricultural land. For example, the Ottawa suburbs of Kanata and Orleans are both outside the cities' greenbelt, so residents of these communities have a longer commute to work and worse access to public transport. It also means that commuters have to travel through the greenbelt, an area not designed to cope with more commuter traffic. Greenbelts can also simply act as a land reserve for future freeways and other highways. This has already happened along sections of Highway 407 north of Toronto and Hunt Club Rd./Richmond Rd. south of Ottawa.

Aside from the greenbelt legislation, under the current system in Ontario, even when regions and municipalities set limits on growth and deny approvals for development on farms, the ultimate arbiter of land use is the provincially appointed Ontario Municipal Board, which can overrule municipal decisions and generally allows development to proceed.

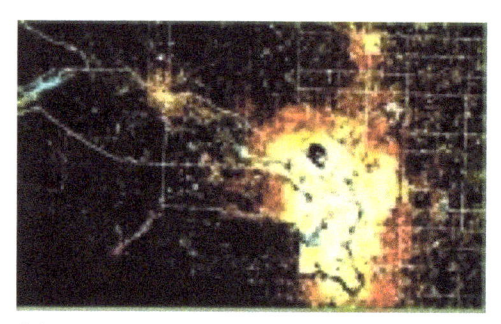

Calgary

Alberta: Pressure on prime farmland is most pronounced around Calgary (see illustration), Red Deer, and Edmonton and in the Edmonton-Calgary corridor, a corridor that covers about 100 km (60 miles) to the east and west of the Queen Elizabeth II Highway and 300 km (185 miles) north-south, an area slightly smaller than Nova Scotia. Between Edmonton and Calgary, the amount of land used for urban and industrial purposes grew by 52 percent between 2000 and 2012.[197] The land converted from agriculture to urban or industrial use was more than 625 km² (62,500 ha; 154,000 acres).

British Columbia: Urbanization around Vancouver is rapidly intruding on the lush Frazer River Delta system, home to 65 percent of the agricultural industry in British Columbia. Urban sprawl areas will eventually become enmeshed with metropolitan Vancouver. With continued immigration (including wealthy Chinese), housing prices have climbed rapidly, forcing many lower-and-middle income Vancouverites to flee the city for the ever-expanding suburbs, particularly Surrey, Richmond, and Delta. At the same time, industrial development along the Fraser River has gradually consumed yet another 4,000 ha (10,000 acres) of prime farmland.

So much land was already being lost along the Fraser River Delta pre-1970 that the province was forced to pass the Land Commission Act of 1973; the origin of the Agricultural Land Reserve (ALR) which, in total, covers approximately 47,000 km² (4.7 M ha; 11.6 M acres) and includes private and public lands that may be farmed, forested or left vacant; large tracts covering thousands of hectares, as well as small pockets of only a few hectares. It was intended to permanently protect valuable agricultural land, is some of the most fertile soil in the country, from being lost. It included about 31,200 ha (77,000 acres) of agricultural land within the Greater Vancouver area (see illustration). Yet despite having been in existence for over 45 years, the ALR continues to be threatened by urbanization and private land developers. Additionally, since it does not protect non-agricultural land, this leads to a substantial, and highly visible, leapfrog-type hillside sprawl. Land fragmentation and absentee ownership also means that only about 75 percent of the ALR land in the Greater Vancouver area is still being actively farmed. Some wealthy residents also buy land in the ALR, build large homes on it, and then just casually "farm" the land simply to satisfy ALR's minimum requirements.

10.4.3   AN ON-GOING ISSUE

Ultimately, nostalgia for our agricultural heritage isn't going to keep farmland in production. Eventually, this gradual Pac-Man©-like absorption of prime farmland could threaten Canadian agricultures' continued capacity to adequately provide domestic and international consumers with safe, nutritious food. Perhaps equally or even more importantly, it also threatens the sustainability of our fragile ecosystem and our capacity to successfully address climate change, as well as the preservation of the rural amenities that urban residents also treasure.

10.5 REGENERATIVE ORGANIC AGRICULTURE (ROA)

Regenerative organic agriculture (ROA) is a holistic farming system that maximizes carbon fixation while minimizing the loss of that carbon once it is returned to the soil. The ROA farming principles and practices increase biodiversity, enrich soils, improve watersheds, enhance ecosystem services, and potentially transform agriculture from a carbon pump (+CO_2e) into a carbon sink (-CO_2e).

In practical terms, regenerative organic agriculture is foremost an organic system that does not employ synthetic pesticides and inputs, which disrupt soil life, and does not use fossil-fuel dependent nitrogen fertilizer, which is responsible for a significant portion of man-induced N_2O emissions. ROA is comprised of organic practices including, at a minimum, conservation tillage, no bare soil exposure, cover crops, residue mulching, composting (manure), and extended crop rotations. Regenerative agriculture is, essentially, sustainable agriculture on steroids—a longer-term approach that seeks to get microorganisms to do the work of supplying the growing plant with the required nutrients.

Figure 10.5 GHG Emissions, Organic vs Conventional

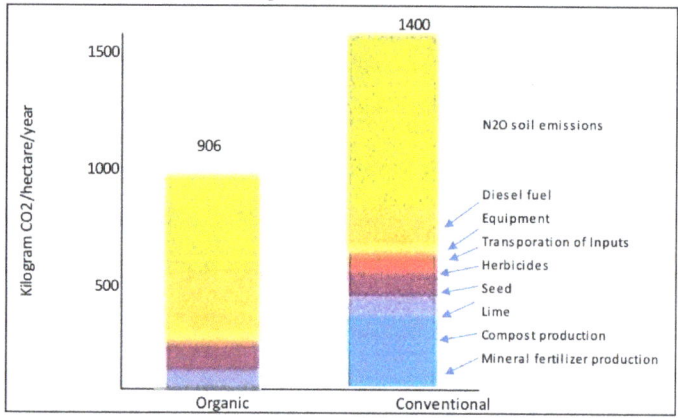

Source: Rodale Institute, Farming Systems Trials

Regenerative organic agriculture (ROA) can have innumerable other components: holistic managed grazing, animal integration[198], ecological aquaculture, silvopasture,[199] agroforestry, strip cropping, alley cropping, intercropping, medicinal plant production, boundary systems (e.g. windbreaks), and many others. ROA is, by definition, holistic. But the components are unique to each farm and farming environment; it is not a one-fits-all package.

At the same time, ROA compared to conventional production practices often has a higher labor requirement/acre, a higher total cost/acre, and a lower crop yield/acre (Table 10.5). As such, a significant product price premium (i.e., a good ROI) is very often mandatory to make ROA economically feasible.[200] For extensive cropping systems where labor is already a severe constraint, widespread adoption of the ROA paradigm seems unlikely. It should, however, have widespread application for high-valued organic crops that are already relatively labor-intensive but in high demand (e.g., some fruits and vegetables).

Table 10.5 HRS Wheat, Organic vs Conventional, Manitoba, 2019

Item	Conventional	Organic	Difference
Revenue:	$ 371.25	$ 630.00	170%
Yield/acre	55	35	64%
Price/bushel	6.75	18.00	267%
Costs:	$ 384.00	$ 455.00	118%
Operating	224	260	116%
Fixed	134	125	93%
Labour @$22/hr.	26	70	269%
Margin:	-$ 12.75	$ 175.00	1373%

Source: Manitoba Agriculture, 2019

What is indisputable is that ROA does increase biodiversity and it does represent a more ecologically sustainable production system. Equally important, a conversion from conventional production to ROA production can substantially reduce CO_2e emissions in agriculture (Figure 10.5), as well as potentially serving as a major carbon sink for total CO_2e emissions from all sources. To quote one enthusiastic and knowledgeable proponent:

> On-farm soil sequestration can potentially sequester all of our current annual global greenhouse gas emissions of roughly 52 gigatonnes of carbon dioxide equivalent. Indeed, if sequestration rates attained by exemplar cases were

achieved on crop and pastureland across the globe, regenerative agriculture could sequester more than our current annual carbon dioxide (CO_2) emissions. Even if modest assumptions about soil's carbon sequestration potential are made, regenerative agriculture can easily keep annual emissions within the desirable lower end of the 41-47 $GiCO_2e$ range by 2020 which is identified as necessary if we are to have a good chance of limiting warming to 1.5°C.[201]

More and more people are taking note. For example, an upstart private company, Boston-based Indigo AG, currently (2019) plans to sign up 3,000 farmers globally (farming more than 1 M acres) and pay them $15 per ton of carbon using venture capital raised by the company.[202]

In short, regenerative organic agriculture could be integral to the climate solution. It may also be possible to eventually merge ROA with sustainable conventional agriculture once more environmentally benign synthetic fertilizers, pesticides, and energy technologies are developed.

Some ROA proponents also emphasize the role that ROA can play in enriching rural communities as well as the spiritual relationship we have with nature. (Also see Chapter 13, Section 13.1.3.)

10.6 PUBLIC LAND MANAGEMENT INITIATIVES

Provincial Ministries (usually Natural Resources or Environment) are generally responsible for all programs concerned with the public land, water, trees, fish, and animals of the province. Typical activities include the administration and protection of provincial Crown lands and waters, regulation of the fish and wildlife resources of the province, management of the production and harvesting of timber, forest fire control, development and administration of provincial parks and recreation areas, and the provision of financial assistance to regional entities in support of local conservation initiatives.

Over the years, Canadian provinces have also utilized various land management tools in an effort to try to accommodate multiple users while, at the same time, assuring that a sustainable ecological system can be maintained. These include:

- **Regional Land Use Plans** – detailed documents that consider multiple land and resource values at a regional level. They provide goals and strategies for the allocation and use of those lands and resources. Aside from the Agricultural Land Reserve (ALR), BC has four regional land use plans. These plans are generally prepared under the auspices of ministries other than the Ministry of Agriculture, often the Ministry of Municipal Affairs. Thus, this type of plan is typically down-loaded to municipal authorities for implementation, particularly with respect to **zoning** for specified land uses.

- **Integrated Resource Plans (IRPs)** – comprehensive land use plans for both public and private-sector lands. These were essentially proposed land use (**zoning**) documents that did not look at the inherent environmental amenities embedded in each designated land use.

They were popular in various provinces in the mid-1980s, but their implementation and impact has been very limited.[203]

- **Conservation Authorities** – community-based natural resource management agencies. Long-established in Ontario (1946), conservation authorities represent groupings of municipalities on a watershed basis who work in partnership with other agencies to carry out natural resource management activities within their respective watersheds. They carry out programs in natural hazard management, nature education, land conservation and management, recreation, and research, and have successfully developed thousands of programs to further the conservation, restoration, development, and management of Ontario's natural resources. Management programs generally occur in lands known as conservation areas, restoration areas, or wilderness areas. (Wikipedia)

- **Specific programs** to protect and enhance the agricultural land base. Two major initiatives on the prairies were both administered by the federal government but provincial initiatives are becoming increasingly important:

  1. The **Community Pasture Program**, originally created in the 1930s by the PFRA to reclaim land that was badly eroded during the prairie drought, has been highly successful. It has now returned more than 145,000 ha (350,000 acres) of poor-quality cultivated lands to grass cover, significantly improving the ecological value of these lands. A prime objective of the program is to maintain a healthy, diverse landscape that is representative of the natural functional prairie ecosystems.[204]

  2. The **Shelterbelt Program** was also introduced by the Prairie Farm Rehabilitation Administration (PFRA) to reduce wind erosion on the Prairies. Under this program, PFRA provided millions of free saplings to prairie farmers prior to the termination of this program in the 1990s. This program was very successful but farmers now sometimes remove these shelterbelts to more easily accommodate large farm machinery.

  3. The **Watershed Management Program**, initiated in Manitoba in 2017, is mandated to establish nutrient runoff targets to keep track of water quality, and provide financial incentives to farmers who adopt practices that restore wetlands, retain water, better manage riparian areas, and increase the penalties for illegal drainage. This program could gradually become a more comprehensive Alternative Land Use (ALUS) model and, perhaps with an institutional framework similar to Ontario's Conservation Authorities (see above), become a template for other similar provincial initiatives. (Also see Section 10.9.3.)

  4. The **Environmental Farm Plan (EFP) Programs** in Ontario, Saskatchewan, and Alberta, designed to assure more sustainable on-farm resource management, are also having their desired affect. Recently promoted nationally, preparation of an EFP is now often a prerequisite to qualify for other government program support. Much smaller provincial initiatives (e.g., **Bucks for Wildlife** in Alberta) have generally had a positive but very limited impact

10.7 WATER MANAGEMENT

Here we look at five different aspects of water management, including the legal framework, supply-demand, irrigation, drainage, and flood control. Irrigation (Section 10.7.3) is addressed in considerable detail.

10.7.1 LEGAL FRAMEWORK

Surface water law in Canada is generally based on English common law and riparian rights, which simply dictates that users have an obligation not to infringe on the rights of downstream users. Astute surface water management has long been a major concern of virtually every provincial government because it is a public resource that must be sustainable yet also be equitably and efficiently rationed among different uses. The users are numerous: minimum flow requirements to sustain aquatic life, domestic and municipal potable water requirements, water for irrigated agriculture, water for livestock, industrial water requirements, water for recreation, water for hydropower, interprovincial flow requirements, as well as non-market considerations (e.g., esthetics). In the absence of a market-based price-allocation procedure for **all** these various uses, provincial governments generally rely on a prescriptive **regulatory framework**.

With respect to **groundwater**, the rules were different. Groundwater use was generally governed by the Rule of Absolute Capture, which essentially allowed users to extract water without any regard for the impact on their neighbors. The general rule was that the first person to capture such a resource owned that resource. Thus, a landowner who extracted or captured groundwater within the sub-surface of his land acquired absolute ownership, even if it was sub-surface drainage from contiguous land. Now, in most provinces, for all uses other than domestic groundwater use, impacts on neighboring groundwater users must be established prior to the issuance of a groundwater license.

Total groundwater use varies by province, but it is generally the industrial sectors—primarily manufacturing, mining, and thermal power generation—that are the biggest consumptive users. In Saskatchewan and Alberta, the oil and gas sectors are also major users. Next come the municipal and agricultural sectors, particularly cattle feedlots. The beverage industry, including the bottled water industry, also depends on potable water, which it typically obtains from municipalities or directly from groundwater sources. Groundwater is the source of potable water for about 30% of all Canadians so protection of its quantity and quality is paramount.

10.7.2 SUPPLY & CONSUMPTIVE USES

Canada has more renewable fresh water resources per person than any other industrialized country in the world.[205] Almost 9% of Canada's total area is covered by freshwater, with an estimated 32,000 lakes.[206] However, about 60% of this water flows north and is not readily accessible to the majority of Canada's population, which is concentrated along its southern border. Urban growth, expanded industrial activity (including mining), increasing use of water for irrigating crops, and changing weather patterns are placing increasing pressure on Canada's freshwater supply, particularly in portions of southern Ontario, southern Alberta, southern Saskatchewan, SW Manitoba, and the Okanagan Valley in BC.[207]

Agriculture accounts for about 44% of total water consumption in Canada, by far the largest single user (Figure 10.6). The oil and gas sector, thermal power generation (net), and manufacturing each utilize about 11 percent of fresh water in Canada. Mining (8%), households (9%), and commercial/institutional (6%) make up the remainder. It is clear that increased efficiency in agricultural water use is where the greatest opportunity lies to reduce water consumption to make more water available, either for other uses or for expanded irrigated agriculture.

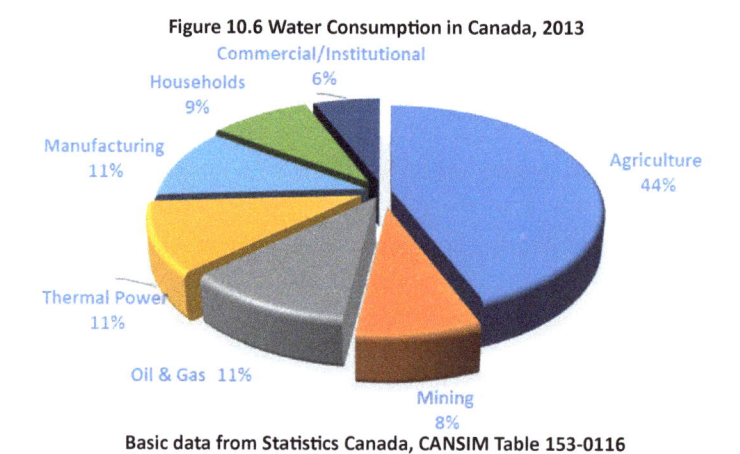

Figure 10.6 Water Consumption in Canada, 2013

- Agriculture 44%
- Mining 8%
- Oil & Gas 11%
- Thermal Power 11%
- Manufacturing 11%
- Households 9%
- Commercial/Institutional 6%

Basic data from Statistics Canada, CANSIM Table 153-0116

10.7.3 IRRIGATION

There are approximately 610,000 hectares (1.5 million acres) of irrigation in southern Alberta, which accounts for about 70% of Canada's total irrigated area. Most of the remainder is located in south-central Saskatchewan, SW Manitoba, southern Ontario, and the Okanagan Valley in BC.

Members of the Alberta Irrigation Projects Association oversee thirteen irrigation Districts in Alberta, which draw virtually all their water from surface sources: St. Mary River, Belly River, Oldman River, Bow River, and the South Saskatchewan River (Figure 10.7). Water licenses (both permanent and terminable) issued to the individual Districts specify the priority date (first in time, first in right), licensed volume, diversion rate, return flows, and source and location(s) of diversion. To access water, farmers pay the Irrigation District a set annual fee which averages $40/ha/year ($16/acre/year) plus various surcharges (for pipelines, additional water, etc.), plus their share of on-going irrigation rehabilitation in their District.[208] A cost-sharing arrangement exists with Alberta Agriculture & Forestry for major rehabilitation projects, including headworks infrastructure, buried pipelines within Districts, and other reparations/efficiency enhancing measures. The cost-sharing arrangement is usually 75% provincial and 25% District, which translates into perhaps an additional $5/acre/year cost to the irrigator.[209]

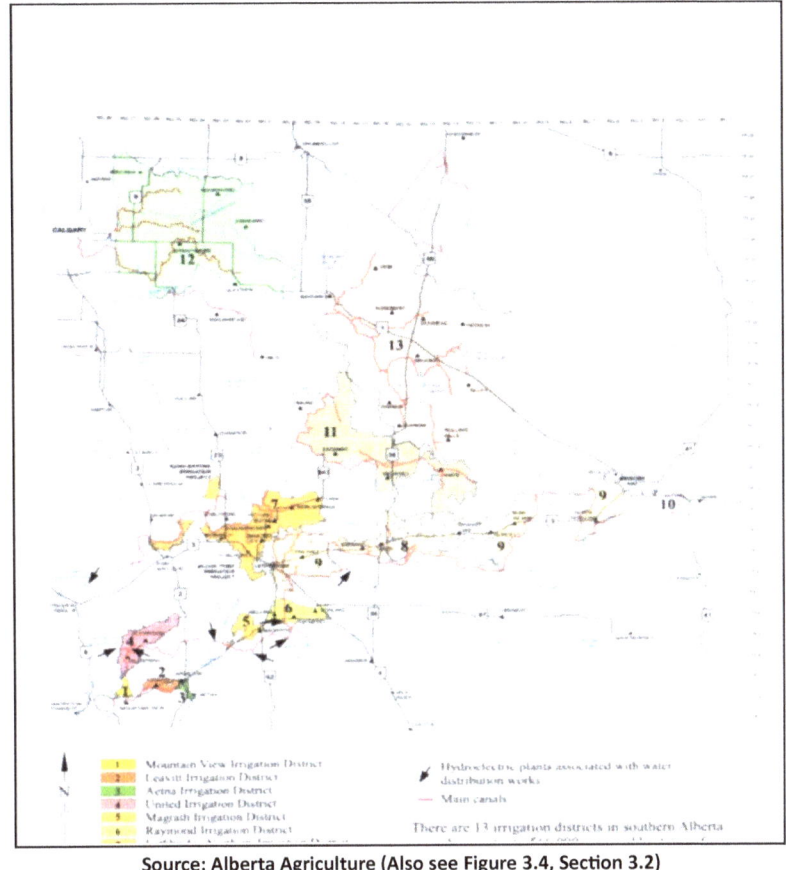

Figure 10.7 Irrigation Districts in Alberta

Source: Alberta Agriculture (Also see Figure 3.4, Section 3.2)

The gross irrigation water requirement (the total diversion) is estimated based on existing conveyance and on-farm irrigation efficiencies, as illustrated in Table 10.6.[210]

Table 10.6 Illustrative Irrigation Water Requirement Calculation

Item	Acre-Feet/Acre	Inches	Meters
Total Diversion	2.55	30.6"	0.78m
Reservoir Evaporation	0.08		
Conveyance Loss (E = 87%)*	0.34		
Return Flow	0.48		
Farm Gate Demand	1.64	20"	0.51m
On-Farm Loss (E = 72%)*	0.46		
Crop Application	1.18	14.2"	0.36m

* Efficiency coefficient

Existing licensed water allocations in the South Saskatchewan River Basin and instream flow requirements already exceed the potential surface water available and no new licenses have been issued since 2001.

In Saskatchewan, the Saskatchewan Water Authority oversees all water management in that province, including the operation of Lake Diefenbaker (supplied by the South Saskatchewan River, which is still not fully allocated).

With irrigation, crop yields approximately double and the crops that can be grown in this semi-arid region are more extensive (Figure 10.8). The principal specialty crops grown are dry beans and potatoes.

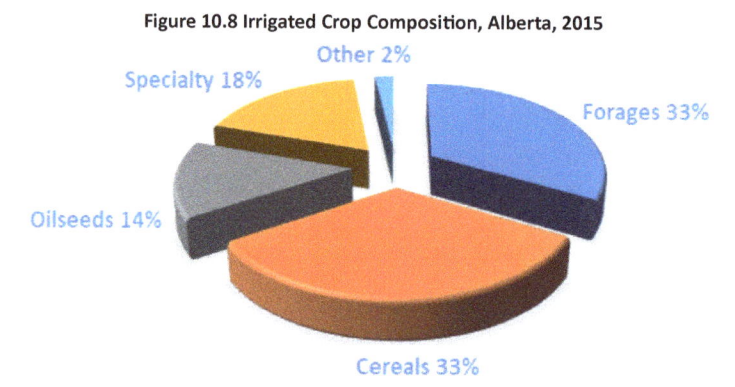

Figure 10.8 Irrigated Crop Composition, Alberta, 2015

- Other 2%
- Specialty 18%
- Forages 33%
- Oilseeds 14%
- Cereals 33%

Source: Alberta Irrigation Information, AAF, 2018

The composition of on-farm irrigation systems in Alberta is now (2018) approximately[211]:

Low-pressure pivot sprinklers	73%
High pressure pivot sprinklers	8%
Side role sprinklers	11%
Surface irrigation	8%

The more efficient low-pressure center pivots became increasingly popular in the mid-1990s. Better on-farm water management (e.g., better scheduling & application rates) also greatly improved on-farm water use efficiencies since the 1990s.

At the Irrigation District level, water use efficiencies have also gradually improved, principally through the use of more efficient conveyance systems (i.e., buried pipelines and more secondary and tertiary canals with a concrete or membrane [plastic] lining). Improved scheduling has also helped. Efforts to address a growing salinity problem in the 1970s and 1980s were also beneficial.

In Alberta, Alberta Environment & Parks owns and operates the primary on-stream dams and diversion structures and some headworks. Procedures for water-sharing during periods of low flow have also been institutionalized. In Saskatchewan, the Saskatchewan Water Authority owns all of the principal structures and manages the entire storage and delivery system.

On-Stream Reservoirs:

On-stream reservoirs can and generally do serve a multitude of functions: irrigation, hydro-electric power, flood control, potable water supplies for households and industry, reliable water supplies for livestock, aquatic protection (i.e., instream flows), and recreational requirements. However, they do flood otherwise pristine land, irreparably change the environmental landscape, alter the recreational mix, and irreversibly change the dynamics of the river itself (e.g., reservoir silting). Consequently, they are advocated and derided in about equal measure.

In North America and Europe, the proposed development of any newly proposed on-stream reservoirs have met with increasing resistance from the general public, as well as from international lending agencies, such as CIDA and the World Bank. This is not the case in many (if not most) developing countries, where on-steam dam construction generally continues unabated—usually for hydro-electric power, irrigation and/or flood control.

From a purely socio-economic perspective, even a **direct** benefit versus cost comparison (i.e., B/C ratio) often indicates that the construction of on-stream reservoirs is only marginally viable, particularly if the principal objective is expansion of irrigated crop production. Numerous Alberta-based studies support this general finding (Table 10.7). The criterion is straight-forward: if the B/C ratio is <1, the proposed initiative is not economically feasible.

Table 10.7 The Economics of On-Stream Reservoirs for Irrigated Agriculture, Alberta

LOCATION	DIRECT B/C RATIO*	YEAR	SOURCE
1. Oldman River	1.1 to 1.3 .94	1978, 1986	MAA, **Oldman River Basin Study, Phase II: Economic Analysis of Water Supply Alternatives**, 1978. MAA, **Oldman River Dam, Economic Analysis**, AE, 1986.
2. Milk River	.48 to .93 .66 to .88 .50 to .68 .77 to .82 .99 to 1.00 .83 - .99	1980, 1981, 1986, 1987, 2004, 2009	MAA, **Milk River Basin Studies: Socio-Economic Component, Part 1**, Alberta Environment, 1980. MAA, **Milk River Basin Studies: Socio Economic Analysis, Part 2**, AE, 1981. MAA, **Milk River Basin Studies: Economic Evaluation of Alternate Reservoir Sizes - Site 2**, AE, 1987. Klohn-Crippen, **Milk River Basin: Preliminary Feasibility Study**, AE, 2004. MAA, **Economic Analysis of Alternative Ridge Reservoir Diversions**, MRWCC, 2009.
3. Blood Band**	.82 to 1.13	1982	UMA/MAA, **Blood Indian Irrigation Feasibility Study: Economic Study Component**, AE, Edmonton, 1982.
4. Clear Lake	1.1	1984	MAA, **Economic & Financial Analysis of Clear Lake Irrigation Development**, AE, 1984.
5. Pine Coulee	1.05	1994	Nichols Applied Management, **Economic & Socio-Economic Evaluation of the Pine Coulee Project**, APW, 1993.
6. Little Bow	.95 to 1.47 .78 to .84	1995	MAA, **Little Bow River Basin: Economic & Financial Analysis of Irrigation Development**, AE, 1986. Foster Research, **Little Bow River Project/Highwood River Diversion Plan: Socio-Economic Assessment**, AE, 1992.
7. Acadia**	0.76	2004	MPE Engineering Ltd., **MD of Acadia Irrigation Study**, AE, 2005.
8. Special Areas**	.64 to .74 .45 to .56	2004	Acres Int. Ltd., **Water Supply to the Special Areas, Phase 1 Study, Final Report**, AE, 1987. Watrecon Consulting, **Socio-Economic Assessment of the Special Areas Water Supply Project**, Special Areas Board, 2004.
9. Meridian	.33 to .35 .34 to .38	2001, 2013	Golder Assoc., **Meridian Dam Preliminary Feasibility Study: Socio-Economic Benefit-Cost Analysis**, AE, 2002. Golder Assoc., **AESRD Economic Analysis: Meridian Dam**, AE, 2013.

* Direct benefit-cost ratio. Does not include spin-offs (multipliers).
  B/C @5% >1 = economically feasible. Studies not strictly comparable. Study #1 (1978) = regional.
** Off-stream diversions.

This one-dimensional assessment, however, has some severe limitations:

- Not all costs and benefits are easily quantified in monetary terms. This is especially true of environmental impacts (positive and negative), social disruption, and unique archeological and cultural values. Quantifying intangibles such as option, existence, and bequest values are even more intractable (Section 10.8).

- There are very real methodological limitations to this type of analysis, particularly to what extent the projected costs and benefits in future years should be discounted and whether relative price relationships will always remain the same (e.g., the value of potable water will always have a fixed relationship to the value of, say, a scrumptious steak). In reality, social-cultural values change over time, as do technologies.

At the same time, extending this type of analysis to encompass the entire regional, provincial, and national economy results in a much less ambiguous picture: the net economic impact, especially regionally, is **irrefutably positive**. These generally positive impacts include:

- Initial construction expenditures and subsequent O&M expenditures in the region (essentially a financial transfer)

- Spin-offs that generate more long-term regional employment and income opportunities and, thus, a more sustainable regional growth trajectory

- Reduced risk, which stimulates greater investment in the regional economy and, subsequently, a lesser dependence on existing safety nets (i.e., other transfer payments to the region)

- Reduced regional income disparities (i.e., less income inequity)

- Water security and the future (higher) option value of the water itself (e.g., water is life, possession is 9/10th of the law, use it or lose it).

The difference between a dryland farming area (say Hanna-Oyen) and an irrigated farming area (say Lethbridge-Taber-Bow Island) is patently obvious. Irrigation generates a regional dynamic that supports the growth of manufacturing, wholesalers, retailers, service industries and public infrastructure, which would otherwise simply not exist. Employment and incomes climb and sustainable socio-economic growth is more easily assured.

About 5% of the cultivated land in Alberta is under irrigation and contributes approximately 20% of the total agriculture GDP in the province.[212] With the spin-offs generated by irrigation development (i.e., secondary net benefits) the total net benefit can be 2, 3, or more times as large as the direct net benefits to producers. Alberta Agriculture & Forestry now claims that "...almost 90% of the benefits accrue to the region and province and only 10% to irrigation producers."[213] In short, even fully recognizing that similar spin-offs arise with most investments in physical infrastructure, on-stream reservoir initiatives can still be a very effective **regional** development tool.

In a broader context, the logic for more water storage (on-stream or off-stream) in Alberta rests, in particular, on its location on the eastern slopes of the Rocky Mountains; the head of major drainage basins where 60% of the annual flow occurs over a period of just six weeks in late spring. More water storage would facilitate better flood management, as well as allowing Alberta to potentially reduce its existing heavy reliance on fossil fuels for electricity (with its correspondingly relatively large CO_2 footprint).[214] Hydro power is clean, continuous, has a long life, and is relatively inexpensive.[215]

At the same time, there are factors that are largely unfavorable to future on-steam reservoir development, particularly: a) a growing public demand to better protect virginal environmental amenities, unimpeded waterways, archeological resources, and (Indigenous) cultural values;[216] b) increasingly stringent environmental and construction standards resulting in higher construction costs; c) construction costs out-pacing the general rate of inflation, and d) diminishing returns to increasingly marginal development proposals.

Water Rights & the Value of Water:

Water has long been considered free, like the air we breath. This, in turn, has given rise to the dictum "possession is 9/10ths of the law." Many jurisdictions have what is the equivalent of a Water Act, which prioritizes access to fresh water: first potable water; second industry, third irrigation, and so on, all subject to environmental constraints. Other jurisdictions (e.g., Alberta) prioritize the allocation of surface water based solely on the date when the license was issued. Jurisdictions that allocate surface water by giving water rights to potential users also generally prioritize these in terms of when they were issued: first in time, first in right. This naturally leads to a morass of accompanying rules, regulations, and attendant regulators. This is not unlike a triage station at your local hospital. Today's costs for water are essentially payments for the services provided to treat and deliver the water; not the water itself.

The central problem with this legacy approach is that over time it potentially becomes increasingly inefficient and socially sub-optimal. A rigid regulatory framework simply cannot quickly and easily reallocate water to its highest and best use. Consequently, over time, water use is almost inherently mis-allocated from a socio-economic perspective. The limited water available, for example, may end up being utilized to produce low-valued feed grains (for livestock) on marginal irrigated lands, instead of being made readily available to nearby households and municipalities. The typical bureaucratic response is to just try to find another (higher-cost) potable water source for households and municipalities.

But what if the water itself was **not** considered "free"? Anything that is scarce does have a socio-economic value and that can be translated into a monetary value. So why can't we buy and sell water like any other commodity?[217]

Clearly, in terms of water allocation, the value of water for different uses is very difficult to determine because it has numerous dimensions,[218] is location-specific, and is situation-specific. Thus, the measurable economic value of water is usually considered synonymous with a **net willingness to pay** for raw water for a particular use under specified conditions (i.e., the difference between the average willingness to pay for delivered water and the average cost of delivery). Willingness-to-pay estimates are similar to *shadow prices*; prices that would emerge **if** there was a competitive market for water. This value is generally greater than the marginal value of water as potentially determined in the marketplace.

The **average value** of water is an appropriate guide when allocating large blocks of water among mutually exclusive uses and some approximate values are indicated in Table 10.8:

Table 10.8 Range of Average Values for Alternative Water Uses

Water Use Purpose	Low Value/dam3	High Value/dam3	Average Value/dam3
Municipal	$ 100	$ 2,430	$ 1,220
Residential	$ 5	$ 3,356	$ 1,681
Irrigation	$ 51	$ 104	$ 69
Livestock Watering	$ 27	$ 682	$ 355
Food and Beverages	$ 10	$ 124	$ 67
Petroleum/Chemicals	$ 17	$ 130	$ 74
Hydro or Thermal Electricity*	$ 7	$ 18	$ 13
Sports Fishing*	$ 20	$ 74	$ 47
Waste Assimilation*	$ 1	$ 4	$ 3

Notes: 1. Irrigation average value is a weighted average for Alberta.
2. Sports fishing estimates bracket most in-stream value estimates exclusive of non-use values which are usually in the range of Cd$ 60 to Cd$ 120 per household (Colby, 1987).
3. One dam3 = .811 acre-feet. One dam3 = 1,000,000 litres. One acre-foot = 271,000 gallons.
* Principally on-stream and main canals.

Basic Sources: Muller (1985), Kulshreshtha et. al. (1986), and Colby (1987).
Reproduced in MPE Engineering Ltd./Russell Consulting, *Southern Regional Stormwater Management Plan - Economic Analysis*, Lethbridge, 2016. Originally compiled by Dr. M. S. Anderson, 2002.

Not unexpectedly, the value of water for domestic purposes greatly exceeds that of any other use. Most other uses are often fairly competitive with one another, particularly irrigation, industry, and in-stream demands. (Here reflected in the value of sport fishing.)

At the same time, when water can be shared, it should ideally be allocated so that the **marginal value** is the same in each use. But water pricing at its marginal value as an allocative basis for different uses is not generally accepted as a valid allocative criterion by the Canadian public.[219] Oblivious to more acute future scarcity (since Canada has ¼ of the Earth's freshwater, a dangerous misconception in a regional context), it is still generally considered to be "free" in most jurisdictions in Canada. BC, however, does impose a water charge based upon the type of use and the licensed amount.

The past decade has heralded a shift in approaches to both groundwater and surface water across Canada, marked by the emergence of new watershed-based governance models (see below), a demand for higher standards of drinking water, and the increasingly active voice of citizens in environmental policy and management. Additionally, there has been a modest and very gradual shift toward a more market-based approach to water pricing. The Water Act in

Alberta, for example, now allows water licensees with water in excess of their needs to actually sell it at its market value, subject to government approval, either temporarily or permanently.[220]

10.7.4   DRAINAGE

Water drainage issues arise both on-farm and regionally.

On-Farm Drainage:

Generally, there is no Common Law right to the drainage of stationary water. A down-slope owner does not have to receive augmented natural surface runoff unless a prescriptive right has been acquired. The general rule is a farmer cannot negatively affect contiguous lands.

From a farmer's perspective, drainage of temporary and even permanent wetlands on cultivated land is often profitable. There are at least six potential private (on-farm) benefits:[221]

1. Annual wetland crop production revenue
2. The benefits of improved timing of farm operations
3. Improved weed control
4. Lower farm production costs because of improved field patterns
5. Improved crop quality in the area immediately surrounding previous wetlands
6. Less crop damage by wildlife (waterfowl, deer, etc.)

The immediate drainage costs are largely a function of the topography and soil type. Additional costs can also be incurred because of resultant erosion, flooding, sedimentation, and water quality deterioration, as well as subsequent remediation.[222]

Nevertheless, even this on-farm financial assessment fails to take into account potential **externalities**—potential off-farm public costs, particularly the probable loss of wildlife habitat and possible external changes to the local hydrological regime (e.g., reduced groundwater recharge) and micro-climate. The central issue is determining who pays and who benefits. Where, for example, is the financial incentive to compensate farmers for not draining temporary prairie wetlands that impede farming operations but support public amenities (e.g., waterfowl production, groundwater recharge)? Provincial governments have, from time to time, instituted programs to develop regional drainage programs. Ontario and Quebec have a long history of supporting on-farm drainage for cropland to lower water tables. Tile drainage, often to deal with salinity problems, is widespread and much less controversial. Yet on the prairies, current government regulations prevent on-farm drainage of temporary and permanent wetlands, primarily to protect wetland habitat.

Off-Farm (Public) Drainage:

Drainage initiatives by provincial governments to intensify agricultural production have generally focused on entire drainage basins or sub-basins (Water Management Units – WMUs), typically 15,000–25,000 ha (40,000–60,000 acres) and on the prairies they are usually in more northerly regions of the respective provinces. These areas generally have gray-wooded and peat soils accompanied by a myriad of related surface water issues: temporary sheet water, sloughs and marshes; seasonal sloughs and marshes; and permanent sloughs, marshes, bogs (muskeg), fens (swamp), lakes and ponds.[223]

The socio-economic rationale for these initiatives has often been rather dubious. Historically, the benefit-cost ratio for five such basins in northern Alberta was estimated to be about 0.75 and, consistent with this finding, wetland consolidation was considered more economic than partial drainage. In turn, partial drainage was considered more economic than total drainage.[224] Nonetheless, large-scale drainage in the northern prairies can sometimes (like irrigation elsewhere) be an effective **regional** development initiative.

10.7.5 FLOOD CONTROL

Contrary to the drainage of stationary surface water, there is a Common Law right pertaining to the drainage of water flowing in a natural watercourse. This watercourse must have a bed, visible confining banks, and a substantive flow.

Flood control measures usually takes the form of river training structures and alterations to bed morphology—dredging, realignment, dykes, dams, weirs, and related structures (e.g. rip rap and gabions). Rarely is the focus of the proposed watercourse changes exclusively directed toward agriculture. Proposed changes to a watercourse are usually intended to protect people, homes, industry, and related infrastructure in densely populated urban areas situated on flood plains or low-lying areas adjacent to rivers with a highly variable stream flow (e.g., High River [hint!], Calgary. and Winnipeg).

The tangible direct and indirect damages, as well as the intangible damages, from floods can be calamitous.[225] Most provinces have a provincial flood forecasting and warning system and they invest heavily in flood prevention. Since 1997, Manitoba has spent more than $1 billion in flood mitigation, including more Red River Valley dykes, Assiniboine River Dykes, and an improved Red River Floodway around Winnipeg. Flooding is still the leading cause of public emergencies in Ontario. Alberta also continues to invest in flood control to better manage the rapid runoff from snowmelt and heavy rains in the mountains and foothills.

10.7.6 OTHER WATER-RELATED GOVERNMENT INITIATIVES

Other water-related public initiatives over the years have included the following:

- Conservation Authorities – initiated in the 1940s in response to severe flooding and erosion problems in Ontario, there are now thirty-six conservation authorities that promote an **integrated watershed management approach** to try to balance human, environmental and economic needs. Over time, conservation authorities in Ontario have become involved in a wide range of activities and responsibilities, depending upon the environmental concerns of local residents, member municipalities, and the province (Section 10.6). More recently, **participatory watershed-based delegated governance models** have become more common elsewhere in Canada. These are, similarly, locally led initiatives to try to resolve local resource use conflicts by seeking a human, environmental and economic consensus.

- Integrated **Watershed Management Plans** – principally plans to direct the planning and management of Crown land, giving priority to the protection of water supplies (British Columbia). Alberta also developed a similar but more comprehensive South Saskatchewan River Basin Plan. It focuses on surface water allocation.

- Provincial **groundwater registries**, provincial/federal **environmental farm plans** (Section 10.6), and **pollution standards** are also employed to protect the quality of both surface water and groundwater across Canada. (See Chapter 9.)

10.8 THE VALUE OF ENVIRONMENTAL AMENITIES

Farmers and ranchers both protect and enhance the environmental amenities of our land. But this generates an inherent social problem. Something that is free benefits everyone but the costs of maintaining those environmental amenities are borne solely by the private property owner. These amenities are "externalities" and the inability of owners of those amenities to capture their full value in the marketplace is referred to as a "market failure." These externalities refer, in particular, to the following:

- **Option Value** – is a measure of what people might pay for the option of being able to experience an environmental amenity at some point in the future.

- **Existence Value** – is the value that is placed on private willingness to pay for maintaining or preserving a public asset or service even if there is little or no likelihood of the individual actually ever using it. It is the benefit (or intrinsic value) people receive from knowing that a particular environmental resource simply exists.

- **Bequest Value** – the value of the satisfaction received from preserving a natural or historical environment in perpetuity. It is what people might pay to maintain an environmental feature for potential use by future generations.

Together, they make up the non-use value of an environmental service or good. But to maintain those amenities, he/she must forego its **opportunity cost**; the cost of not utilizing that landscape for a potentially more profitable private use.

And these amenities do have a very real monetary value, however difficult they are to measure in terms of dollars and cents.[226] Table 10.9 provides some indicative values.

Table 10.9 Indicative Non-Market Values for Habitat & Related Recreation

Habitat/Related Recreation	Item	Monetary Value, $2016*
HABITAT:		
Protection of endangered or threatened terresstrial species (USA)	Grizzly bear, bald eagle, bighorn sheep, whooping crane	$2 to $39 per species/household/yr
	Northern Spotted Owl	$48 - $541/household/year
Habitat preservation (Australia)	all	$96 - $225/person/yr
Wetland Habitat**	Fish	$163 - $336/ha ($66-$136/acre)
	Waterfowl	$850/ha ($344/acre)
	Sediment accretion	$15/ha ($6/acre)
	Flood control	$1,843/ha ($746/acre)
	Water quality	$5,637/ha ($2,282/acre)
	Waste assimilaton	$31,675/ha ($12,824/acre)
	Recreation	$193 - $388/ha ($78 - $157/acre)
RELATED RECREATION:		
Hiking & Horseback Riding		$65/person/day
Big Game Hunting		$105/person/day
Small Game Hunting		$76/person/day
Migratory Waterfowl Hunting		$70/person/day
Non-Consumptive Fish/Wildlife		$57/person/day
Wilderness		$52/person/day

* Adjusted 1993 US$ values. USA CPI change = 2.18. Cd$ = 0.78 US$.
** Adjusted for 1995 US$. US CPI change = 2.06. Exchange rate = .78. 1 hectare = 2.47 acres.

Sources: Asian Development Bank, *Economic Valuation of Environmental Impacts: A Workbook*, ADB, Manila, 1996. (Also see docs. by Phillips & Pattison)
Wetland habitat values from: Weibe, K.D. and R.E. Heimlich, "The Evolution of Federal Wetland Policy," *Choices*, 1995.

This is entirely consistent with a growing consensus that the value of environmental amenities should also be included in measures of our Gross National Product (GNP), a conventional measure of wealth. One analyst has estimated that the value of natural resource services is maybe twice our conventional measure of GDP.[227] Including the value of these services in the GDP calculations, it is argued, would more accurately capture the notion of time preference, non-market pricing, and sustainability.[228] Numerous governments have already recognized the need to bring depreciation of forests, minerals, wetlands, and other resources into a national accounting framework and a number of OECD countries (including Canada) have started the process of compiling accounts on natural resource stocks.[229] If they are being enhanced, it enriches us; if they are being depleted, it impoverishes us.

Additionally, we now know that the natural green world around us not only improves our health but it may also prevent an early death. Natural environments, including parks, forests, lakes and open water are now widely recognized as having the potential to mitigate the adverse effects on health associated with urban living, such as traffic congestion, noise, and air pollution. Experiencing a green, outdoor environment—ecotherapy—can reduce anxiety and depression, attention deficit disorder, and chronic illnesses such as diabetes.[230]

10.9 PAYMENT FOR ALTERNATIVE LAND USE SERVICES (ALUS)[231]

Most public initiatives fail to fully recognize that farmers and ranchers are a constituency that could be more aggressively recruited to pro-actively enhance and maintain the eco-diversity of the agricultural land base. This must change.

More and more analysts and commentators are also starting to say that simply using non-targeted direct and indirect subsidies to farmers/ranchers to produce more and better food is misguided. They argue that protecting the fragile ecosystem in Canadian agriculture should take precedence. The European Union, Japan, United States, and Canada have all gradually moved toward **agri-environmental policies** (AEPs), sometimes called **agri-environmental schemes** (AES's). Each promotes payments to farmers/ranchers for environmental services that reduce the negative externalities of agricultural production, while also serving as a means of transferring public funds to farmers.

10.9.1 EUROPEAN UNION

EU environmental programs address a wide range of externalities, and are focused on paying for a particular farming activity as well as serving as a driver for rural development. This objective is achieved by compensating farmers for the private delivery of positive public goods, such as attractive landscapes produced by agriculture. And there is ample empirical evidence that Europeans are willing to pay for these positive externalities. An example is the United Kingdom's Countryside Stewardship and Organic Farming Scheme, which cost-shares, subsidizes, or otherwise pays farmers for providing environmental benefits from land that remains in production. With BREXIT, the current government is considering replacing all EU CAP subsidies with subsidies awarded for delivering public goods, the most important of which is environmental protection and enhancement, such as planting woodlots, restoring peat bogs or maintaining hedgerows.[232]

10.9.2 UNITED STATES

AEPs (or AES's) in the United States have focused almost entirely on reducing agriculture's footprint, such as soil erosion. US AEPs also tend to be more targeted and more often take into account the opportunity cost of the activity. In particular, the US Conservation Reserve Program (CRP) pays agricultural producers an annualized fee to obtain a stream of environmental benefits over a predetermined contractual period. Under the CRP selection process, bids from eligible applicants are ranked using an environmental benefit index (EBI) that measures and weights multiple environmental benefits from specific parcels of potential reserve land, as well as bid parameters that approximate the opportunity cost.

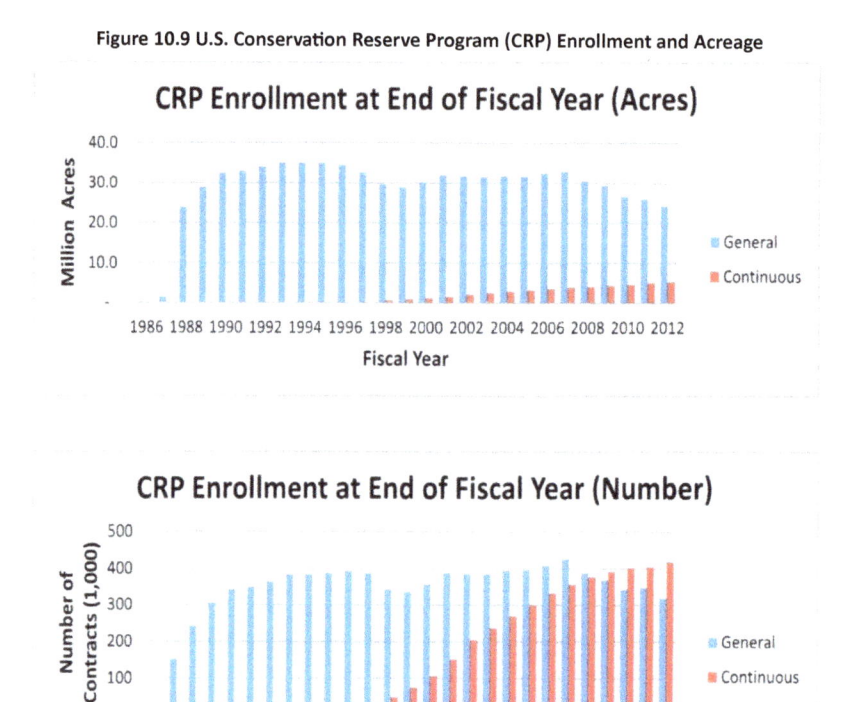

Figure 10.9 U.S. Conservation Reserve Program (CRP) Enrollment and Acreage

Source: USDA/FSA, *Conservation Reserve Program, Annual Summary & Enrollment Statistics*, FY 2012, Washington, D.C., 2013

The CRP (Figure 10.9) involves about 400,000 US farmers and about 12 M ha (30 M acres); about 20% of all US farmers and 3% of all US farmland.[233] (About two times the number of farmers in Canada and about 19% of all the farmland in Canada.) Annual payments are about US$124/ha (US$50/acre) for a general CRP agreement and US$222–$321/ha (US$90–$130/acre) for a continuous CRP agreement. Maintaining farm wetlands under CRP pays about US$274/ha (US$111/acre). The total annual cost of the CRP is about US$1.7 billion (2012 data).

10.9.3 CANADA

Since 2003/04, Canada has had a National Farm Stewardship Program with similar objectives, that is, to reduce the risk to water and air quality, improve soil productivity and enhance wildlife habitat. More recently (2019), a voluntary national Alternative Land Use Service (ALUS) Program is intended to further help farmers and ranchers restore wetlands, plant windbreaks, install riparian buffers, create pollinator habitat, and establish other ecologically beneficial projects on their properties. Under this umbrella program, PEI has now established a province-wide Alternative Land Use Service (ALUS) Program, and some municipalities and communities in Ontario, Saskatchewan, and Alberta have also established ALUS projects. Approved projects must represent a Best Management Practice (BMP) and be part of an Environmental Farm Plan (EFP) (Section 10.6). And modest per-acre annual payments are available to some participating farmers and ranchers.

But the federal government and most provinces have so far studiously avoided making any major financial commitments to an ALUS-type program, believing that it is a slippery slope. Once the principal of paying farmers/ranchers for maintaining or enhancing environmental amenities is accepted, so they say, where does it end?

Yet in Manitoba, as far back as January 2002, the Keystone Agricultural Producers (KAP – Manitoba's largest general farm organization) and Delta Wildlife proposed a province-wide ALUS program for that province that would have done exactly that. They proposed a provincial crop insurance program for farmers to convert part of their land to green manure crops or grazing, wildlife habitat areas, or woodlots for carbon sequestration. They argued then that:

> There are many pressures on farmers to be environmentally responsible, ensure safe water, care for the land [and] a policy like this [would] help us deal with these issues and, rather than be reactive, we could be proactive.[234]

This was envisioned as providing a broader source of income for rural communities; income in addition to the income derived from the then-existing farm safety nets. Recently proposed legislation in Manitoba (Sustainable Watersheds Act, November 2017) would extend the ALUS model and provide financial incentives to farmers who adopt practices that restore wetlands, retain water, and better manage riparian areas. The Act would also increase protection of seasonal, semi-permanent and permanent wetlands.[235] This would represent one more step toward compensating landowners, at the very least, for ecological services that preserve things like water quality and natural habitats. (Also see Section 10.6.)

10.9.4 THE FUTURE

The public payment to farmers/ranchers for maintaining or enhancing environmental amenities could have at least five advantages over an almost exclusive dependence on the existing suite of "safety nets," which are explicitly designed to include disguised production-based subsidies (Chapter 9):

- It could completely eliminate all subsidies to farmers, rich and poor; big and small, east and west. It would, instead, favor farmers/ranchers who have environmental amenities to sell to the public according to the public's willingness to pay.
- It could allow the existing safety nets to be re-calibrated to reflect actual risk and, thus, much-needed public transparency and a much lower public cost.
- All major agricultural programs in Canada would immediately become WTO "green"; not ambiguously "amber," as sometimes presently designed.

- It would also unambiguously link payments to sustainable environmental management and facilitate the more widespread adoption of Regenerative Organic Agriculture (ROA) (Section 10.5).

- It would make farmers and ranchers integral, proactive partners in the on-going national and international effort to reduce CO_2 emissions.

The tremendous diversity and expanse of Canadian agriculture makes any major change to public policy excruciatingly difficult. But a whole host of socio-economic factors (e.g., climate change, on-going environmental degradation, health concerns, international trade patterns, etc.) are virtually all trending toward the widespread acceptance and, eventually, the widespread adoption of this ALUS paradigm.

11. *Climate Change & Agriculture*

11.1 OVERVIEW

The research on climate change (or global warming) is voluminous and, to some extent, still somewhat controversial. The objective of this chapter is simply to provide the reader with a basic understanding of how this very exhaustive literature all fits together, as illustrated in Figure 11.1

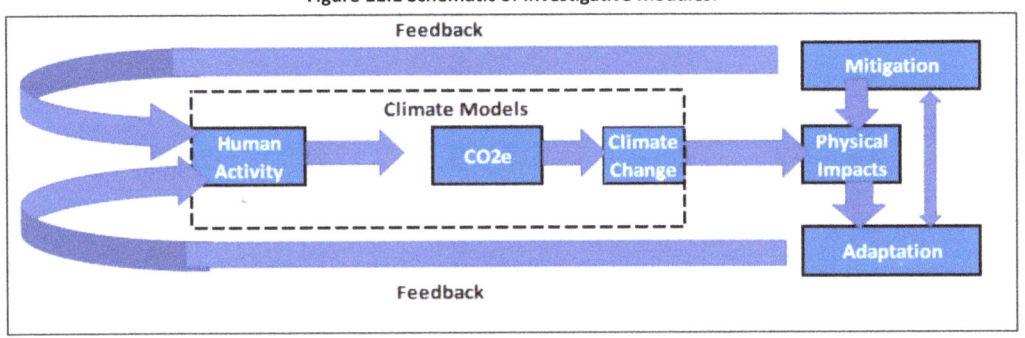

Figure 11.1 Schematic of Investigative Modules:

This is a gross oversimplification but it captures its essence. Human activity increases greenhouse gas emissions which, in turn, change our climate which, in turn, affects how human activity can be conducted (i.e., impacts). We then respond with specific mitigative measures to reduce greenhouse gas emissions, while, simultaneously, human activity is necessarily adapting on its own. Some adaptation overlaps desired mitigation measures.

A climate model is essentially a complex system of dynamic equations that historically links inputs (human activity) to intermediate outputs (CO_2e) and the ramifications of those intermediate outputs (climate change). Atmospheric CO_2e concentrations are projected to climb from about 275 ppm in the pre-industrial age to about 550 ppm (2X) by the year 2050 if future human activity remains unaltered. (The CO_2e level in 2017 was already 405 ppm.) Thus, with an established empirical relationship between atmospheric CO_2 concentrations and a multitude of related variables, the model then simulates (extrapolates) the expected implications for the future (e.g., an average +2C degree temperature change). But these are projections; not predictions. This also assumes causality; not just a correlation. Thus, crucial to these projections is how these simulations also account for technological change and human adaptation to anticipated climate change over time. Anticipated climate change is less speculative than the anticipated ramifications and how to address the negative ramifications. More distant projections are increasingly speculative.

Greenhouse gases (GHG) consist of carbon dioxide (CO_2), methane (CH_4), nitrous oxide (N_2O), and by-product emissions (hydrofluorocarbons, perfluorocarbons, and sulfur hexafluoride). For simplicity, the principal greenhouse gases are all expressed in terms of a carbon dioxide equivalent (CO_2e) where methane * 21 = CO_2 equivalent and nitrous oxide * 310 = CO_2 equivalent. The national composition of greenhouse gases is approximately 82% carbon dioxide, 10% methane, 5% nitrous oxide, and 3% HFCs, PFCs, and SF6.[236]

Projected climate change (veracity, causes, consequences, appropriate response, etc.) is all subject to considerable (and sometimes very heated and passionate) debate and discussion. Ultimately, society's global response is highly dependent upon the socio-economic-environmental criteria employed. An economic perspective,[237] which typically equates discounted incremental costs to discounted incremental benefits, would call for less onerous adjustments than would an environmental perspective.[238]

Aside from identifying the most efficient and effective policy instruments to employ, an additional concern is the **distribution** of costs and benefits. Who pays and who benefits?

11.2 A GLOBAL PERSPECTIVE[239]

In a global context, Canada contributes about 1.6% of greenhouse gas (GHG) emissions, as measured in terms of its CO_2 equivalents. China, the USA, Europe, India, and Russia are the major CO_2 contributors. (Figure 11.2)

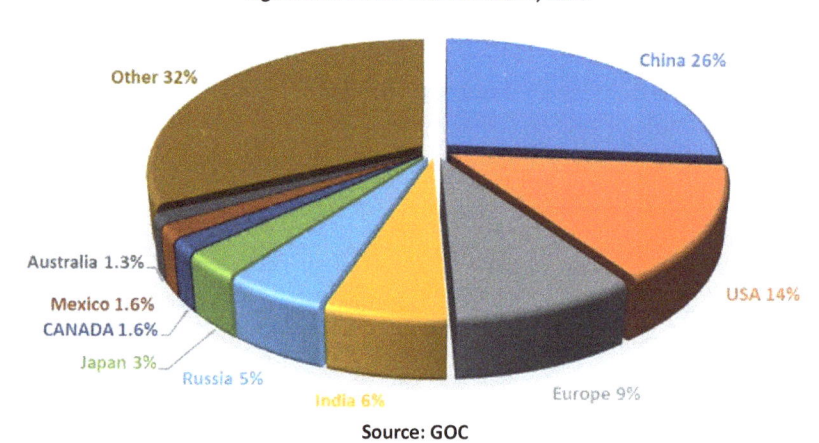

Figure 11.2 Global GHG Emissions, 2013

China 26%
USA 14%
Europe 9%
India 6%
Russia 5%
Japan 3%
CANADA 1.6%
Mexico 1.6%
Australia 1.3%
Other 32%

Source: GOC

On a per person basis, however, Canada's contribution is still relatively large; about 15 tonnes of CO_2 per year.[240] Even comparing countries with a similar standard of living, Canada still does relatively poorly, only slightly better than the United States and Australia, but more than twice as bad as the European Union and Scandinavia.

Table 11.1 Per Capita CO_2 Emissions, Selected Countries*

Country/Region	Per Capita tonnes/yr. 2014	World Ranking, 2011 (1 = worst)
United States	16.5	12
Australia	15.4	11
CANADA	**15.1**	**14**
Russia	11.9	23
China	7.5	55
European Union**	6.9	49
Scandinavia***	6.6	50
India	1.7	133

* Only includes fossil fuels and cement manufacture.
** Seven countries, simple average.
*** Denmark, Sweden, and Norway, simple average.

Basic Source: World Bank; U.S. Department of Energy

Canada's GHG emissions in 2015 amounted to about 722 mega tonnes (mega=million) and out of this 722 mega tonnes, the oil and gas sector and the transportation sector contributed about 26% and 24%, respectively. Agriculture contributed a relatively meager 54 mega tonnes.[241] (Figure 11.3). How these contributions have changed over time is illustrated in accompanying Figure 11.4.

The electricity sector's share has gradually declined in recent years, while the oil and gas sector and the transportation sector have witnessed gradual increases. Agriculture's contribution, about 7% of the national total, has remained relatively constant over time.

Figure 11.3 GHG Emissions, Canada, by Sector, 2015 (Total = 722MT)

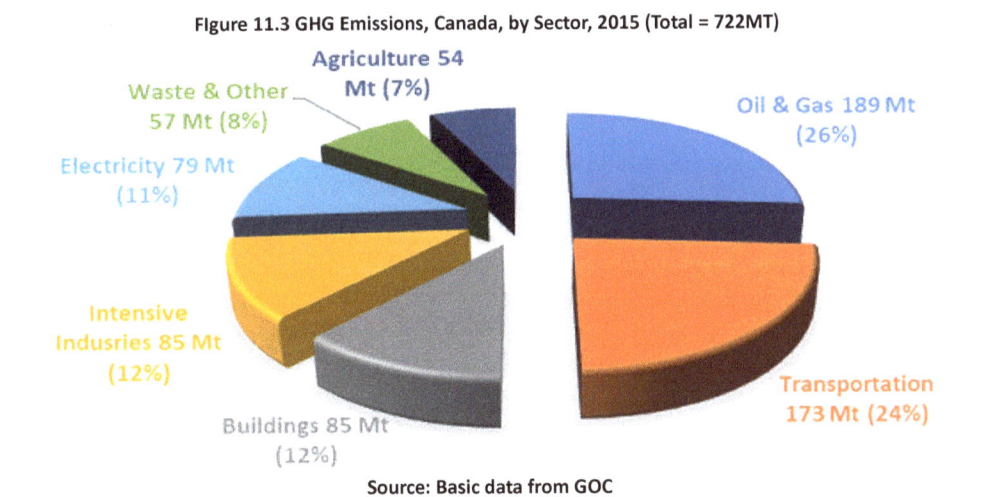

- Agriculture 54 Mt (7%)
- Waste & Other 57 Mt (8%)
- Electricity 79 Mt (11%)
- Intensive Industries 85 Mt (12%)
- Buildings 85 Mt (12%)
- Oil & Gas 189 Mt (26%)
- Transportation 173 Mt (24%)

Source: Basic data from GOC

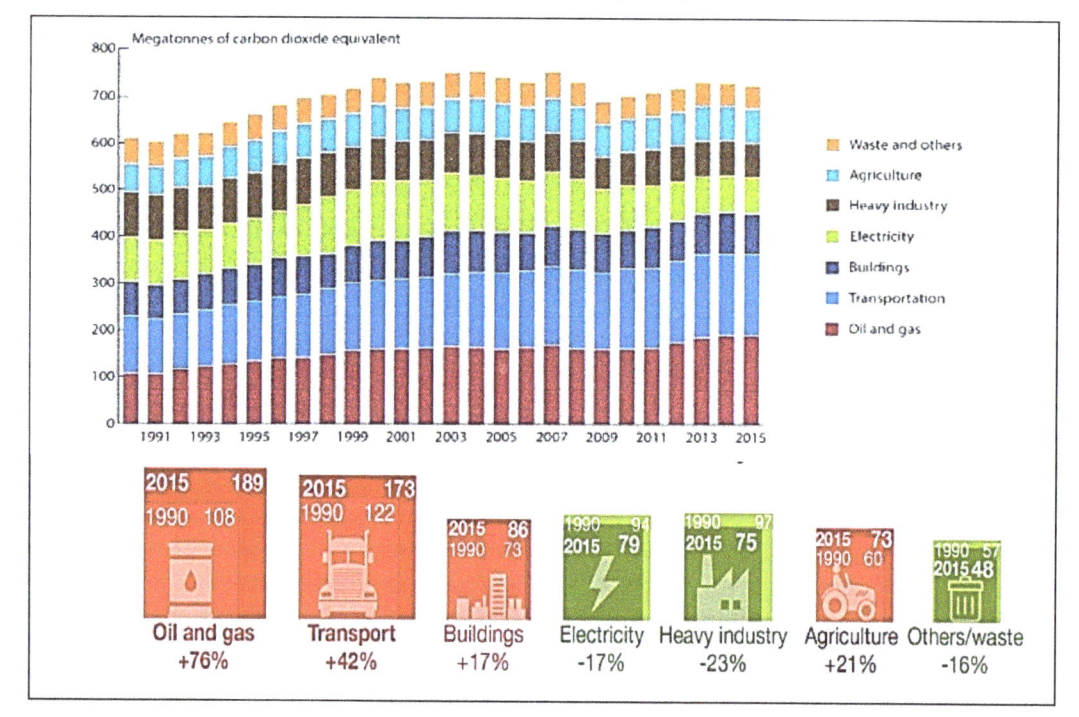

Figure 11.4 GHG Emissions, Canada, by Sector, 1990 - 2015

Sector	1990	2015	Change
Oil and gas	108	189	+76%
Transport	122	173	+42%
Buildings	73	86	+17%
Electricity	94	79	-17%
Heavy industry	97	75	-23%
Agriculture	60	73	+21%
Others/waste	57	48	-16%

Source: OECD, *Environmental Performance Reviews - Canada 2017*

In short, as a sparsely populated country, Canada's contribution to global GHG emissions is, similarly, also relatively small. At the same time, our heavy dependence on fossil fuels, both domestically and for export, translates into a relatively large per-person CO_2e footprint. And since 1990 (the Kyoto Accord baseline) Canada's total GHG emissions have gradually ratcheted upward, especially with respect to the oil and gas sector, the transport sector, and, yes, even in the agricultural sector.

11.3 CO_2E SOURCES IN AGRICULTURE

The sources of GHG emissions (CO_2 equivalent) in Canadian agriculture, by province and region, are divided into seven major categories in Table 11.2 and illustrated (also by source) in accompanying Figure 11.5.

Table 11.2 GHG Emissions, by Source, Canadian Agriculture, by Region, 2011* Thousand Tonnes

CO2 SOURCE & SINK CATEGORIES	MARITIMES	QUEBEC	ONTARIO	PRAIRIES	B.C.	CANADA
1. Enteric Fermentation	391.2	2270.8	2724.9	11679.2	917.1	17983.2
Dairy Cattle	159.4	936.3	884.2	442.7	209.4	2632.1
Non-Dairy Cattle	217.4	1152.7	1655.4	10803.8	666.2	14495.5
Other Livestock/Poultry**	14.4	181.8	185.3	432.7	41.5	855.6
2. Manure Management	91.5	721.5	661.8	1077.3	168.2	2720.4
Dairy Cattle	43.5	150.7	173.7	113.5	90.1	571.5
Non-Dairy Cattle	24.6	112.2	127.4	379.1	45.7	689.0
Other Livestock/Poultry**	23.5	458.6	360.7	584.7	32.3	1459.8
3. Manure Management Systems***	89.7	535.3	805.5	2072.5	171.9	3674.9
4. Direct Soil Emissions	311.9	2151.2	3230.7	9950.2	311.1	15955.1
Synthetic Fertilizers	158.0	1006.3	1488.4	6392.5	123.0	9168.2
Manure Applied as Fertilizer	63.5	525.8	522.0	691.4	87.9	1890.7
Crop Residues	77.5	481.4	1003.0	3448.2	51.8	5061.9
Other****	12.9	137.7	217.2	-582.0	48.5	-165.7
5. Manure on Pastures	44.6	244.1	254.2	2021.0	160.4	2724.3
6. Indirect Soil Emissions*****	223.3	1299.3	1879.7	7179.9	257.9	10840.1
7. Field Burning of Crop Residues	0.1	0.2	0.3	26.4	0.0	26.9
TOTAL AGRICULTURE	1152.2	7222.5	9557.1	34006.5	1986.5	53924.9

* Methane (CH4) X 21 = CO2 equivalent. Nitrous oxide (N2O) X 310 = CO2 equivalent.

** Hogs, poultry, sheep, goats, buffalo, horses, and llamas/alp *** Includes solid storage/dry lot, anaerobic lagoons, and liquid.

**** Summerfallow, tillage, irrigation, & histosols. ***** Atmospheric deposition, leaching and runoff.

Source: Basic data from Environment Canada (2012), *National Inventory Report: Greenhouse Gas Sources and Sinks in Canada, 1990-2011,* Parts 1-3, GOC, Ottawa

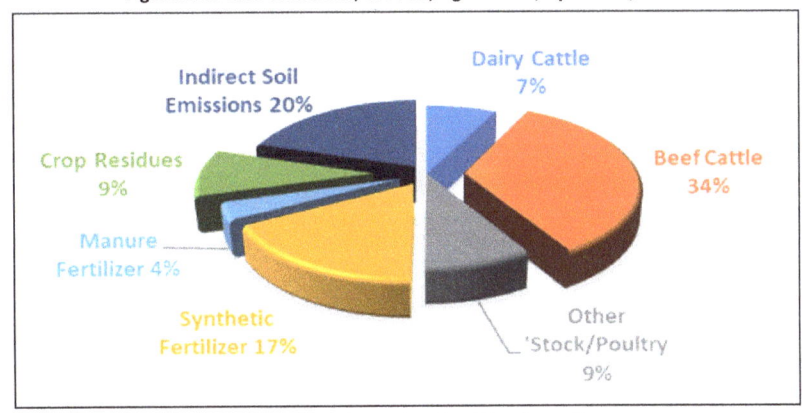

Figure 11.5 GHG Emissions, Canada, Agriculture, by Source, 2011

- Dairy Cattle 7%
- Beef Cattle 34%
- Other 'Stock/Poultry 9%
- Synthetic Fertilizer 17%
- Manure Fertilizer 4%
- Crop Residues 9%
- Indirect Soil Emissions 20%

Indirect soil emissions = atmospheric deposition, leaching & runoff.
Dairy cattle includes 21% of manure management systems.
Beef cattle includes 25% of manure management systems
+ 90% manure on pastures.
Other livestock & poultry includes 54% of manure management

Source: Basic data from Table 11.2

The principal CO_2e sources are beef cattle (34%), indirect soil emissions (20%), and synthetic (chemical) fertilizer (17%). The livestock sources are basically from ruminants: beef cattle, dairy cattle, sheep, and goats. Ruminants ferment their food in a specialized stomach, regurgitate it (as cud), and then chew it again prior to digestion. This generates gases, which they then release, mostly methane (CH_4). Both methane and nitrous oxide (N_2O) is also released from manure. (Figure 11.6)

Figure 11.6 GHG Emissions from Livestock

Indirect soil emissions (20%) includes atmospheric deposition, leaching, and runoff. Synthetic fertilizers (principally N, P, and K) are derived from natural gas and phosphate/potash rock. Not included as a source of agriculture pollution in the above calculations is the CO_2e generated from either related transportation or agri-industry. Clearly, in a vast country like Canada, truck and rail movement of both agricultural products to market and inputs to farms also contribute to CO_2e pollution. The extensive use of other carbon-based products (especially pesticides, and farm fuels) is also deleterious. (This adds about 20 mega tonnes/year. See Section 11.3.)

Relative GHG agricultural emissions across Canada generally reflect relative livestock numbers and the relative amount of cultivated cropland. Almost 2/3rds of all agricultural GHG emissions arise on the Canadian prairies—Manitoba (11%), Saskatchewan (21%), and Alberta. (31%). Ontario contributes about 18 percent of national agricultural emissions, followed by Quebec at approximately 13 percent. BC and the Maritimes emit a relatively insignificant 4% and 2%, respectively (Figure 11.7).

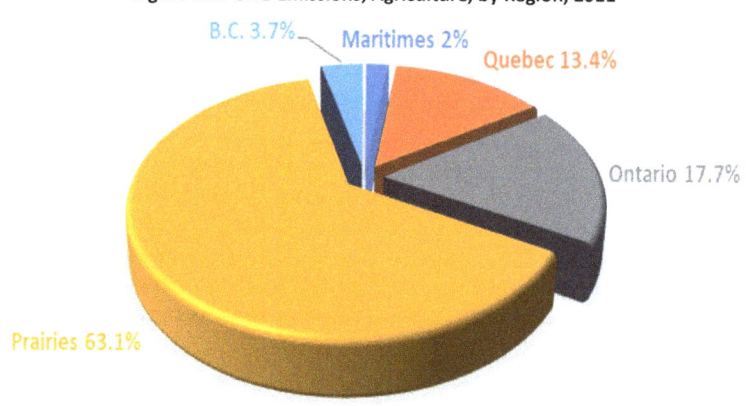

Figure 11.7 GHG Emissions, Agriculture, by Region, 2011

B.C. 3.7%
Maritimes 2%
Quebec 13.4%
Ontario 17.7%
Prairies 63.1%

Source: Basic data from GOC

Finally, two additional points should be highlighted. Firstly, despite a growing agricultural sector, GHG emissions from Canadian agriculture have not changed much in the last 15 years. They remain in the range of 50–60 mega tonnes per annum (Figure 11.8)[242], about 7 percent of all GHG emissions in Canada (Figure 11.3).

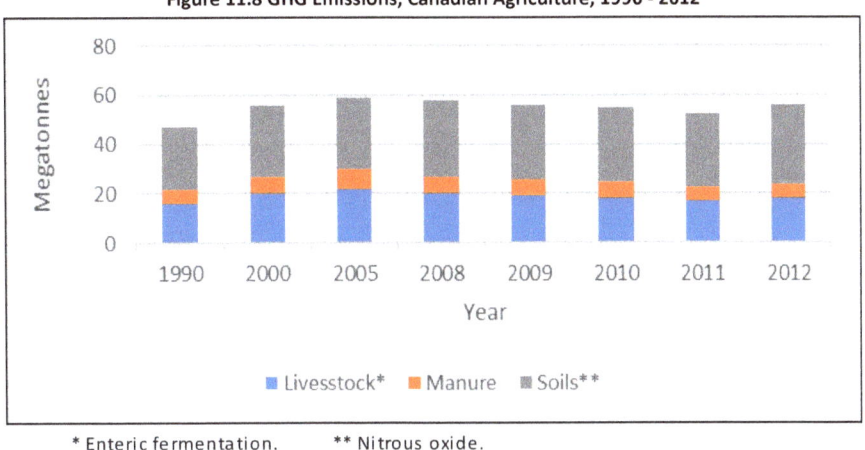

Figure 11.8 GHG Emissions, Canadian Agriculture, 1990 - 2012

* Enteric fermentation.   ** Nitrous oxide.
Source: AAFC

Secondly, from an international perspective, this is still relatively minute on a per-acre basis. A comparison with the performance of other developed countries with an aggressive environmental agenda, Table 11.3, is instructive.[243] In an international context, Canadian agriculture ranks very well.

Table 11.3 Arable Land and CO_2e Emissions from Agriculture, 2016

Country	Arable Land (M acres)	GHG Emissions	Emissions (kg/acre)
CANADA*	113.6	56	0.49
Spain	30.9	37.7	1.22
France	45.0	89.0	1.98
Italy	16.8	35.0	2.08
Germany	29.4	70.0	2.38
U.K.	16.3	52.0	3.19
Ireland	2.7	18.0	6.67

* Differs slightly from Table 3.2.

Source: Basic data from the World Bank, European Commission, and Agriculture and Agri-Food Canada

11.4 CO_2E AGRICULTURE SECTOR BALANCE

The Canadian landscape consists of two major CO_2e sinks: Crown lands and agriculture. Provincial Crown lands, in particular, consisting of more than 1 billion acres (400 M ha) of land, primarily boreal forest and wilderness (ocean aside) are easily the largest CO_2 sink in Canada (Section 10.1). But the agriculture sector is number two, totaling some 159 million acres (64 M ha) (Table 3.2). Thus:

A conservative estimate of Canada's existing carbon-absorption capacity, based on land area and the global carbon-absorption average, indicates that Canada could already be absorbing 20 to 30 percent more CO_2 than we emit. Using the same calculation, the 'Big Four' polluters of China, the U.S., the European Union, and India, which together are responsible for a whopping 60 percent of global emissions, release ten times more CO_2 than their combined land area absorbs.[244]

Our very approximate estimate of carbon absorption by the existing agricultural landscape (Table 11.4) is about 52 million tonnes/year; almost identical to the estimated CO_2e generation (which includes summer fallow) of 54 million tonnes/year (Figure 11.3).

Table 11.4 Carbon Sinks in the Canadian Agricultural Landscape

Farm/Ranch Land Use	Farm/Ranch Area*	Carbon Sequestration**	TOTAL CO2e (million tonnes)
Pasture (seeded & natural)	93.4	0.2	18.7
Annual Crops	47.8	0.4	19.1
Woodlands/Wetlands	11.5	1.2	13.8
TOTAL	152.7		51.6

* Excludes summerfallow (2.2M acres), a CO2 source, and "all other".
** Tonnes of CO2e per acre per year. Author estimates are very approximate.
  CO2e = 3.67 SOC

The estimate of 54 million tonnes/year of CO_2e generated, however, still does **not** include the GHG emissions from field machinery operation, on-farm transport, heating, electricity, or for those GHG emissions attributed to machinery manufacture and agrochemical (mostly N fertilizer) manufacture and this totals, say, an additional 20 million tonnes of CO_2e per year.[245] Thus, the grand total of the CO_2e Canadian agriculture generates (**sources**) is, say, 75 mega tonnes per year—well in excess of the CO_2e Canadian agriculture currently absorbs (**sinks**).[246]

Expanding this **sink potential** can make Canadian agriculture CO_2e neutral. But perhaps, even more importantly, expanding this sink potential could be an integral part of the solution of how to cope with existing or increasing national CO_2e emissions.

11.5 ROLE OF AGRICULTURE IN CO₂E MITIGATION

It is now generally recognized that **agriculture has the greatest near-term (i.e., by 2030) GHG mitigation potential among all the economic sectors**, largely by soil organic carbon (SOC) sequestration.[247] Three principal ways to do this are:

- Land use changes
- Land management changes
- Crop mix & yield changes

11.5.1 LAND USE CHANGES

Gradual land use changes can be a very important means of increasing carbon sequestration. Very approximately (Table 11.4), woodlands and wetlands have about six times the capacity to sequester carbon compared to barren pastureland; annual crops have about twice the capacity to sequester carbon compared to barren pastureland. Planting windbreaks, developing woodlots, and reclaiming wetlands can obviously make a big difference. Reducing the farmland area, increasing agricultural yields, and actively restoring natural habitats on the land saved (called *sparing*) would achieve significant reductions in the net emissions from agriculture[248]

11.5.2 LAND MANAGEMENT CHANGES

Table 11.5 highlights the potential sequestration benefits of even modest changes to existing agricultural land use practices.[249]

Table 11.5 Land Management Change & CO_2e Savings

Land Management Change (LMC)****	Dry Prairie* SOC Mg/ha/yr***	Dry Prairie* CO2e Savings *****T/year	Parkland** SOC Mg/ha/yr***	Parkland** CO2e Savings *****T/year
IT to NT	0.04	0.15	0.06	0.22
IT to RT	0.02	0.07	0.02	0.07
RT to NT	0.02	0.07	0.03	0.11
Decrease Fallow	0.12	0.44	0.12	0.44
Increase Perennials	0.23	0.84	0.22	0.81

* Dry Prairie = Brown & Dark Brown Soil Zones.
** Parkland = Black, Grey Wooded South, and Grew Wooded Peace Soil Zones.
*** SOC = Soil Organic Carbon. Mg = million grams = 1 tonne. Twenty years.
**** IT = intensive till   NT = no till   RT = reduced till
***** CO2e = 3.67*SOC   3.67

Source: Environment Canada, *National Inventory Report, 2013*, p. 128

The benefits of less summer fallow are well-known and the widespread adoption of no-till and minimum-till technologies have already made a big difference. No-till and minimum-till are now the predominant tillage systems in Canada and, in 2015, this alone resulted in a net CO_2e sink of about 3.8 mega tonnes.[250]

11.5.3 CROP MIX & YIELD CHANGES

On cultivated land, shifting to crops with a greater root depth (called a root-to-shoot (R/S) ratio), especially canola and oats, also increases agriculture's capacity to sequester SOC (soil organic carbon). The increase in the yield and area planted to canola has already greatly increased the overall capacity of crops to act as a carbon sink (Figure 11.9). During 1990–2015, annual crops offset about 1/3rd of all the CO_2e emissions generated by the agricultural sector.[251] Other studies have also emphasized the importance of increasing crop yields to achieve reductions in GHG emissions.[252]

Figure 11.9 Carbon Sink Capacity, Selected Crops, 1971 - 2015

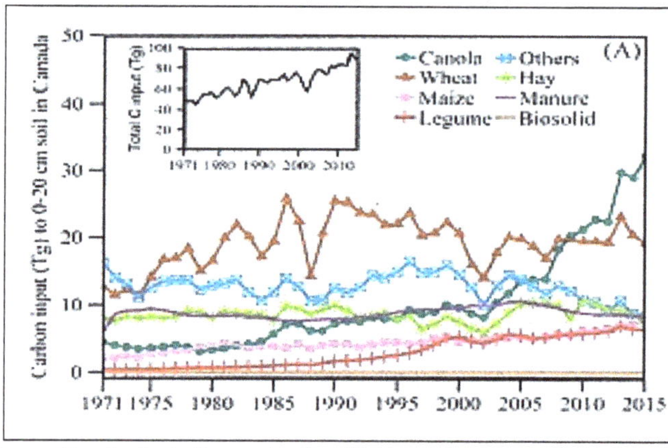

Source: Fan, J., et. al., "Increasing Crop Yields and Root Input Make Canadian Farmland a Large Carbon Sink," *Geoderma*, Vol. 336 (2019), p. 53

In short, what is required is a more sustainable system of farming principles and practices that simultaneously increase biodiversity, enrich soils, improve watersheds, and enhance ecosystem services. This is sometimes referred to as **regenerative agriculture** (Chapter 10, Section 10.5) and, if practiced on the planet's 3.6 billion tillable acres, could possibly sequester up to 40% of current man-made CO_2e emissions worldwide.[253] Set-aside land management initiatives and their rationale are discussed in greater detail in Chapter 10, Section 10.9.

Clearly, however, there are literally hundreds of potential agriculture-based mitigation measures that could be adopted, in whole or in part, to reduce CO_2e emissions (Table 11.6).[254]

Table 11.6 Potential CO_2e Mitigation Measures in Agriculture

Measure	Examples	Mitigative effects[a] CO_2	CH_4	N_2O	Net mitigation[b] (confidence) Agreement	Evidence
Cropland management	Agronomy	+		+/-	***	**
	Nutrient management	+		+	***	**
	Tillage/residue management	+		+/-	**	**
	Water management (irrigation, drainage)	+/-		+	*	*
	Rice management	+/-	+	+/-	**	**
	Agro-forestry	+		+/-	***	*
	Set-aside, land-use change	+	+	+	***	***
Grazing land management/ pasture improvement	Grazing intensity	+/-	+/-	+/-	*	*
	Increased productivity (e.g., fertilization)	+		+/-	**	*
	Nutrient management	+		+/-	**	**
	Fire management	+	+	+/-	*	*
	Species introduction (including legumes)	+		+/-	*	**
Management of organic soils	Avoid drainage of wetlands	+	-	+/-	**	**
Restoration of degraded lands	Erosion control, organic amendments, nutrient amendments	+		+/-	***	**
Livestock management	Improved feeding practices		+	+	***	***
	Specific agents and dietary additives		+		**	***
	Longer term structural and management changes and animal breeding		+	+	**	*
Manure/biosolid management	Improved storage and handling		+	+/-	***	**
	Anaerobic digestion		+	+/-	***	*
	More efficient use as nutrient source	+		+	***	**
Bio-energy	Energy crops, solid, liquid, biogas, residues	+	+/-	+/-	***	**

Source: IPCC, 2013

With respect to any of these mitigation measures, however, the central question is: who pays and who benefits? Current C0₂e levels in primary agriculture are largely determined by market-based economic incentives and prevailing technologies. This is unlikely to radically change unless or until governments fully acknowledge that greater biodiversity, healthier soils, and improved watersheds on private lands have a provincial, national, and international value. And then commit to reimbursing the custodians of these resources accordingly.

11.6 CLIMATE CHANGE & CANADIAN AGRICULTURE

With the anticipated climate change in Canada, how is this expected to impact Canadian agriculture and to what extent can the industry successfully adapt to this change?

11.6.1 CLIMATE CHANGE IMPACTS

The IPCC[255] has predicted average temperature increases of 2°C for Canada by 2050; 3° to 5°C over the next 100 years. Most scenarios of climate change in Canada also predict longer, warmer and slightly wetter (about 5%) growing seasons, accompanied by more extreme weather events.

The consensus of professional opinion seems to anticipate that this projected climate change, holding all other variables constant, should have a **net positive impact on Canadian agriculture**, at least until 2050.[256]

Figure 11.10 The Carbon-Oxygen Cycle

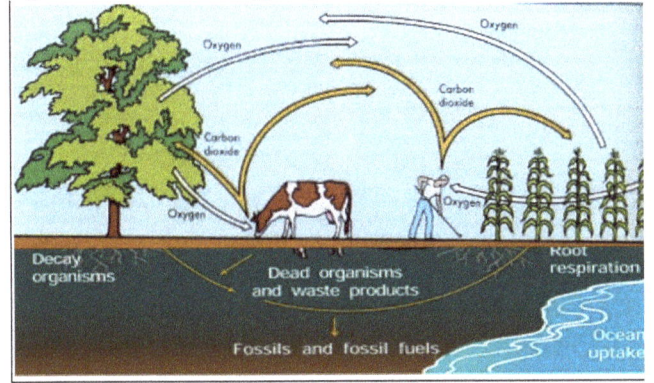

The underlying rationale for this is the expected **CO_2 fertilization effect;** the anticipated increase in the rate of photosynthesis in plants that results from increased levels of carbon dioxide in the atmosphere.[257] Since plants absorb carbon dioxide from the air, combine it with water and light, and make carbohydrates, it follows that as CO_2 levels in the atmosphere increase, the rate of photosynthesis should also increase. The effect varies depending on the plant species, the temperature, and the availability of water and nutrients. Thus, slightly warmer average temperatures with slightly more moisture, along with slightly higher CO_2 levels should increase crop yields, accelerate crop maturation rates, and possibly also allow for the growing of new crops. This is particularly true in the many parts of Canada where crop production is currently constrained by low temperatures and short growing seasons.

By 2050, a study by McGinn et al.[258] predicts positive GHG effects on canola and wheat yields on the Prairies of 21% and 124%, respectively. Yet another report suggests cereal yields could increase by 37% to 60% by mid-21st century, albeit accompanied by higher nitrous oxide emissions.[259] A more recent report by Smith et al. anticipates crop yield increases for southern Saskatchewan by 2050 of 41%–74% relative to 1961–90 yields. Very extensive USA research suggests that if atmospheric CO_2 concentration levels double (from 275 ppm to 550 ppm), crop yields would likely increase by 20 to 30 percent. Wheat, barley, canola, soybean, sugar beet, and clover (so-called C3 crops) would be particularly responsive; corn and sorghum (C4 crops) less so.[260]

Figure11.11 The Anticipated Climate Change Transposition, Alberta, 2050

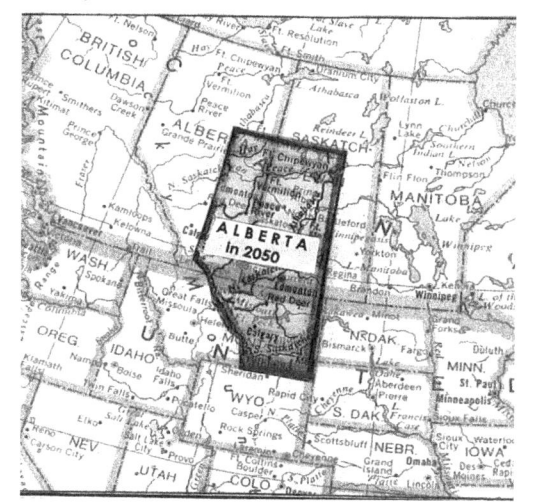

Source: *Policy Implications of Climate Change on Alberta Agriculture, AAFRD Agriculture,* AAFRD, Edmonton, 2008

With anticipated climate change, Williams[261] determined that the resulting shift in crop boundaries for wheat and barley in western Canada would probably be about 175 km. for every +-1°C change. Carter and Saarikko estimate that +1 °C would probably shift everything northward 100–130 km.[262] And various other studies[263] suggest that a 1°C increase in the mean annual temperature (MAT) would tend to advance the thermal limit of cereal cropping in the mid-latitude northern hemisphere by about 150–200 km, while also raising the altitudinal limit to arable agriculture by about 150 to 200 meters. Our own Alberta calculation is a northerly shift of 154 km for every +-1°C change.[264]

Thus, projected climate change is likely to give the Canadian prairies the weather that is now experienced in Iowa, Nebraska, and the Dakotas[265]; "Nebraska summers in Regina,"[266] and there would be similar shifts throughout Canada. Numerous other studies draw similar conclusions.[267] This transposition, for Alberta, for example, is illustrated in Figure 11.10.

At the same time, more **extreme weather events** are also anticipated and are, perhaps, the greatest source of anxiety in the agricultural community. Most Global Climate Models (GCMs) suggest a tendency toward greater late-summer aridity, less surface water and soil moisture, as well as greater variation from season to season and year to year. Using data from the US Oceanic and Atmospheric Administration, the Canadian Institute of Actuaries has found that between 1961 and 1990, extreme weather fell outside the range of normal variability only five times, but in the last ten years it happened twelve times.

Extreme events of most concern include extended droughts, violent rainfall and attendant flooding, severe hail storms, high winds and tornados, and violent temperature changes. While detailed regional projections remain illusive, the most likely agricultural regions in Canada to experience an extended drought are:

- Southwestern Ontario
- SW Manitoba, southern Saskatchewan, & SE Alberta (essentially the Palliser Triangle)
- BC's Okanagan region

One hydrological simulation anticipates a reduction in stream flows in the South Saskatchewan River Basin in twenty-five years of about 5 percent,[268] which could negatively impact, in particular, prevailing irrigation regimes.

Finally, with climate change, there could also be **more pests and disease**. Under a warmer climate with milder winters, there probably would be an increase in the over-wintering range and population density of various insects, perhaps 50% more grasshoppers. The number of weed species could also increase. This, in turn, might require the use of more pesticides, augmented further by the possible concurrent decline in pesticide efficacies. This would impose an additional ecological cost on society.

11.6.2 POTENTIAL FOR AGRICULTURE ADAPTATION[269]

Reflecting upon the incredible capacity of Canadian farmers/ranchers to adapt to all types of change, especially during the last 100 years, they will almost certainly be able to also successfully adapt to gradual climate change during the next thirty-two years (and longer). Change is a constant.

Technological change (biological/genetic, chemical, mechanical, electrical, managerial, etc.) has an internal dynamic that should automatically respond to the changing needs of a market-driven economy:

- Technological change in agriculture during most of the last century averaged between 1 and 2 percent *per year* and a continuation of this long-term trend implies a crop production doubling time of between 35 and 70 years. By 2050 (in 32 years), even a very modest annual 1% increase in yields would translate into about a 38% cumulative yield increase; an increase that would be about 2.5 times as large as our projected yield losses under the most adverse of climate change scenarios (say -15%).[270] An annual yield increase of 2% would translate into about a 88% cumulative yield increase, almost 6 times as large a the worst-case climate change scenario. During 1946–1980 output per person-hour in Canadian agriculture increased over 400 percent.[271]

- A large number of on-the-shelf technologies could be adopted within, say, 5 – 10 years, much more quickly than the impacts of climate change (however defined) are projected to become apparent. There should be time.

In terms of potential **land use (cropping pattern) adjustments**, a hypothetical regional transposition which conceptually shifts the Prairies into the northern USA (Figure 11.10) also suggests, in particular:[272]

- With a hotter-drier scenario, the production of traditional prairie cereals (particularly

HRS wheat) would move farther north. A hotter-wetter (Mediterranean) scenario might also encourage more barley production.

- In southern parts of the prairies, barley production would probably decline; wheat increase.

- In southern parts of the prairies, other more heat- and drought-resistant crops would be introduced (e.g., *millet* and *sorghum*).

- More southerly parts of the prairies could also see more and more *soybeans* and *corn*. This would follow a century-long trend.

- Other oilseeds could also become more widespread, probably more mustard, sunflower, and safflower.

- Other than chickpeas, it is unlikely that the acreage in pulses would decrease much, and this acreage might even increase.

- "Specialty crops" (e.g., sugar beets) would remain highly dependent upon irrigation and irrigation water availability.

- Hard red *winter wheat* could also eventually replace some of the existing HR spring wheat acreage. This would help avoid potential late-summer heat blasts and possibly reduce pesticide costs as well.

- Without additional irrigation, more land in the southern prairies is almost certainly going to switch to perennial crops, mainly improved grasses, forages, and rangeland. Less cultivation of ecologically fragile and already economically marginal lands on the prairies, particularly in the Palliser Triangle, is likely.

- Migration of crop production into more northerly areas of the prairies, mostly relatively unproductive gray-wooded soils, is problematic because of existing constraints. The principal existing constraints are soil quality, terrain, and distance from markets, none of which would be greatly affected by climate change.

- Economic activity and people will probably continue to gradually migrate from hotter-drier areas to cooler-wetter areas. This is also a century-long trend.

At the same time, the resources supporting agriculture should not be further depleted by farmers' various autonomous adaptation initiatives. For example, the more efficient use of fertilizer and more minimum tillage would mitigate or store emissions of nitrous oxide and CO_2. The exception is water, a resource that will become less available to agriculture whether the climate changes or not.

While the industry needs to remain vigilant with respect to its role in climate change, agriculture is not a major contributor to net GHG. Further, it can be a big part of the solution to the GHG emission problem. Advocates for the industry should be more proactive in letting others know that their industry compares well with agriculture in other countries and with many other sectors in Canada.

12. *Related Pollution Issues & Mitigation Measures*

Ever since 1962, when Rachel Carson published *Silent Spring*, a damning condemnation of chemical companies and pesticides, the eco-warriors have had a spring in their step. Primary agriculture has not entirely escaped the onslaught. Here we will briefly review the sources of agricultural pollution of the air, soil, and water, and highlight some of the actual and potential mitigative measures that are being pursued (Table 12.1).

Table 12.1 Agricultural Pollution: Sources & Impacts

Sources	Impacts		
	Air**	Soil	Water
ABIOTIC:			
Fertilizer	17%	x	x
Pesticides	x	x	x
Irrigation		x	x
ID Soil Emit.	20%		
Crop Residue	9%		
Auto Emissions	x		
BIOTIC:			
Manure	4%		x
Livestock*	50%		

* Beef, dairy, poultry, & other.
** Percentages represent the CO2 contribution.

(Table 11.2 and Figure 11.5)

12.1 SOURCES & IMPACTS[273]

We focus on six potential sources of air, soil, or water pollution in the agricultural industry:

- Synthetic (chemical) fertilizer
- Pesticides and disease
- Irrigation
- Machinery emissions & discharges
- Agricultural processing wastes

12.1.1 SYNTHETIC FERTILIZER

Since 1950, the production and utilization of chemical fertilizer has increased almost 1000% (i.e,. 10X)—about 1/3rd of it nitrogen-based. These much higher application rates have environmental consequences.

Some nitrogen-based synthetic (chemical) fertilizer escapes into the atmosphere as ammonia (NH_3) and some of this becomes nitrous oxide (N_2O) which, in terms of its CO_2 equivalent, is 310 times as bad. Anhydrous ammonia and urea cause the highest N_2O emissions, estimated in some soils to amount to about 1.7% of the N applied.[274] This translates into about 17% of all the CO_2 emissions in agriculture. (See Chapter 11, Tables 11.1 and Figure 11.4.)

Additionally, in heavily populated/industrialized areas, there is another potential negative impact. These fertilizer emissions can combine in the air with pollutants from combustion—mainly nitrogen oxides and sulfates from vehicles, power plants, and industrial processes—to create very tiny solid particulates called **aerosols**. These can then penetrate deep into the lungs, causing heart or pulmonary disease and even death (3.3 M worldwide). A global study suggests that more than one-half the aerosol ingredients in the eastern United States come from farming and in Europe and China their contribution is even greater.[275]

Perhaps an even larger concern is the impact of unused nitrogen and phosphorus fertilizers on groundwater and streams/water bodies. Since not all of nitrogen-based fertilizer is utilized by plants, the remainder accumulates in the soil, leaches into groundwater, or is lost as runoff. The nutrients, especially nitrates, in fertilizers can cause problems for natural habitats and human health (especially gastrointestinal issues) if they are washed off soil into watercourses or leached through the soil into groundwater. Natural soil nitrification and the nitrification of sewage effluent and animal wastes also contribute to this problem. Downstream, **eutrophication** can also occur in a body of water when an increase of mineral and organic nutrients reduces the dissolved oxygen content, producing an environment that favors algae blooms. This is increasingly common in mid-summer on the prairies and is especially bad where carbon and phosphorous are particularly plentiful; the result of both soil erosion and nutrient-loaded runoff.

12.1.2 PESTICIDES & DISEASE

Most pesticides serve as plant protection against weeds (herbicides), fungi (fungicides), or insects (insecticides). Agriculture-related diseases can originate in association with either plants or animals.

Pesticides:

There are numerous environmental concerns about the use of pesticides in agriculture:

- Pesticides can contaminate the soil when they persist and accumulate. This alters microbial processes, which can be toxic to soil organisms. In turn, pesticides can accumulate in animals that eat contaminated pests and soil organisms which, in turn, can be ingested by humans. The extent of this chemical residual in the soil depends upon the compound's unique chemistry.

- Pesticide leaching can also occur when the pesticides mix with water through rainfall or irrigation and move through the soil, potentially contaminating groundwater. Pesticide leaching can arise not only on treated fields but also from pesticide mixing areas and pesticide application machinery.

- Pesticides may also be harmful to beneficial insects, especially aquatic insects such as midges and mayflies which, in turn, are important in the food chain for fish and birds.

- Pesticide use, accompanied by other deleterious agricultural practices, can similarly cause a dramatic decline in flying insects, which in turn can reduce bird populations. For example, bird numbers across France have declined by 1/3rd in just the past 15 years and this is attributed, at least in part, to on-going pesticide use.[276] The decline in the bald eagle population (largely attributed to DDT) is yet another example.

- In terms of air quality, some pesticides and fumigants, such as methyl bromide, are toxic substances that can escape as gases or as particulate matter carried by the wind.

Pesticides include **herbicides**, **insecticides**, and **fungicides**. The typical annual applications of various pesticides are: field crops 1.2 kg/ha; fruit 27.9 kg/ha; and vegetables 3.6 kg/ha.[277]

The public and the scientific community in Canada have recently been particularly concerned about the potentially harmful effects of **neonics,** a widely used seed treatment.[278] Because neonics do not break down quickly or easily in the soil and because they are water soluble, they have the potential to easily run off into water courses and negatively impact aquatic insects such as midges and mayflies which can, in turn, negatively affect insect-eating fish and bird populations. A recent concern was their possible link to the collapse of bee populations.

In mid-2015, new regulatory requirements for the sale and use of neonicotinoid-treated seeds in Ontario came into effect to reduce the use of neonics on seeds and beans to 20% within two years. The new rules helped ensure that neonicotinoid-treated corn and soybean seeds were used only when there was a demonstrated pest problem. Health Canada's Pest Management Control Agency now (early 2019) proposes totally prohibiting the use of the well-known and widely used neonics (produced by Bayer and Syngenta) by 2021 for crops with alternatives available and by 2023 for uses that do not yet have viable alternatives. The European Union has already restricted their use.

Total pesticide use, as reflected in data for the USA (Figure 12.1), has remained relatively constant during the past 30-40 years.[279] With gradually increasing yields, herbicide use on a per unit-of-output basis has actually been declining, particularly since the mid-1980's (Figure 12.2).[280]

Figure 12.1 Pesticide Use in U.S. Agriculture, 1960 - 2008 (21 crops)

Source: USDA/ERS and proprietary data

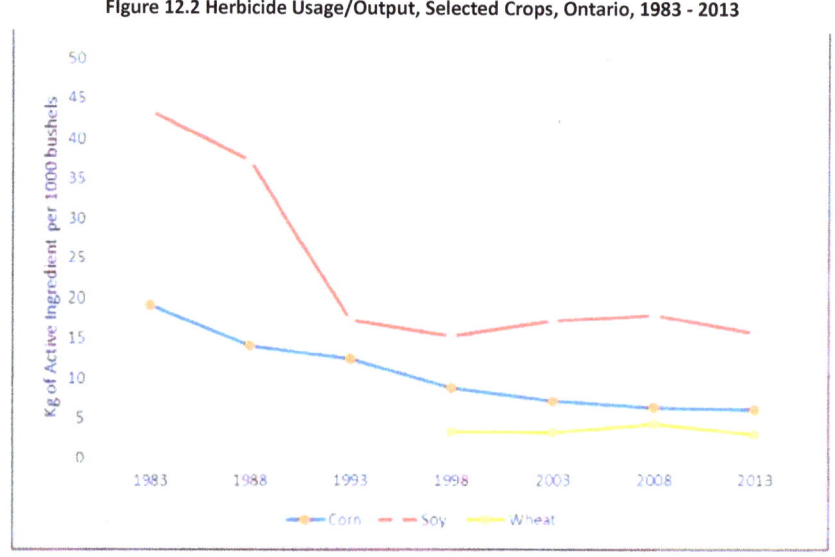

Figure 12.2 Herbicide Usage/Output, Selected Crops, Ontario, 1983 - 2013

Source: FFC Ontario, *Survey of Pesticide Use in Ontario, 2013/14,* November 2015

The US data (Figure 12.1) also indicates a gradual decline in insecticide use, offset by a fairly recent national increase in the use of herbicides. Much of this increase in total herbicide use can probably be traced to the increasingly important role of glyphosate for glyphosate-resistant GM canola, corn and soybeans. It is also increasingly being used as a desiccant to synchronize crop maturity. For weed control, a glyphosate is better than an atrazine mix at pre-emergence because there is no soil residue limiting future crop rotations.

At the same time, there is a public perception, widely held, that exposure to some pesticides can and do lead to numerous human health issues, especially various cancers, neurological problems, reproductive and fertility problems, and long-term respiratory problems. And in many cases, there is an acknowledged statistical **correlation**. But establishing **causality** (i.e., A actually causes B) and under what conditions (strength, duration, frequency, etc.) is decidedly more difficult. The Canadian Cancer Society website, for example, still says that "for most pesticides studied, research does not show a definite link with human cancer."

The widespread use of glyphosate became especially controversial in 2015 after the International Agency for Research on Cancer, a division of the World Health Organization (WHO), classified the herbicide as probably carcinogenic to humans. In fact, glyphosate has a 40-year history of safe use in 160 countries around the world and has one of the lowest environmental hazard ratings in the industry.[281] And after a rigorous eight-year re-evaluation, Health Canada and the PMRA (Pest Management Regulatory Agency) again said:

> "Glyphosate is not genotoxic [damaging DNA and causing mutations] and unlikely to pose a human cancer risk", adding

> "No pesticide regulatory authority in the world currently considers glyphosate to be a cancer risk to humans at the levels at which humans are currently exposed."[282]

Very recently (mid-2019), the US Environmental Protection Agency (EPA) was even more categorical:

> Glyphosate is not a carcinogen…EPA continues to find that there are no risks to public health when glyphosate is used in accordance with its current label.[283]

And the US National Cancer Institute also totally concurs.

Still, this has not deterred lawyers from filing class-action lawsuits, particularly against Bayer-Monsanto, makers of the very popular glyphosate Roundup©. (There are other glyphosate manufacturers.) There is at least one on-going class-action case in Saskatchewan and literally hundreds in the US, often claiming a "substantial" link between Roundup© and non-Hodgkin's lymphoma (NHL).

For decades, farmers have actually been transitioning to safer, more environmentally friendly compounds to combat pests and disease. In California, between 1998 and 2009, there was a 66% decline in the use of older, traditional broad-based chemicals in favor of more targeted, softer, and often organically approved pest and disease control alternatives. By 2009, two out of the top three pesticides utilized (sulfur and mineral oil) were already organic, and about 39% of all pesticides applied were organic. Thus, pesticide toxicity levels have also been gradually trending downward.[284]

The future should be even more environmentally friendly:
- New, less toxic chemicals will emerge. For example, a diamide-based insecticide[285] may turn out to be a more economical and more environmentally friendly alternative

compared to existing neonics. Treating canola seed with Lumiderm™ to control flea beetles and wireworms, for example, is already commonplace on the prairies.

- The nature of pesticides will also likely change. New technologies, such as RNA interference, which involves the use of compounds designed to block the action of specific genes and enzymes in various pest organisms, are being developed.

- Regenerative agriculture, which utilizes crop rotations to break insect and disease cycles, will also become increasingly important, thus reducing the need for the chemical control of pests and disease.

- The development of new non-GMO crops utilizing genomic-splicing and other genetic methodologies to make crops resistant to many pests and diseases is on-going (Section 12.2.4).

Disease:

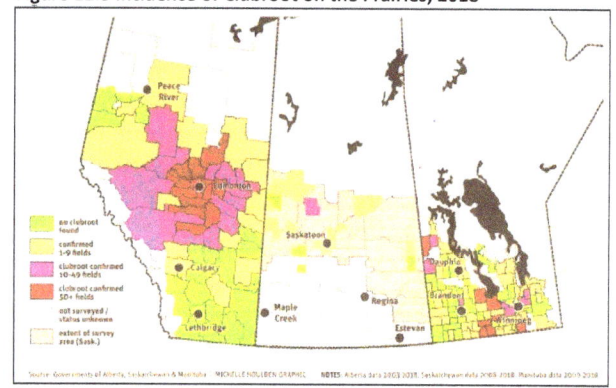

Figure 12.3 Incidence of Clubroot on the Prairies, 2018

Our Source: *Western Producer,* January 17, 2019, p. 4

Disease can also be a serious environmental issue. Clubroot, a prime example, is a microscopic soil-borne plant pathogen that can have a devastating impact on crop yield and quality. Visible signs of clubroot are galls or clubs that form on roots, stunted growth, and wilted plants. It affects canola, mustard, and other crops in the cabbage family. Cole crop vegetables, for example, broccoli, Brussels sprouts, cabbage, cauliflower, Chinese cabbage, kale, kohlrabi, radish, rutabaga and turnip, are all susceptible to clubroot, as are many cruciferous weeds; for example, wild mustard, stinkweed, and shepherd's purse. Clubroot was first identified in canola crops in Alberta in 2003, and by 2014 it was found in forty-two municipalities. More recently, there have been occurrences in Saskatchewan and Manitoba (Figure 12.3).

Clubroot is being spread mainly through soil infested with resting spores. Infested soil can be carried from field to field by farm machinery, especially tillage equipment, and can also be moved by wind and water erosion. Seed of various crops, as well as hay and straw, can also become contaminated with resting spores via dust or earth tag when they are grown in clubroot-infested fields.

Key strategies for clubroot prevention include[286]:

- Using clubroot-resistant seed varieties (note that resistant varieties are susceptible to emerging strains of clubroot)

- Practicing a four-year crop rotation
- Controlling volunteer canola and susceptible weeds
- Practicing good sanitation, specifically cleaning equipment by removing soil by pressure washing and disinfecting with a 1–2% bleach water mixture.

Landowners and other occupants such as oil and gas and utility rights-of-way, are legally required to take measures to prevent clubroot, control and destroy existing infestations, and control the spread of clubroot.[287]

There are also a multitude of animal diseases that can originate in the farming/ranching community. Animal diseases, however, are diseases that typically do not transfer to human beings and, as such, are generally not a major concern to the general public. Still, although generally not fatal to humans, there are numerous parasites that can be transmitted from domestic animals to humans, including tapeworms, roundworms, hookworms, fleas, ticks, mites, and fungal organisms. And some diseases transmitted from animals can, indeed, be deadly (e.g., rabies).

12.1.3 IRRIGATION

Although irrigation typically increases crop yields by 200–300% (with even greater increases in net farm income), salinity[288] can be a major problem in irrigated agriculture. Excess salinity reduces crop yields by at least 25 percent and, worldwide, it is not unusual for 20–30% of irrigated land to be affected by excess salinity. The estimate for Canada is lower, but still an on-going concern.

The primary cause of man-made salinization is the salt brought in with irrigation water. Through evapotranspiration, salts in the irrigation water become more concentrated in the soil and the drainage effluent, thus reducing the quality of return flows. The second most important cause is waterlogging in irrigated land. With irrigation efficiencies of, say, 60%, the remaining water must efficiently drain from the irrigation land. Yet another cause is seepage from irrigation canals.

The primary method of controlling salinity in irrigated agriculture is to permit 10–20% of the water to leach the soil and then be discharged through an effective drainage system. Canal seepage salinity can be greatly reduced by utilizing canal linings (concrete or plastic membranes) and cut-off curtains or entirely eliminated utilizing pipelines.[289]

Dryland salinity is also a major degradation problem on the Canadian prairies, but this is generally caused by groundwater redistributing salts and accumulating these on the surface. It is not man-made. Dryland salinity affects 0.65 M ha (1.6 M acres) in Alberta, 1.3 M ha (3.3 M acres) in Saskatchewan, and 0.24 M ha (0.6 million acres) in Manitoba (Chapter 10). Planting alfalfa, which has a high transpiration rate and long growing season, lowers the water table and this gradually helps make saline dryland somewhat more productive.

12.1.4 MACHINERY EMISSIONS & DISCHARGES

Hundreds of thousands of tractors, combines, swathers, haybines, and other self-propelled farm machinery emit gases from petroleum-based fuels and this equipment generally lacks the stringent emission control devices now mandatory on newer cars and trucks. Unsafe farm fuel storage facilities, spills, and discarded petroleum-based lubricants are also potential soil pollutants.

12.1.5 LIVESTOCK

Livestock emit large amounts of both methane and nitrous oxide (Figure 11.6) and, as already noted in Chapter 11, methane emissions are 21 times and nitrous oxides 310 times more damaging as greenhouse gases than CO_2. Twenty-five hundred dairy cows produce as much waste as a city of 400,000 residents. Worldwide, approximately 37% of human-induced methane production comes from livestock.[290] In Canadian agriculture, in terms of CO_2e, beef production is responsible for 34% of total emissions, dairy cattle 7%, other livestock and poultry 9%m and manure 4 percent (Table 11.2), Related issues can also arise with respect to air quality (odor) and surface and/or groundwater contamination, especially from large intensive livestock operations (e.g., beef feedlots).

12.1.6 AGRICULTURAL PROCESSING WASTES

The wastes from the processing of agricultural products can be yet another source of pollution. These include runoff or effluent from fruit and vegetable processing, cleaning of dairies, slaughtering of meat animals, oilseed crushing plants, ethanol production, corn processing (starch, etc.), sugar refining, and numerous others. The runoff from agricultural processing facilities can contain disease organisms and other infectious agents. Insects associated with agriculture can potentially transmit diseases while plant diseases can also potentially move from agriculture to urban greenery.

12.2 MITIGATIVE MEASURES[291]

Following, we highlight seven initiatives that could result in more reductions in agriculture-related pollution:

- More nitrogen-fixing plants in crop rotations
- Increased efficacy of nitrogen-based chemical fertilizers
- Biological pest control

- Genome editing
- Lower residual feed intake in beef
- Increased manure treatment
- Alternative agricultural fuels

12.2.1 NITROGEN-FIXING PLANTS

Nitrogen fixation by legumes can be in the range of 28 to 84 kg per ha (25–75 pounds per acre) of nitrogen per year in a natural ecosystem and several hundred pounds per acre per year in a cropping system. (Table 12.2)

Table 12.2 Soil Nitrogen Contribution, Legumes

Crop	lb./Acre/Yr.	$ Value/Acre*
Fababean	112-300	$66
Field Pea	60-180	$38
Lupine	60-160	$35
Alfalfa	50-150	$32
Soybean	37-150	$30
Lentil	50-120	$27
Chickpea	50-120	$27
Bean	30-70	$16

* Based on an actual N cost of $0.32/lb.
Basic Source: Entz, P., *Canola Guide,* Jan. 1995, p. 19

Perennial and forage legumes such as alfalfa, sweet clover, true clovers, and vetches can fix 280–560 kg per ha (250–500 lb per acre) of nitrogen. An herb native to North America, white prairie clover, fixes over 180 kg per ha (160 pounds per acre) of nitrogen per year. This means that these crops (after inoculation) require little or no additional nitrogen-based chemical fertilizer. Further, nitrogen is also returned to the soil via the vegetation (roots, leaves, fruit) when it is harvested and the residue decomposes. And this, in turn, also reduces the chemical fertilizer requirement of the subsequent non-leguminous crop.

Soybean Roots

Ontario and (to a lesser extent) Manitoba already grow soybeans very extensively. Saskatchewan is already a major producer of both lentils and chickpeas.

Breaking the wheat-canola rotation (neither of which can fix nitrogen), typical of much of the prairies, could greatly reduce chemical fertilizer requirements and associated CO_2e emissions. If wheat and oilseeds could also be bred to fix nitrogen (apparently complex, if not impossible), this would really be a game-changer.

12.2.2 INCREASED EFFICACY OF NITROGEN-BASED CHEMICAL FERTILIZERS

Figure 12.4 Expected Benefits of N-Stabilizers

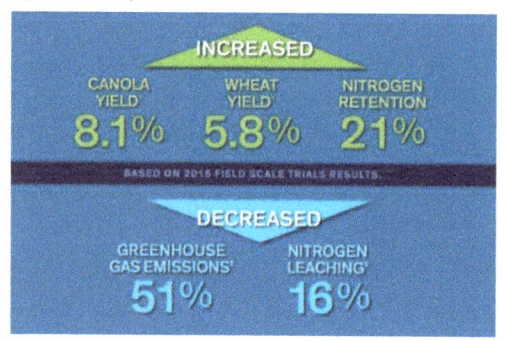

INCREASED
CANOLA YIELD 8.1%
WHEAT YIELD 5.8%
NITROGEN RETENTION 21%
BASED ON 2015 FIELD SCALE TRIALS RESULTS

DECREASED
GREENHOUSE GAS EMISSIONS 51%
NITROGEN LEACHING 16%

Source: As advertised

The major chemical companies (Chapter 6) are now promoting nitrogen stabilizers to protect against nitrogen losses when a producer applies a large portion of their nitrogen fertilizer in the fall or early spring. These stabilizers reduce the volatilization and denitrification of UAN (a solution of urea and ammonium nitrate in water), urea, and manure. By maintaining the fertilizer in ammonium form the nutrients are released more slowly, which also enhances crop yields. This also reduces leaching (Figure 12.4).

12.2.3 BIOLOGICAL PEST CONTROL

The use of biological pest control agents, or using parasites and pathogens to control agricultural pests also has the potential to reduce agricultural pollution by reducing chemical pesticide use. International development agencies (e.g., CIDA) have long advocated biological pest control in developing countries, but with mixed results.

The principal advantages are: (1) it is self-sustaining; (2) it can be cost-effective; and, (3) it is an environmentally friendly alternative. Still, it is not always easy to identify, propagate, and disseminate the appropriate pest control agent, and it may not be possible or practical to apply on extensive farming operations in Canada.

12.2.4 GENOME EDITING

Genome editing is a process that tweaks existing DNA *in situ* by adding, subtracting, or substituting a piece of DNA (called SNPs or "snips"), not unlike (but much quicker than) the natural process of mutation and conventional plant breeding.

This will allow us to develop crops that are resistant to certain herbicides, pests, or disease that are **not** genetically modified organisms (GMO), mutations that relied upon the transfer of desired characteristics from one organism to another (utilizing bacteria). New gene-editing tools such as CRISPR and GRON can now quickly tweak crops to add new traits for specific end-use qualities, protection against new disease pathogens, or better weather tolerance.

In crop production, we have already seen a 30% reduction in herbicide use after the introduction and widespread adoption of glyphosate-resistant crops (Table 13.1 following). New gene-edited crop varieties should generate the same economic and environmental benefits without the much-maligned GMO designation. (For details, see Chapter 13.)

With respect to animal agriculture, the development of swine that have a greater digestive efficiency could also both reduce the cost of production and reduce this industries' environmental footprint. The Enviropig is a genetically modified Yorkshire pig that has phytase in its saliva to allow it to break down the phosphorus (in the form of phytic acid) in feed grains such as wheat and corn. The ability of the Enviropig to digest the phosphorus from the grains eliminates the waste of that natural phosphorus (20–60% reduction) while also eliminating the need to supplement the nutrient in the feed. [292]

12.2.5 ALTERNATIVE FUELS

The major farm machinery companies (see Chapter 6) are already developing farm tractors that operate on methane rather than diesel. The cited benefits are:[293]

- 10% lower CO_2 emissions and 80% overall emissions compared to a standard diesel tractor
- Virtually zero CO_2 emissions when fueled by biomethane produced from manure or crop residues
- 30% saving in running costs with the same service intervals as its diesel equivalent
- Emission controls are not required
- 50% reduction in drive-by noise

This initiative envisions a farm that not only produces food, but also produces the material it needs to generate the energy it uses to run its operations and power its tractors and other machinery; a more self-sufficient and environmentally friendly future.

12.2.6 LOWER RESIDUAL FEED INTAKE (RFI) IN BEEF[294]

The development of a RFI beef animal uses a conventional selective breeding technique to produce progeny that can both reduce feed requirements and reduce methane emissions. This is done by repeatedly selecting bulls with the reduced feed intake trait.

Animals with the reduced feed intake trait can, reportedly, produce a reduction (gain) of 0.4% to 0.8% each year, which can then build over two to four generations to a cumulative 25% reduction in enteric (methane) emissions and a 15% reduction in manure (methane and nitrous oxide) emissions.[295] At the same time, both the cow herd and the offspring (feeders) require less feed. This fully realized genetic gain accrues incrementally over about a twenty-five-year period.

12.2.7 MANURE TREATMENT

There are three already widely employed manure treatment options:
- Anaerobic digestion – the biological treatment of liquid animal waste using bacteria in an area absent of air to promote the decomposition of organic solids. The remaining

nutrient-rich liquid can then be applied to fields as a fertilizer while the methane gas can be used for heating, generating electricity, or as an engine fuel. (See 12.2.5 above.) Aerobic digestion is superior to current anaerobic lagoons in reducing GHG emissions. This is also the best way to control the odor associated with manure management.

- Solid-liquid separation – mechanical separation into a solid and liquid portion for easier manure management. The liquids (4–8% dry matter) can be easily pumped and spread on fields while the solid fraction (15–30% dry matter) can be either spread on fields or composted.

- Composting – sourced directly or from a liquid manure separator. Passive composting (w/o churning) has lower GHG emissions due to incomplete decomposition and lower gas diffusion rates.

12.3 SECTOR METAMORPHOSIS

Canadian agriculture has constantly evolved, buffeted by the marketplace, rapidly changing technologies, government initiatives, and more. Change is engrained in the sector's DNA and it has always successfully responded to rapidly changing times. Today is no different.

The leaders of tomorrow's agriculture, however, will be different. It won't be a lethargic, cash-strapped government. And it won't be new technologies. It will be consumers, both domestic and international.

Now the consumer is increasingly demanding healthier food, safer food, alternative foods, organic foods, convenient foods, and other iterations. Now the consumer is increasingly demanding that food be produced in a more humane and ecologically sustainable manner. And now the consumer is increasingly demanding more field-to-fork transparency.

Different forms of regenerative organic agriculture (ROA) will become the norm. Agricultural technologies will gradually re-align and government support for agriculture will probably re-align under the Ministry of Health and the Ministry of the Environment. Input suppliers will adapt, processors will adapt, and food retailers will adapt. And this evolution will go hand-in-hand with private and public initiatives to successfully adapt to climate change and a more ecologically healthy, sustainable agricultural industry.

13. *Organic Crops, Functional Foods, & GMO's*

This Chapter provides a brief description and assessment of organic foods, functional foods, and genetically modified organisms (GMOs or GM'). It provides a legal definition, a brief description of the regulatory framework, an abbreviated market profile, and an objective third-party assessment of each.

13.1 CERTIFIED ORGANIC PRODUCTION[296]

What is organic food and how is it regulated? And what is its market potential and how might it evolve?

13.1.1   LEGAL DEFINITION

Organic food production in Canada is certified organic as specified in the Canadian Organic Standards: **Organic Production Systems – General Principles and Management Standards**; and **Organic Production Systems – Permitted Substances Lists**.

Organic production is a holistic system designed to optimize the productivity and fitness of diverse communities within the agri-ecosystem, including soil organisms, plants, livestock, and people. The use of molecular DNA manipulation (i.e., GMOs) in the production of organic food is prohibited. Pesticides are allowed so long as they are not synthetic.

Processed organic food usually contains only organic ingredients. If non-organic ingredients are present, at least 95% of the food's total plant and animal ingredients must be organic. Foods claiming to be organic must be free of artificial food additives and must not involve chemical ripening, food irradiation, or genetically modified ingredients.[297]

Organic meat certification dictates that the animals only be fed certified organic food that contains no animal by-products, receive no growth hormones or antibiotics, and must spend as much time outdoors as is possible and reasonable.

13.1.2 REGULATION

Any agricultural product that is labeled organic (including food for human consumption, livestock feed, and seeds) sold in Canada (including imports) is regulated by the Canadian Food Inspection Agency (CFIA). The CFIA oversees, monitors, and enforces the requirements of the Canada Organic Regime using a third-party that includes conformity verification bodies, certification bodies, and organic operators. The federal regulations apply to products that are exported, sold across provincial borders or that carry the Canada Organic logo.

Products grown and sold as organic within the same province and which do not carry the Canada Organic logo are only subject to provincial regulation. Six provinces have organic regulations, but this list does not yet include Ontario and Saskatchewan. Ontario is the biggest consumer of organic goods in Canada, and Saskatchewan is the largest producer of organic products in Canada.

13.1.3 MARKET PROFILE

Organic food sales are becoming increasingly popular and continue to outpace sales of non-organic food. In the United States, organic products account for about 4% of total food and beverage sales and 12% of all fruit and vegetable sales (2012); about 5–10% of the total food market. Organic food processors are rapidly being acquired by multinational companies (e.g., in 2017 e-commerce giant Amazon.com bought US-based Whole Foods Market Inc. for US$ 13.7 billion).

In Canada, by 2012, organic food sales reached some $3 billion, about 1.7 percent of the total food and beverage market. By 2017, organics accounted for approximately 2.6 percent of Canadian grocery sales for food and beverages.[298] British Columbia purchases a disproportionate share of organic food products.

By 2017, on the prairies, there were 1,605 organic producers farming some 1.8 million acres.[299] In 2011, organic corn was already produced by 1.1% of all corn farmers while sweet corn was produced by 6.5% of all sweet corn producers in Canada.[300] In 2016, organic milk represented 1.3% of total milk production[301] (Figure 13.1).

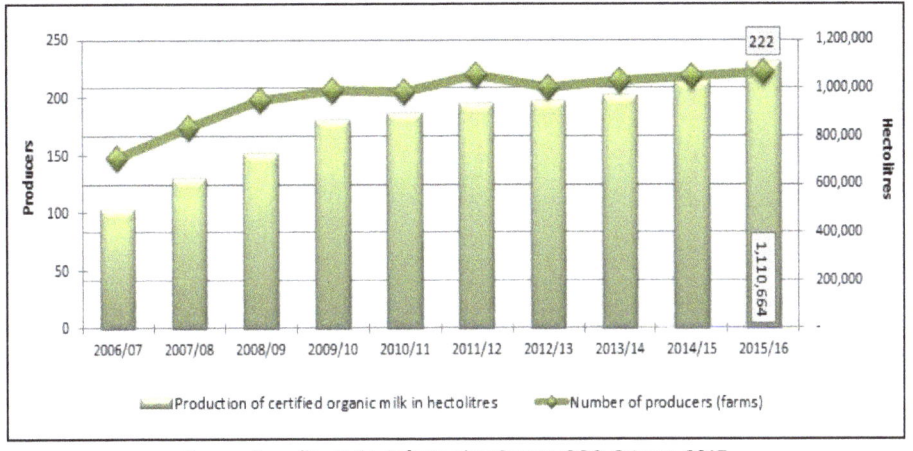

Figure 13.1 Production of Certified Organic Milk & Number of Producers, Canada

Source: Canadian Dairy Information Centre, GOC, Ottawa, 2017

At the farm level, there is a strong price incentive to switch to organic production. Some 2018 price differentials on the Prairies are indicative:[302]

Table 13.1 Organic vs Conventional, Relative Prices, 2018

Item	Organic	Conventional	% Premium
Wheat	$17/bushel	$6.50/bushel	260%
Flax	$36/bushel	$13/bushel	280%
Lentils	$0.84/lb.	$.20/lb.	420%
Oats	$5.75/bushel	$2.85 bushel	200%
Milk			197%

Sources: *Western Producer* (11/18) and *Washington Post* (5/17)

At the retail level, organic food products typically sell at a 30–40% price premium. Organic milk, for example, typically has a retail price premium of 9%–53%, depending upon the country.[303]

It is, however, still difficult to generalize about the relative (private) profitability of organic versus conventional productions systems.[304] It takes as little as 15 months (but up to 36 months), free of prohibited substances, to transition from conventional farming to organic farming. (Also see Chapter 10, Section 10.5.)

13.1.4 ASSESSMENT

Public perception and scientific analysis of organic versus non-organic foods are generally not consistent. Organic Alberta, for example, advertises the estimated nutritional benefits of some organically grown vegetables (Table 13.2). And some consumers are adamant that organic food contain less of the compounds that adversely affect their health. Yet numerous meta-analyses[305] of various food properties (nutrients, vitamins, pesticide residues, bacteria) have generally failed to identify statistically significant differences between organic and non-organic foods.

The health risk reduction of organic versus conventional foods is, therefore, debatable because both contain pesticide residues, but both are well below government established guidelines. It is generally acknowledged, however, that the perceived taste can be different.

Table 13. 2 Organic versus Conventional, Mineral Content, Selected Vegetables

Type of Soil Management	Minerals in milliequivalents						
	Calcium	Magnesium	Potassium	Sodium	Manganese	Iron	Copper
Snap Beans							
Organic	40.5	60.0	99.7	8.6	60.0	227.0	69.0
Conventional	15.5	14.8	29.1	0.0	2.0	10.0	3.0
Cabbage							
Organic	60.0	43.6	148.3	20.4	13.0	94.0	48.0
Conventional	17.5	15.6	53.7	0.8	2.0	20.0	0.4
Lettuce							
Organic	71.0	49.3	176.5	12.2	169.0	516.0	60.0
Conventional	16.0	13.1	53.7	0.0	1.0	1.0	3.0
Tomatoes							
Organic	23.0	59.2	148.3	6.5	68.0	1938.0	53.0
Conventional	4.5	4.5	58.6	0.0	1.0	1.0	0.0
Spinach							
Organic	96.9	293.9	257.0	69.5	117.0	1584.0	0.0
Conventional	47.5	46.9	84.0	0.8	1.0	19.0	0.5

Source: Research conducted by F. E. Bear, Rutgers University, 1995. (Reproduced in *Ag Advisor,* July 1, 2018)

Although all food products have to adhere to government established guidelines, some food products have higher pesticide residues than others. The worst products according to one source are: apples, peaches, nectarines, strawberries, grapes, celery, spinach, sweet bell peppers, cucumbers, and cherry tomatoes.[306]

Finally, food labeling can also be deceptive. For example, because of strict withdrawal rules, all milk and meat are essentially antibiotic free. Similarly, since all animals have hormones in them, meat and milk will never be hormone-free. Conventional pork and poultry production are free of added hormones, but beef cattle are often given a hormone implant to stimulate growth.

Readers are encouraged to consult the voluminous scientific research on organic vs non-organic foods to make their own determination. Various CFIA websites and organic food market reports (see website) are also helpful.

13.2 FUNCTIONAL FOODS[307]

Another rapidly growing food market is in functional foods. What are they and how are they regulated? And what does their future look like.

13.2.1 DEFINITION

In Canada, the terms "functional food" and "nutraceutical" (or natural health product) have no legal differentiation. A nutraceutical can either be marketed as a food or as a drug. Agriculture and Agri-Food Canada makes the following distinction:

- **Functional foods** are foods enhanced with bioactive ingredients and which have demonstrated health benefits, such as probiotic yogurt, or breads and pasta with added pea fiber.
- **Natural health products** (nutraceuticals) are extracts derived from natural sources and which have demonstrated health benefits, such as omega-3 capsules or beta-glucan supplements. These are generally sold in medicinal forms and have a demonstrated physiological benefit, a potentially positive effect on health beyond basic nutrition, or provide protection against chronic disease.

13.2.2 REGULATION

Responsibility for the regulation of functional foods and natural health products is divided between Health Canada's Food Directorate and the Natural Health Products Directorate. The Canadian Food Inspection Agency (CFIA) monitors and enforces the safety of functional foods and nutraceuticals.

13.2.3 MARKET PROFILE

More than 750 Canadian companies specialize in functional foods and natural health products, with 2011 sales in excess of $11 billion. The worldwide market for functional foods in 2018 was about US $161.5 billion and this is expected to almost double to US $275.7 billion by 2025; an implied growth rate of about 10% per year.[308] Functional ingredients such as probiotics, which help fight bad gut bacteria, are increasingly popular and probiotic yogurts are the most widely consumed products with a proven health claim.

13.2.4 ASSESSMENT

Claims linking the consumption of functional foods or food ingredients with health outcomes require sound scientific evidence and significant scientific agreement. Some functional food classes, sources, and purported health benefits are indicated in Table 13.3. We obviously cannot attest to the veracity of the purported health benefits of specific products.

Functional foods are basically any food that enhances your physiological well-being while nutraceuticals are essentially the concentrated form of the active ingredient in functional foods. Presumably, this could include foods with an organic or GMO designation.

Numerous websites are easily accessed and readers are encouraged to make their own determination.[309] The **Journal of Nutrition** and the Cision PR Newswire are both helpful websites.

Table 13.3 Examples of Common Functional Components, Sources, and Potential Health Benefits

Class/Components	Source	Potential Benefit
Carotenoids Beta-carotene	Carrots, pumpkin, sweet potato, cantaloupe	Neutralizes free radicals that may damage cells; bolsters cellular antioxidant defenses
Dietary Fiber Insoluble fiber	Wheat bran, corn bran, fruit skins	May contribute to maintenance of a healthy digestive tract; may reduce some cancers
Fatty Acids PUFAs—omega-3 fatty acids—DHA/EPA	Salmon, tuna, marine and other fish oils	May reduce risk of coronary heart disease (CHD); may contribute to maintenance of mental and visual function
Flavonoids Anthocyanins—cyanidin, delphinidin, malvidin	Berries, cherries, red grapes	Bolsters cellular antioxidant defenses; may contribute to maintenance of brain function
Isothiocyanates Sulforaphane	Cauliflower, broccoli, broccoli sprouts, cabbage, kale, horseradish	May enhance detoxification of undesirable compounds; bolsters cellular antioxidant defenses
Phenolic Acids Caffeic acid, ferulic acid	Apples, pears, citrus fruits, some vegetables, coffee	May bolster cellular antioxidant defenses; may contribute to maintenance of healthy vision and heart health
Plant Stanols/Sterols Free stanols/sterols	Corn, soy, wheat, wood oils, fortified foods and beverages	May reduce risk of CHD
Prebiotics Inulin, fructo-oligosaccharides (FOS), polydextrose	Whole grains, onions, some fruits, garlic, honey, leeks, fortified foods and beverages	May improve gastrointestinal health; may improve calcium absorption
Probiotics Yeast, Lactobacilli, Bifidobacteria, and other specific strains of beneficial bacteria	Certain yogurts and other cultured dairy and non-dairy applications	May improve gastrointestinal health and systemic immunity; benefits are strain-specific
Phytoestrogens Isoflavones—daidzein, genistein	Soybeans and soy-based foods	May contribute to maintenance of bone health, healthy brain and immune function; for women, may contribute to maintenance of menopausal health
Soy Protein Soy protein	Soybeans and soy-based foods	May reduce risk of CHD
Vitamins C	Guava, sweet red/green pepper, kiwi, citrus fruit, strawberries	Neutralizes free radicals that may damage cells; may contribute to maintenance of bone health and immune function

Source: Adapted from the International Food Information Council

13.3 GENETICALLY MODIFIED ORGANISMS (GMO'S)[310]

The research, development, and field application of GMOs in North America and elsewhere has had a somewhat checkered history. What are GMO's, how are they regulated in Canada, to what extent are they being embraced by the marketplace, and what is their probable future?

13.3.1 DEFINITION

The traditional process of selective breeding in which organisms with desired traits/DNA were retained (>12,000 years) is a precursor to the modern concept of genetic modification (GM).

Genetic engineering is a form of biotechnology where specific genes are added or removed from an organism to change its genetic makeup.[311] Genetically modified foods are foods produced from organisms that have had changes introduced into their DNA using the methods of genetic modification as opposed to traditional cross-breeding (mating or natural recombination). The most common modification is to add one or more genes to an organism's genome. This relies upon the transfer of desired characteristics (utilizing bacteria) from one organism to another. The result is a genetically modified (GM) product or a genetically modified organism (GMO).

13.3.2 REGULATION

In 1983, the Canadian government established the Federal Regulatory Framework for Biotechnology. The GMO approval process is based on numerous regulations that are enforced by Health Canada for "novel" foods, the Canadian Food Inspection Agency (CFIA) for seeds and livestock feed, and Environment Canada for "new substances intended for environmental release." The Canadian consumer may not know if there are GM ingredients in a food, since the mandatory labeling of foods derived through biotechnology is only required if there are significant composition differences or documented health impacts.

The Canadian regulatory process focuses on the product rather than the process. The focus is on the traits expressed in products and not on the method used to introduce those traits. The approach applies to both traditional breeding and genetic engineering.

13.3.3 MARKET PROFILE

A profile of the various GM crops that are already in commercial production is provided in Table 13.4. The level of market penetration is only indicative of farmer acceptance, not consumer acceptance.

Table 13.4 A Profile of Selected GMO Food Products

Food Product	GMO Characteristic	Market Penetration**
Vegetable Oils	The oil content of canola was modified (1995) and subsequently* made glyphosate-resistent* for weed control. Also developed BT corn (maize) with glyphosate-resistance for weed control, as well as glyphosate-tolerant soybeans that produce healthier oils. Most rancid-resistant vegetable oil used in the US and Canada is now produced from GM crops canola, corn, cotton, and soybeans. It is sold directly to consumers as cooking oil, shortening, and margarine. Corn also produces starches and syrups. GM meal is a protein supplement for animals.	Canola = 93% US Soybeans = 94% US Corn = 92%
Sugar Beets (2005)	Glyphosate-resistance for weed control.	
Cotton	Developed to be immune to *Bacillus thuringiensis* (Bt). Named Bt cotton. Also made resistent to the herbicide bromoxynil for weed control.	US Cotton = 94%
Rice (2000)	**Golden rice** is a variety of rice produced through genetic engineering to biosynthesize beta-carotene, a precursor of vitamin A, in the edible parts of rice. It is intended to produce a fortified food to be grown and consumed in areas with a shortage of dietary vitamin A.	
Tomato (1994)	Delayed ripening for a longer shelf life.	
Potato	Developed to be immune to *Bt* and, later, late blight. Also a GM potato that does not easily bruise and has better fry/protein characteristics.	
Squash	Virus resistance.	
Apple (2015)	Gene silencing to prevent the fruit from browning.	
Papaya	Developed to resist ringspot virus.	Hawaii Papaya = 80%
Zucchini	Genetically modified to resist three viruses.	US Zucchini = 13%
Salmon (2015)	Added a growth hormone regulating gene so as to grow year-around.	

* Dominated by the Roundup Ready © brand.   **USA estimates.

Basic Source: Wikipedia

By 2010, twenty-nine countries had already planted commercialized GM crops and a further thirty-one countries had granted regulatory approval for transgenic crops to be imported. The USA is the leading country in the production of GM foods, with 25 GM crops having received regulatory approval. Canada is the third largest producer of GM crops and in 2017 planted 32.4 million acres of genetically modified crops: canola 21.8 M acres; soybeans 6.2 M acres; and corn 4.4 M acres. This is essentially all of the land devoted to these three crops and about 30% of all the arable land in Canada.[312] In terms of their total arable land base, the United States and Brazil lead the way with 50 and 60 percent, respectively (Figure 13.2). Canada is the largest producer of GM canola seed in the world.

Figure 13.2 Area Planted to GM Crops

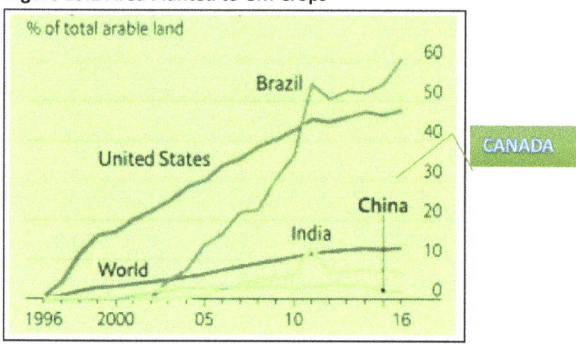

Sources: International Service for the Acquisition of Agri-Biotech Applications; FAO, StatsCanada

Some countries restrict GM production and GM imports. In response to consumer anxiety over the possible implications of transferring DNA material from one organism to another, the original GMO methodology, the European Union still does not allow either the production or import of GM crops.

13.3.4   ASSESSMENT

There seems to be a scientific consensus that currently available foods derived from GM crops pose no greater risk to human health than conventional (non-GM) food. Scientists with individual expertise in molecular biology, toxicology, chemistry, nutritional sciences, and microbiology assess every potential GM food prior to its potential registration. On this basis, the Canadian government generally views transgenic organisms as not very different from non-GM food and crops.

The financial benefits to producers as well as improvements to human health and the environment, both nationally and internationally, are decidedly positive[313]:

- The net benefit to Canadian canola producers is about $350–$400 million/year
- With Canadian canola production, there are major reductions in pesticide use (35%), soil tillage, soil erosion, fossil fuel use, and greenhouse gas emissions
- In India, the introduction of GM cotton has increased poor farmer incomes by 134%, reduced pesticide use to control insects by 41%, and reduced pesticide application poisoning by 21%
- In China, GM cotton increased farmers annual income by $200 per acre and reduced pesticide use to control insects by 90%
- In Burkina Faso, GM cotton farmers receive $30 per acre more than non-GM cotton

farmers while, at the same time, reducing pesticide application poisoning by 30,000 cases per year

- Golden rice, with enhanced vitamin A characteristics, has also improved human health throughout Asia

In this context, it is difficult to disagree with Calestous Juma, an advocate for innovation in Africa, who argued throughout his life that GM seeds that were resistant to drought and pests were just what Africa needed.[314] Perhaps most importantly, people throughout the world have now been consuming GM crops for a quarter century without any ill effects.

Nonetheless, opposing viewpoints are held by many, and members of the public are much less likely than the general scientific community to perceive GM foods as safe. Readers are encouraged to consult the voluminous scientific research on GMOs (some 151 references in *Wikipedia* alone) to make their own determination.[315]

The related issue regarding the safety of glyphosate (Roundup©, other) herbicide, which is used for weed control in many GM crops is addressed in Section 12.1.2.

13.4 IMPLICATIONS FOR FARMERS

The consumer will ultimately rule. If **organic** and **functional** foods are perceived to be healthier, irrespective of any scientific evidence to the contrary, the demand for these products will continue to climb more quickly than conventional food products. Some consumers may also reason that if organic and functional foods are more expensive, they must also be better. This may open up market opportunities for farmers who wish to take advantage of markets that provide, at least for a time, improved pricing margins.

Similarly, if the consumer perceives **GMOs** to be riskier, irrespective of any scientific evidence to the contrary, the demand for products with a GMO attribute will weaken. This will, however, lead to the development of crops with essentially the same desired attributes utilizing alternative genomic methodologies. The most promising is **genome editing**, a process which tweaks existing DNA in situ by adding, subtracting, or substituting a piece of DNA (called SNPs or "snips"), and basically resembles the natural process of mutation and conventional plant breeding.

14. The Future Of Agriculture In Canada[316]

No one knows what the future will really look like, especially a lowly economist. As John Kenneth Galbraith (a noted economist) once dryly observed, "The only function of economic forecasting is to make astrology look respectable." Reasonably accurate predictions about next year or even the next five years are possible (however unlikely) but predictions about the next 10–20 years are little more than professional "best guesses." And, clearly, extrapolations into the far-future (say 30–50 years) are little more that drug-induced hallucinations.

Our imagination is earth-bound; constrained by prevailing social, economic, environmental and political parameters. Nevertheless, we will fearlessly try to identify the mega-variables that we believe will drive continued change in Canadian agriculture in the challenging years ahead.

14.1 THE MEGA-VARIABLES

The identified mega-variables are six:

- Internationalization
- Consumer-Driven
- Regulation
- Technological Change
- Water Availability
- Structure of Agri-Industry

We believe that these six mega-variables will largely determine the structure, conduct, and performance of primary agriculture in the next 10 to 20 years.

14.1.1   INTERNATIONALIZATION

There are numerous dimensions to this: a) population growth; b) international trade agreements; c) off-shore sourcing of agricultural inputs; d) internationalization of R&D;

e) globalization of financial markets; and f) increasing food self-sufficiency in "developing" countries.

Between now and 2050, the planet's population is forecast by most global agencies to increase by almost one-third; from 7.3 billion people to 9.7 billion people. This will mean agricultural production will have to increase very considerably, perhaps by as much as 70 percent (FAO, 2009). The negative consequences of projected climate change on agriculture production in equatorial and other drought-prone areas, could put additional strain on agricultural production in more northerly latitudes, including Canada. Despite current protests against unrestrained "globalization," it will ultimately prevail because it must.

Canadian agriculture may develop a comparative advantage because it has a relatively benign production system versus all other export-oriented countries. Canada uses fewer pesticides, less tillage, and has a rapidly diminishing carbon footprint on a comparable competitive basis. As our agricultural industry continues to develop high-quality niches, rather than standard (generic) commodities, differentiation on the basis of the production environment, carbon footprint, animal care, and other perceived preferences of end users will be a driving market force.

The umbrella trade agreement, the General Agreement on Tariffs and Trade (GATT) is really a trade rule-book. And consistent with GATT, we belong to a growing number of multinational and bilateral trade agreements (CUSMA, CPTPP, CETA, etc.). All of these agreements impact our competitive position and establish trade parameters. Trade-related matters will continue to challenge all jurisdictions and many historic government functions and roles will be replaced by global protocols, the responsibility of non-government organizations.

The off-shore sourcing of agricultural inputs is also increasingly widespread. We produce our own potash and most of our own nitrogen, but virtually all of our phosphorus fertilizer is imported. All the major chemical companies have a foreign address. And most farm machinery is manufactured everywhere. Virtually all <100 hp tractors are manufactured outside of North America. This will continue.

At the same time, research and development will continue to be internationalized. This is driven by both the multinational input providers and by international research organizations. The Consultative Group for International Agricultural Research (CGIAR), for example, has fifteen research centers around the world, which collaborate with hundreds of partners, including national and regional research institutes, civil society organizations, academia, development organizations, and the private sector.

Additionally, the agricultural sector will continue to be impacted by an increasingly seamless international financial market. Fewer and fewer countries have a pegged (artificial) exchange rate. Money quickly moves to countries with the most buoyant economy and the most advantageous interest rate and investment climate. Most major banks and credit institutions are already international in scope.

Finally, all of these trends will coalesce to generate a synergy that will make more and more countries strive toward increasing self-sufficient in food production, especially in Latin America and SE Asia. Some historically food-deficient regions will even become large food exporters (e.g., Russia-Ukraine). For generic food exports (e.g., wheat), the competition will become increasingly fierce. Value-added products and niche (segmented) markets will most likely continue to grow more rapidly than (generic) commodity and intermediate product trade, as consumers force transparency and new values/benefits in what they consume.

All of this will be greatly facilitated by lower (unit) costs of transport, and sophisticated logistics and communication systems.

14.1.2 CONSUMER-DRIVEN

Increasingly, health and environmentally conscious consumers are now beginning to set the agenda and this will continue unabated. Tomorrow's consumers will not only want food products that have the desired nutritional and health characteristics, but they will also be increasingly concerned about the environmental impacts of production, convenience, flexibility, and variety. Consider the menus and the food now consumed versus two or three decades ago. Truly an international and ethnically diverse food system for all. Worldwide, consumers will increasingly demand natural, wholesome food products which, in turn, will increasingly impact on such issues as animal welfare, production practices, and biotechnology. The market for **organic foods**, **functional foods,** and **nutraceuticals** will continue to grow and the aversion to so-called genetically modified (GM) foods will persist. An aging population, and a more intense focus on healthier more diverse diets simply reinforces this trend.

At the same time, environmental issues will become increasingly important to more and more people. This is largely a function of relative income levels. Environmental sustainability will be increasingly paramount. And this, in turn, will also impact such issues as animal welfare, production practices (including agricultural pollution and resource degradation), response to climate change, biotechnology, and soon on the share of fresh water resources. This could also generate more complex trade arrangements, as well as an accelerated demand for market-differentiated food products which are clean, wholesome and safe.

By addressing these consumer concerns, emerging technologies will increasingly find a way into agricultural and food systems through real-time sensing, tracking, measuring quality changes en route to consumers, and creating instant feedback loops from consumers to farmers. Transparency of the system will be as easy as swiping built-in bar codes. Blockchain as part of the agricultural food chain should be welcomed by all.

14.1.3 REGULATION

Often overlooked by the futurists is the almost inevitable rise of more regulation over time, whether government-imposed or simply quality characteristics demanded by the food chain.

Partially in response to the renewed consumer focus, more and more regulation can be expected with respect to:

- Farm safety – farm labor working conditions (hours, Workmen's Compensation, holiday pay, seat belts, clothing [e.g., masks and impervious clothing to protect against toxic chemicals], etc.)

- Food safety – tracking from farm to fork, stricter guidelines for animal and plant GMO development, etc.

- On-farm environmental dictates with respect to both groundwater and surface water maintenance, species maintenance, tree cover, petroleum and chemical storage and disposal, etc.

- Increased restrictions and protocols on operator qualifications (e.g., strict operator licensing for sprayer operations and who can access more advanced inputs).

The resulting regulatory regime will, no doubt, have the widespread support of the general public. But, for farmers and ranchers, each and every additional regulation will almost invariably increase their unit cost of production, thus making them less competitive in the international marketplace if competing countries lack a similar regulatory framework.

14.1.4 TECHNOLOGICAL CHANGE

Technological change will continue to be a powerful presence in our lives. It is organic and incessant. Dramatic technological change in Canadian agriculture and elsewhere will continue, particularly with respect to:

a) Precision agriculture

b) Sensors

c) Robotics

d) Genomics

e) Closed ecological systems

f) Alternative food sources

g) Alternative food uses

h) Real-time, high-resolution data

a. Precision Agriculture – GPS & AI

Tractors and other machinery with Global Positioning System (GPS) technology can now be located within a few centimeters anywhere on Earth. This eliminates overlaps or misses, which both reduce fuel bills (by as much as 40%) and improve the uniformity and effectiveness of fertilizers, herbicides, and pesticides.[317] Tractors outfitted with GPS receivers and pulling heavily monitored air-seeders will increasingly drive themselves up and down fields on precise tracks, dispensing just the right amount of all inputs throughout the field.

Hands off the steering wheel, the farmer only needs to watch, simply de-activating the GPS guidance at the end of the field, turning, and re-locking the receiver to make another pass. Combines, programmed for a specified width and height, are increasingly recording information on grain moisture, crop yields, and more, and then automatically referencing that information to GPS coordinates, which, in turn, can be linked to subsequent seeding and spraying operations.

Artificial Intelligence (AI) will also become increasingly pervasive. Farm machinery that has the ability to instantly adapt to changing conditions will be widely adopted. Machinery "reasoning," linked to GPS receivers, is quickly being developed by all the major machinery companies and is not too far away (e.g., a combine that makes adjustments on the go [which people can still override] to adapt to field variability).

b. **Sensors**

The use of multi-spectral sensors in agriculture will become ubiquitous. Successful robotic applications in agriculture largely depend on multi-spectral cameras. Some impending applications:

- Multi-spectral sensors on sprayer booms to determine nutrient needs and apply the appropriate liquid nutrients: nitrogen (N), phosphorus (P), potassium (K), sulfur (S), etc. Airborne multi-spectral cameras will also likely be adopted for nutrient and pest mapping.
- Relatively inexpensive drones or mini-satellites equipped with multi-spectral sensors will facilitate individual weed control, especially in row crops. The future potential use of lasers or microwaves would have the advantage of also making the cultivated crop "organic."
- Multi-spectral cameras will also be increasingly adapted to facilitate the robotic harvesting of fruits and vegetables, discerning when a fruit or vegetable is ready for plucking.

Sensing technologies in real time to measure vital nutrient, microbial, and plant growth factors will strongly influence crop choices and develop data to enable crop rotations to be symbiotic. Regenerative agriculture will have sensing and tracking technologies to clearly account for beneficial practices. Regenerative agriculture, which includes energy self-sufficiency and vegetation to guarantee a negative carbon footprint, will be transformative.

Electronic monitoring to track livestock well-being and grain storage conditions is already in widespread use. Monitoring the nutrition, disease, breeding, and housing conditions and stress levels of every animal in the herd (dairy, beef, and swine) will become commonplace. Digital IP cow cameras can save calves. Ubiquitous rumen monitors facilitate automated ration changes to promote growth and minimize methane emissions. Attached monitors track the health of each individual animal. Imbedded monitors can indicate estrogen levels and signal when animals should be bred. In crop production, inexpensive monitors will record location-specific temperature, moisture, and other grain characteristics, all forwarded in real time to monitoring locations almost anywhere in the world.

c. Robotics

Dozens of gigantic hauler trucks already operate remotely in the Canadian oil sands. Led by Tesla, Apple, Uber, Google and various other car companies, driverless cars are currently being tested on circuitous urban roads. Can driverless farm equipment be far behind?

Agricultural robots (agbots) will increasingly be used to automate irrigation, planting, weeding, soil maintenance, and fruit picking, among others. Robotic weeders (especially on row crops), robotic crop pickers (e.g., strawberries), and autonomous power units to replace people-driven tractors will soon enter commercial production.[318]

Robotics for repetitive farming operations are also likely to become increasingly commonplace. Robotic milking machines are already in widespread use.

Successful robotic **AI** applications will typically depend upon both **GPS** and numerous **sensors**.[319]

d. Genomics

There is little doubt that biotech research will continue to have a profound and worldwide impact on the agricultural sector. The potentially most transformational will arise through **genome editing,** which basically resembles the natural process of mutation and conventional plant breeding. Some possibilities:

- Develop crops that are resistant to certain herbicides, pests, or disease that are not genetically modified organisms (GMO), mutations that relied upon the transfer of desired characteristics from one organism to another (utilizing bacteria). The best known of these technologies is CRISPR-Cas9, which can turn an organism's own genes on and off with the potential to elicit almost any trait desired. This would make existing GMOs (especially GMO canola, corn, soybeans, and cotton) obsolete.
- Develop crops with nutrient enhancements (called **biofortification**)[320]:
    - Rice, beans, sweet potato, cassava, and legumes with enhanced iron
    - Sweet potato, corn, and cassava with higher levels of vitamin A
    - Wheat, rice, beans, sweet potato, and corn with an enhanced zinc content
    - Sorghum and cassava with amino acid and protein biofortification
- Develop crops that are more resistant to drought, heat, cold, and salt.
- Develop crops that no longer self-pollinate (e.g., wheat), thus facilitating hybridization.
- Develop more crops (especially basic food crops like wheat and rice) that can produce their own nitrogen requirements (like legumes). An alternative

technology, the development of more microbe-based products (innoculants) to duplicate nitrogen-fixing legumes is also likely.
- Develop enhanced photosynthetic pathways to greatly increase yields and efficiencies of nutrients, water, and microbial action in the soil. An improved photosynthetic process will create two to five times the energy presently grown and stored in plant material.
- Develop animals, including fish, that grow more rapidly and avoid diseases, thus minimizing the use of hormones, antibiotics, and the like (e.g., growth hormones for beef and hormones to stimulate milk production).

e. Alternative Ecological Systems

We expect three alternative ecological systems to be increasingly scaled up and play an increasingly important role in global food production: **vertical farming, hydroponics,** and **inland fisheries**.

Vertical farming in urban areas is similar to a typical greenhouse environment and is generally not an entirely closed ecological system. Protected from adverse weather conditions, it allows for the production of crops (especially vegetables) all year round. This also more easily provides the consumer with more control over his/her crop species selection, nutrition, and safety (i.e., consumer autonomy). It has the added advantage of reducing the cost of transporting food from rural areas to urban centers (i.e., it can be grown nearby to where it is consumed). More underground farming, where sensors control everything—temperature, humidity, fertilization and LED-light illumination at needed wavelengths in the light spectrum—is also possible.

Hydroponics is a method of growing plants without soil by using mineral nutrient solutions (from manures or chemical fertilizers) in a water solvent, with or without a supportive perlite or gravel. Canada already has hundreds of acres of large-scale commercial hydroponic greenhouses, producing tomatoes, onions, lettuce, radishes, peppers, and cucumbers. As of 2018 and beyond, marijuana production may represent yet another important hydroponic opportunity.

Both vertical farming and hydroponics have the advantage of minimizing water use and ensuring contamination-free water in production.

Inland aquaculture, a completely closed ecological system where there is no reliance on outside matter, will also grow in importance. Fish-farming will generate much-needed animal protein. This system constantly recycles the supply of brine and converts fish excrement into nitrogen-based electrical generation, potentially providing nutrients for adjacent crop production and plant capture.

f. Alternative Food Sources

Different food sources will be developed. Examples include:

- Protein pellets generated from natural gas. This will be an industrial process combining methane (from the gas), nitrogen (from the air), and oxygen (from water). This will primarily be used as a fish food for inland aquaculture operations.
- Increasingly, more plant protein (e.g., oilseed meal) may be converted into meat-like substitutes, or simply protein foodstuffs not recognizable today.
- Animal products will also be grown in factory-like facilities. Factory-produced muscle cells plus fat cells will be utilized to produce meat-like cuts. Non-meat burgers and other delicacies are already available at your favorite fast-food outlet.

g. **Alternative Food Uses**

As already noted, the improved photosynthesis of plants is expected to create two to five times the energy presently grown and stored in plant material. This will position plant material as the next competitive generator of energy. Capturing and converting that energy into commercially useful processes and products will equal or surpass the impact of the emerging production of electric vehicles. A competing or complimentary energy source could be hydrogen, particularly for tractors and heavy truck transport.

The current heavy dependence on hydrocarbon-based energy is going to decrease. The major farm machinery companies (see Chapter 6) are already developing farm tractors that operate on methane rather than diesel. This initiative envisions a farm that not only produces food but also produces the material it needs to generate the energy it uses to run its operations and power its tractors and other machinery; a more self-sufficient and environmentally friendly future. Electric motors will provide the capacity for numerous applications, even in concert with internal combustion engines. The twenty parts to build an electric drive are a lot less than the 1000 or more parts required to build an internal combustion engine.

More recently, there has also been a growing interest in trying to develop products that will not only be of higher quality and less expensive, but also more environmentally sustainable with a lesser dependence on non-renewable resources. For example, at the Institute for Nanotechnology, University of Waterloo, they are trying to invent a process that combines wheat stalks with plastic to create lighter car parts. One car company has, reportedly, already used these new parts in some of its vehicles.

Similarly, researchers at the Bioproducts Discovery and Development Centre at the University of Guelph are presently trying to develop and commercialize a technology that can turn plant materials into resins, polymers, and fibers to produce petroleum-free plastics. They have already produced plastic bio-bins, flower pots and bird feeders that contain 25–30 per cent bioproducts that are already being sold in Canadian building centers. This development will grow exponentially.

h. Real-Time, High-Resolution Data

The sale of livestock via video-demos at auction markets is already commonplace. Next, grain growers will be offered real-time access to all grain buyers, giving them more choice and more price transparency so they can more conveniently and effectively sell their grain utilizing on-line cell phones, anywhere and anytime. This will become increasingly transformational, especially in lesser developed countries across the world.

Additionally, high resolution data gleaned from mini-satellites will increasingly be able to provide producers with customized real-time information on anticipated weather conditions.

We also expect the development of wireless farm and rural resident security systems, which are linked, in real time, to on-site monitors, as well as monitors in neighboring rural residences. All movable on-farm assets will be assigned a bar code, which triggers an alarm system if the non-authorized movement of these assets arises.

14.1.5 WATER AVAILABILITY

The supply of freshwater that supports human health and enterprise is basically constant, representing only about 1 percent of the water available worldwide. Good-quality, non-saline water is the global asset most important in satisfying the increasing demand for basic food, fiber, feed, and fuels. People are utterly dependent on water for their lives (humans consist mostly of water) and their livelihoods. Water is life.

With the world's population set to increase by 35 percent (2.4 B) by about 2050, a mere 30 years from now, the additional food required to feed future generations will put further pressure on freshwater resources because irrigation already accounts for about 80% of the total fresh water consumed and about 2/3rds of the total diverted for human uses. Yet, globally, in both irrigated and rain fed agriculture, only about 10–30 percent of the available water (as rainfall, surface or groundwater) is used by plants in transpiration. There is, therefore, great potential for improving water use efficiencies in agriculture.

The focus of future water development will be on increasing water productivity, the ratio of the net benefits from crop, forestry, fishery, livestock, and mixed agricultural systems to the amount of water used to produce those benefits. The is the efficiency with which transpired water produces biomass. Increasing water productivity will allow producers to: (a) better realize the full benefits of other production inputs (fertilizers, seeds, machinery, energy, and labor); (b) meet the rising demands for food and changing dietary patterns of a growing, wealthier and increasingly urbanized population; and, (c) respond to pressures to reallocate water from agriculture to cities and industries and ensuring water is available for environmental uses and climate change adaptation.

This will address both irrigated and dryland agriculture. An estimated 260 M ha ((640 M acres) constitute approximately 17% of the world's total cultivated farmland but produce 40%

of its food and fiber. Drylands cover 83% of the cultivated famland and 2 billion people use it for cropping and grazing. Limited and erratic water availability and the inability of crops to use the available water efficiently are major constraints. Achieving synchrony between nutrient supply and crop demand under various moisture regimes is the key to optimizing trade-offs between yield, profit, and environmental protection. Expect an increasing emphasis on improved dryland cropping systems, more water conservation initiatives (wetland preservation, shelterbelts, etc.), electronic input optimization sensors, and genetic improvements to further enhance crop water use efficiencies and drought tolerance.

14.1.6   STRUCTURE OF AGRI-INDUSTRY

Continued mergers and consolidations in both the product processing industries and farm input suppliers is also anticipated. Largely driven by economies of scale, in the last 25 years we have already witnessed:

- 12–14 major grain companies reduced to 5–6 major grain companies; 5700+ grain elevators shrink to 350+ high-throughput elevators[321]
- 6–8 major meat processors shrink to 2–3 major meat processors
- 30+ major farm machinery companies shrink to 4–5 major farm machinery companies
- 12–15 major fertilizer companies reduced to 6–8 ever-larger fertilizer companies
- 80+ major seed and chemical companies rapidly shrinking; now reduced to 4–5.

And there is no indication that this Pac-Man© phenomena is going to abate any time soon.

The implications of this for primary agriculture are anything but benign and will almost surely mean: a) increased market power for both processors and input suppliers; b) more vertical integration of input suppliers-producers, as well as producer-processors; and, c) increasingly narrow margins for primary producers as more and more services are outsourced and provided by both input suppliers and processors. The ever-increasing use of on-farm proprietary software will gradually shift farm management to off-farm entities, further erode on-farm profit margins, and probably drive even more smaller farmers out of primary agriculture. Tightly specified production contracts regarding the quantity, quality, production technology, time-of-delivery, etc., will become increasingly commonplace.

At the same time, the widespread adoption of on-farm 3-D printers may gradually undermine the traditional machinery business model, which relies on parts replacement and generous mark-ups. A 3-D printer works much like an ink-jet printer except it builds a physical object by stacking one very thin layer of material on top of another, and is already being used in medicine (prosthetic limbs), food (customized candy), and fashion (clothes). The US military is currently experimenting with a mobile 3-D printer that can generate needed tank and truck parts on the battlefield. This technology will eventually make it into farm machinery shops, or at least into the farm machinery pipeline.

14.2 FARM STRUCTURE

Farm structure, conduct, and performance at any point in time will largely depend on on-going technological change and the social-economic-environmental-political milieu.

Figure 14.1 The Principal Drivers of Future Farm Structure, Conduct, and Performance in Canada

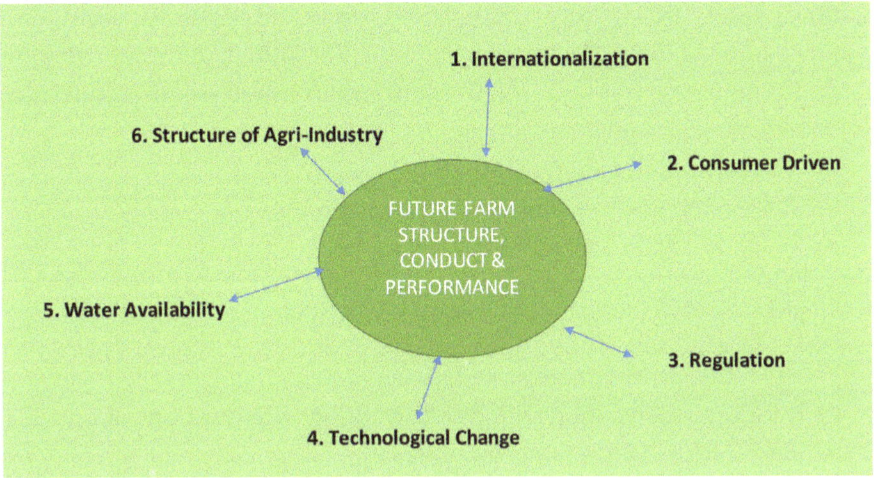

1. Internationalization
2. Consumer Driven
3. Regulation
4. Technological Change
5. Water Availability
6. Structure of Agri-Industry

FUTURE FARM STRUCTURE, CONDUCT & PERFORMANCE

The future we envision for primary agriculture has nine predominant (and some potentially contradictory) features: a) heterogeneity; b) bi-modality; c) increasing commercial scale; d) different capitalization mechanisms; e) vertical integration; f) narrower margins; g) increased self-sufficiency; h) environmental sustainability and enhancement, and, i) increased competition for land and water resources.

14.2.1 HETEROGENEITY[322]

It used to be that averages adequately described primary agriculture. Farms were quite similar and a normal distribution, where many farmers were close to the average, was a reasonably accurate description of primary agriculture. But that is certainly not the case today. Even within a small geographic area, great diversity exists in terms of size, volume, efficiency, financial performance, managerial ability, leverage, production technologies, and so on. And this is unlikely to change in the future.

14.2.2 BI-MODALITY

Bi-modal means that what we have now is a relatively few full-time large farmers and numerous smaller farms, either very specialized farms (e.g., organic) and/or part-time farmers. For its life-style and/or its environmental amenities, part-time farmers, with off-farm employment, will remain an integral part of the rural community. The products and services that large scale compared to small-scale farmers want and need are different. With this increased diversity, market segmentation and niche marketing will become increasingly important.

14.2.3 INCREASING COMMERCIAL SCALE

The conventional wisdom is that as farm size increases the average unit cost of production gradually decreases but must, eventually, increase. This wisdom is gradually being overturned as larger farm operations can, essentially, just clone an existing highly efficient operation. Modules that have a semi-independent capital and management structure can be replicated indefinitely. This is particularly straight-forward with respect to confined livestock and poultry operations. Synergies and efficiencies can also be secured by having various complementary farming operations at one location (e.g., beef and crop production). The economic organization of various agricultural enterprises on a Hutterite colony is illustrative. (Also see 14.2.7.)

14.2.4 DIFFERENT CAPITALIZATION MECHANISMS

The farm business will continue to move from owner-operator toward partnering, joint ventures, more complex shareholder structures, and commercial relationships that stabilize the structure and provide other benefits such as access to new technology, markets, and capital for growth.

Particularly for large-scale farms, innovative financial arrangements are going to be required. Extended incorporated family farms will gradually change their debt structure from bank credit to other credit instruments, particularly input supply or guaranteed product credit. They will also seek alternatives to debt such as leasing and modified debt instruments (e.g., adjustable term loans) that reduce their risk. Securing more financing through private equity firms (e.g., Blackstone) or pension funds can also be expected. Selling shares in a public stock offering is yet another option.

14.2.5 VERTICAL INTEGRATION

The structure of farm businesses will continue its shift toward collaborative commercial arrangements.

Greater vertical integration seems almost inevitable. The impetus for this largely comes from: a) increasingly large input providers and processors who continually want to increase their on-farm value-added component; b) on-farm capital and management constraints; and c) quality and logistical (efficiency) considerations. On-farm out-sourcing, however, is almost inevitably accompanied by narrower margins (i.e., the farmer contributes less so he also captures less net revenue/unit of output).

14.2.6 NARROWER MARGINS

Narrower commercial on-farm net margins are expected. This situation will arise because of: a) on-going vertical integration; b) ever-increasing scale; c) increased capital dilution; d) more operational out-sourcing; and, e) more regulation.

The argument that new technologies will continue to bail farmers/ranchers out is unlikely because many, if not most, of these new technologies are proprietary—meaning that the owners of this technology will largely capture the net financial benefit accruing from that technology.

The implications surrounding more vertical integration, increased capital dilution, and more operational out-sourcing are similar. The more inputs (land, labor, capital, and management) that are outsourced, the less the farmer/rancher keeps. He/she is providing fewer inputs and so the return to those inputs is similarly squeezed.

Ever-increasing scale also puts pressure on farmers/ranchers to constantly grow. Grow or perish. For example, a larger farmer can out-bid a smaller farmer for surrounding land or land rentals simply because they can effectively subsidize these costs with revenue from other operations. This is not unlike the competitive and consolidation/merger process in most other industries.

On top of all of this, government regulation, however desirable from a social perspective, almost always increases unit costs as well which, in turn, also serves to reduce net on-farm margins.

14.2.7 INCREASED SELF-SUFFICIENCY

At the same time, there is some potential for more on-farm integration. Most commercial grain farmers now have their own B-trains to haul their own grain to market. They may also have their own mechanic and service truck. Or they may have their own nutritionist, their own veterinarian, or their own in-house crop specialist. Or, alternatively, keep this in-house by sending the next generation to colleges and universities to secure these necessary skills. The 3-D printing of on-location parts would also be consistent with this trend.

Increased adoption of regenerative agriculture will facilitate both the opportunity for farming/ranching operations to generate their own fuels, as well as decreasing their current dependence on chemicals to control pests and disease. Pesticide use per unit of output will decline.

These in-house initiatives would serve to help offset the impact of otherwise gradually narrower margins (see 14.2.6 above).

14.2.8 ENVIRONMENTAL SUSTAINABILITY & ENHANCEMENT

Canadian agriculture will lead in the movement from resource sustainability to resource enhancement with a reduced carbon footprint as an added bonus. Through knowledge technology and stewardship, land, air and water quality—above and below ground—will develop better symbiotic relationships with the crops grown. Sensing technologies in real time to measure vital nutrient, microbial, and plant growth factors will strongly influence crop choices and develop data to enable crop rotations to be symbiotic. Regenerative agriculture will have sensing and tracking technologies to clearly account for beneficial practices.

Publicly supported land use initiatives to address anticipated climate change will reinforce this realignment. Government financial support through a transition to ALUS programs will belatedly recognize the social benefits of environmental amenities on private lands. This will help farmers and ranchers restore wetlands, plant windbreaks, install riparian buffers, create pollinator habitat, and adopt other ecologically beneficial projects on their properties.

14.2.9 INCREASED COMPETITION FOR LAND-WATER RESOURCES

The arable land base in Canada is essentially exhausted. Thus, with a growing urban population, more intense competition for surrounding arable land can be anticipated. More conflict between different interest groups wanting access to different land attributes on both private and public land is almost a certainty.

Increased conflict between multiple uses (i.e., domestic, industry, irrigation, stock water, hydro, recreation, etc.) for assured access to freshwater (surface and ground) can also be anticipated. This will only intensify as the demand for safe potable water continues to increase and the existing supply is increasingly rationed. The impact of climate change, particularly in the southern prairies, could exacerbate this growing supply-demand imbalance and the attendant allocation issues.

14.3 AGRICULTURE IS THE FUTURE

We must come together and stay together. We must continue to successfully meet the many challenges that our industry constantly faces. And we must continue to not only prevail but prosper in an ever-changing social, economic, environmental, and political milieu.

- The positive edge of most technology brings with it the negative edge related to its ownership and use. There is a growing danger that transparent smart technology and ubiquitous blockchain technologies could eventually result in a loss of control of property, production, and practice in the agricultural community. This must be thwarted. Farmers/ranchers should have the right to own, control, and monetize their own data.

- A growing base of collaborative and cooperative agricultural and food science research and development is essential to protect the integrity and success of Canadian agriculture. Science is the bedrock. Spurious claims most always be vigorously challenged and successfully discredited.

- The components of agriculture and the food system can be limitless and can soon be constructed to essentially end scarcity. The food system has the capacity to replace many manufacturing systems by providing the ingredients/components that can be technologically transformed on-farm or nearby to meet almost limitless demands in a renewable, sustainable way.

- For all its blemishes and constraints, agriculture is part of the complex biological system that creates matter by combining the air, land, water, and sunlight, along with some natural and synthetic inputs that can be more efficient and beneficial than man-made systems. The agriculture and food system can successfully function as both an energy supplier and a carbon sink to ensure recycling of all components of production and perhaps even use.

- Agriculture is the future. Canadian agriculture's unique environment, management, patented technological and scientific processes will surely become a key driver of our future world.

Endnotes

CHAPTER 2

1 This chapter borrows heavily from The Canadian Atlas, Royal Ontario Museum, online. Also see: Dick, L., "History of Agriculture to the Second World War," The Canadian Encyclopaedia, May 2015; and "Agriculture in Canada," Wikipedia, on-line.

2 Harari, Yuval N., Sapiens: A Brief History of Humankind, Signal, Random House of Canada, 2016.

3 Morrison, R. Bruce, and C. Roderick Wilson, Native People: The Canadian Experience, 2nd edition, Toronto, Oxford University Press, 1995, pp. 35.

4 Reaman, G. Elmore, A History of Agriculture in Ontario, Vol. 1, Saunders of Toronto, 1970, p. 13.

5 For Newfoundland and Labrador, see especially an article by Robert D. Pitt in The Canadian Encyclopaedia, Vol. 3 of 5.

6 See especially the article by Virginia P. Miller in Native People: The Canadian Experience, pp. 348-352.

7 See especially, the article by Lyle Dick and Jeff Taylor in The Canadian Encyclopaedia, Vol. 1 of 5.

8 This is when Sir Wilfred Laurier became Prime Minister and the great immigration boom of 1895–1914 (mostly to the prairies and BC) began.

9 The data for 1931, here and elsewhere, can be found in: GOC, Historic Statistics of Canada, 2nd edition, Ottawa, 1983.

10 Reaman, G. E., op. cit., p. 16 and p. 26.

11 Drawn from Dairy Farming in Canada, 2017.

12 Much of this early history is drawn from G. E. Reaman, op. cit., pp. 14–24.

13 Founded in Stoney Creek in 1897, now a community in the city of Hamilton. Subsequently promoted by Macdonald College, University of Guelph.

14 Russell, Peter A., How Agriculture Made Canada: Farming in the Nineteenth Century, McGill-Queen's University Press, 2012, pp. 210–212.

15 The Prince of Wales also owned a 100-acre ranch in the Alberta foothills along the Pekisko Creek, the EP Ranch, 1919–1962.

16 Irving, J. A., The Social Credit Movement in Alberta, University of Toronto Press, 1959.

CHAPTER 3

17   Statistics Canada, Publication 95-640-X, Chapter 1, Figure 6, Total farm area as a proportion of total land area, Canada and the provinces, 2011.

18   OMERS = Ontario Municipal Employees Retirement System. OTPP = Ontario Teachers Pension Plan.

19   Number of colonies from the Census of Agriculture, 2016.

20   Statistics Canada, Census of Agriculture, 2016.

21   Figure 3.18 is based on 2012 data from both Statistics Canada and a Taxation Data Program, and covers both unincorporated farms and communal farming organizations with total farm operating revenues equal to or greater than $10,000, as well as incorporated farms with total farm operating revenues of $25,000 and over.

22   See comments by David Sparling, Chair, Agri-Food Innovation & Regulation, University of Western Ontario, London.

23   Figure 3.19 is also (like Figure 3.18) based on 2012 data from both Statistics Canada and a Taxation Data Program and covers both unincorporated farms and communal farming organizations with total farm operating revenues equal to or greater than $10,000, as well as incorporated farms with total farm operating revenues of $25,000 and over.

24   The Gini Coefficient is most often used in economics to measure how far a sector or country's wealth or income distribution deviates from a totally equal distribution. Perfect equality implies a coefficient of 0; perfect inequality implies a coefficient of 1.

CHAPTER 4

25   Glucosinolates are the compound that gives mustard, cabbage, and horseradish their pungent smell or odor.

26   See, in particular: Brophy, M., "Barley Production and Research Trends in Canada," Brewing and Malting Barley Research Institute, January 2011.

27   See, especially, Back, J., et al., Prairie Oat Growers Manual, University of Alberta, Edmonton, 2010.

28   Previously it was Lucerne Foods, a Division of Safeway, that processed sweet corn and sweet pea in the Lethbridge area. After the merger of Sobeys (Empire) and Safeway, anti-combines legislation dictated that the processing unit be sold to Bonduelle, based in Quebec.

29   Manufactured nitrogen fertilizer is produced by using heat and passing nitrogen and hydrogen gases over four beds of catalysts (i.e,. the Haber-Bosch process). Natural gas or some other fuel is required to generate the required heat.

30   Statistics Canada, CANSIM 001-0017.

31   According to the CCA website (cattle.ca) and based on the 2016 Census of Agriculture there were 59,784 farms and ranches with beef cows in 2017. The total number of cattle and calves on farms in January 2017 was 12.07 million, including 1.40 million dairy cattle. Canada fed 2.7 million cattle to market weight so if those 2.7 million dressed out at 700-800 lb each, that should have yielded about 2.1 billion lb (X1000), or roughly a million tonnes. This excludes cattle shipped to and slaughtered in the USA.

32   This section draws heavily upon: Heminthavong, K., Canada's Supply Management System, Library of Parliament, 2015.

33   The USA has approximately 9 million cows (9X Canada) and about 58,000 dairy farmers (5X Canada), even though the USA has 10X the population. See USDA/NASS, Data and Statistics, 2013.

34 No attribution. From: White, E., "Economists Suggest Overhaul of Dairy System," Western Producer, November 1, 2018, p.13.

35 Worldwide sales of robotic milking machines is about US$ 1.2 billion per year and sales are expected to double in just the next five years. (Guebert, A., "The next generational change in farming is already underway," Alberta Farmer, June 3, 2019, p. 4.)

36 McKenna, B., "Canada's Dairy Industries a Rich, Closed Club," Globe and Mail, June 25, 2015. Ontario and Quebec capped the quota price at $25,000 per kg. in 2010.

37 Very approximate. Calculations assume whole milk is 3.5% BF, the market price for 1 kg of BF/day is $25,000, and one cow produces 9000 litres of milk per annum. Some estimates are even higher.

38 The Montreal Economic Institute, the Fraser Institute, and the Conference Board of Canada argue that supply management is expensive for consumers, as estimated herein. Also see: Carter, C., "Hidden Costs of Supply Management in a Small Market," Canadian Journal of Economics, 2016. Research by the University of Waterloo and the Nielson Company (commissioned by the Dairy Farmers of Canada) disputes these findings.

39 The World Trade Organization (WTO) calls this an MPS (Market Price Support), which is measured as the difference between the reference price (average during 1986–1988) for the two commodities butter and skim milk powder for which prices are administered and the currently administered prices. This is an Aggregate Measure of Support (AMS), typically an amalgam of various measures unique to each country. Thus, it is really an estimate of the value of the implicit consumer-to-producer transfer; the extent of the annual hidden consumer subsidy to dairy producers. Basic data from D. Hedley, 2018; and WTO, Domestic Support, Canada, 2013.

40 See Global Affairs Canada.

41 Very approximate. The broiler estimate assumes a quota cost of $500 per square meter and 13.9 birds per square meter. The turkey estimate assumes a quota cost of $500 per square meter and 2.8 birds per square meter. The egg estimate assumes a quota value of $245 per unit and about 24 M laying hens. Some estimates are even higher.

42 Again, see Global Affairs Canada.

43 Includes donkeys/mules, bison (buffalo), llamas and alpacas, rabbit, mink, fox, deer, and elk. Other fowl (other than duck and geese) includes quail, pheasants, and partridge.

44 There are about 1 billion sheep in about 100 countries. Thus, Canada represents about 0.1 percent of this worldwide total.

45 Farm and Food Care Ontario, "Goat Fact Sheet,," 2018.

46 Sheep (especially sheep with hair) and goats can look very much alike. The tell-tail sign is the tail: a goat's tail turns up; a sheep's tail turns down.

47 Equestrian Canada, 2010 Canadian Equine Industry Profile, 2011.

CHAPTER 5

48 AAFC, An Overview of the Canadian Agriculture and Agri-Food System, 2015, Ottawa, 2016, p. 34.

49 Ibid, p. 34.

50 Ibid, p. 34.

51 Ibid, p. 34.

52 Countries that signed the WTO Nairobi Declaration (2015), including Canada, have agreed to eliminate subsidized exports by the end of 2020 and as such Canada would presumably be out of the dairy export market as of 2021. But

Canada has developed a Class 7 milk product designation, which is presently a way to clear the Canadian skim milk powder market by offering a competitive price in both the domestic and export markets. Canada insists that this is not subsidized and outside WTO limits on subsidized exports.

53   AAFC, An Overview of the Canadian Agriculture and Agri-Food System, 2015, Ottawa, 2016, p. 29.

54   Skogstad, G., The Politics of Agricultural Policy-Making in Canada, U of T, 1987, p.18.

55   Website @FAOWHOCodex.

56   Earlier analysts calculated producer and consumer subsidy equivalents (PSEs and CSEs). See, for example: Webb, A. J., et al., Estimates of Producer and Consumer Subsidy Equivalents: Government Intervention in Agriculture, 1982-87, USDA Statistical Bulletin No. 803, Washington, D. C., 1990.

57   According to the Western Producer, November 22, 2018, p. 10.

58   The reader is encouraged to consult other sources to verify the accuracy of the general descriptions following.

59   Hedley, D. D., "Governance in Canadian Agriculture," Canadian Journal of Agricultural Economics, Vol. 65 (2017), p. 536.

60   Harper, S. J., Right Here Right Now, McClelland & Stewart, Toronto, 2018, p. 34.

61   Murray, M., NAFTA and Agriculture, online, 2016.

62   Class 7 was developed to essentially clear the Canadian skim milk powder market by offering a "competitive" price in both the domestic and export markets, thus under-cutting international market prices. Prior to establishing this Class 7 designation, the US dairy industry had been exporting ultrafiltered milk to Canada because it was not subject to high Canadian tariffs or TRQs.

63   Newsletter, January 2019. albertawheat.com; albertabarley.com.

64   This includes, in particular, frozen potato chips and mashed potato.

65   Newsletter, January 2019. albertawheat.com; albertabarley.com.

66   Still, there is no provision that specifically allows for trade in GM food crops. The EU and Canada maintain zero tolerance for the imports of GM foods that have not been approved by the respective regulatory agencies. But CETA could maintain pro-GMO pressure on the EU regulatory process. Labelling of GMO foods and food ingredients in the EU and consumer resistance to GMOs will likely continue to keep these products out of the EU food system. Canadian GM soybeans are already widely used as an animal feed in the EU.

67   Table 5.5 says 50,000 tonnes. These two numbers have not been reconciled.

68   "Geographic indicators" are like trademarks (e.g., champagne from Champagne; scotch from Scotland).

69   A staple of virtually every basic economic textbook. See, for example: Samuelson, P. A., Economics – Introductory Analysis, McGraw-Hill Book Company, Toronto, various editions.

70   As reported in the Western Producer, November 29, 2018, p. 4.

71   "Income elasticity of demand" refers to the demand response for Product X when incomes rise. Elasticities > 1 are "elastic"; elasticities <1 are "inelastic."

CHAPTER 6

72   Canadian Council of Food Processors.

73   See: Canadian National Millers Association.

ENDNOTES 261

74 AAFC, Consumer Trends-Bakery Products in Canada, Market Indicator Report, Ottawa, January 2013.

75 Canada baked goods market size in 2017 was reportedly US$5.6 billion (Cd$7.5 B). In 2011, the bakery market in Canada, including frozen bakery and desserts, had sales of Cd$8.6 B.

76 See https://www.canolacouncil.org/markets-stats/industry-overview/

77 One hectolitre = 100 litres. See http://cmbtc.com/about/mission/

78 A list can be found on Wikipedia.

79 A popular account can be found in: Bronfman, C., Distilled, HarperCollins, Toronto, 2016.

80 Very approximate. The number 800 comes from Robert A. Bells' website "Wines of Canada" and another website indicated there were 278 licensed grape wineries in B.C. Also see the Wine Business Monthly, which reports 432 wineries in 2011.

81 See especially http://www.pulsecanada.com/producers-industry/

82 See, especially: Christensen, N., and C. Dade, "Pulse Fractionation a Huge Opportunity," Winnipeg Free Press, September 1, 2017. Also see: Canada West Foundation, "How Lessons from Canola Can Help Unleash Pulse Fractionation on the Prairies," November 8, 2017.

83 In February 2018, Simplot announced yet another doubling of potato processing capacity in Manitoba. (Source: The Western Producer, 2-18.)

84 AAFC, Potato Market Information Review, 2016–2017, on-line.

85 This total is thought to include potato processors, discussed elsewhere. See: R. F. Barratt and M. Fafard, "Fruit and Vegetable Industry," March 4, 2015, website.

86 Mussell, A., and K. Grier, Ontario Processing Vegetables: An Economic Analysis, OMAFRA, 2016.

87 Basic information from the Canadian Sugar Institute.

88 This changes frequently, as it is believed, for example, that Pound-Maker Adventures (Lanigan, SK) and Milligan Bio-Tech (Foam Lake, SK) are no longer in operation.

89 Proceedings of the National Academy of Sciences, summer 2015.

90 There are really only three major pork processing plants in Western Canada. Maple Leaf Foods and Olymel dominate. See: Brisson, Y., "The changing face of the Canadian hog industry," Statistics Canada, 2017. Also see: https://www.ridgetownc.com/research/documents/mcewan_

91 GOC, Canadian Dairy Information Centre, 2017.

92 Approximate. Based on their reported purchase of 348 independent retail outlets in the last decade, their recent acquisition of 18 Andrukow Group Solutions outlets (Western Producer, May 2016), and their even more recent acquisition of 210 Viterra retail outlets (Western Producer, June 2017).

93 Approximate. Based on an estimated 77 Ag. Business Centres (Richardson website), the recent purchase of 13 more independent outlets (CHS, Webbs, Agro Guys), as well as on-going construction of at least 2 more outlets in Saskatchewan. (Data from various websites.)

94 Western Producer, December 2018, p. 46.

95 Western Producer, August 2018. BASF subsequently sold the Clearfield system to Corteva Agriscience (DowDupont). (Western Producer, February 28, 2019, p. 16.)

96 The annual investment in canola breeding in Canada is $75–$80 M; the investment in cereal crop breeding is about $25 M. See: Western Producer, November 22, 2018.

97   Pearce, R., Secan at 40, October 2016.

98   Hambre. B. M., and A. Asseltine, Agricultural Implements Industry, on-line, March 4, 2015.

99   https://www.export.gov/article?id=Canada-Agricultural-Machinery-and-Equipment

100   MacDon was recently purchased by Linamar and, consequently, is now expected to also quickly expand its reach in Europe.

101   See the summary of Canadian Industry Statistics – Agricultural Implement Manufacturing.

102   https://www150.statcan.gc.ca/n1/pub/96-325-x/2014001/article/14084-eng.htm

103   https://www.anacan.org/about-our-industry/canadian-feed-industry-statistics.html. There are, however, only 165 members in the Animal Nutrition Association of Canada and it is estimated that they represent 90% of the total commercial feed volume in Canada.

104   Based on 2007 data from The Canadian Veterinary Medical Association (CVMA), estimated (in 2007), there were about 8,000 veterinarians in private practice (and 2,000 other) broken down as follows: 59% specializing in companion animals (CA), 29% in mixed animals (MA), and 12% in food animal practices (FA). We assume that most of the MA practices would be in rural areas. https://www.canadianveterinarians.net/about/statistics

105   http://www.dairyinfo.gc.ca/pdf/genetics_publication_2018_

106   Also see: Gaston, C., and M. Bedard, Highlights of the 1974 Interprovincial Input-Output Model, Statistics Canada, Structural Analysis Division, Ottawa, 1975; Statistics Canada, The Input-Output Structure of the Canadian Economy, 1971–80, Catalogue 15-201E, annual; as well as earlier studies by Gordon McEachern and Tim Josling. Now, most provinces also have their own I/O models.

107   The indirect industry effect is also sometimes referred to as a "spin-off." Note that a multiplier of, say, 2.5 implies a direct impact of 1 and an indirect impact of 1.5.

108   Statistics Canada and AAFC calculations.

CHAPTER 7

109   See especially: Bollman, R. D., Rural Canada 2013: An Update: A Statement of the Current Structure and Trends in Rural Canada, Federation of Canadian Municipalities, Ottawa, 2014. Also see: Lauzon, A., R. Bollman, and W. Ashton, State of Rural Canada; Introduction, CRRF, Ottawa, 2016.

110   Statistics Canada, Rural and Small Town Canada, Analysis Bulletin, Vol. 3, No. 3, Catalogue #21-006-XIE.

111   Reimer, B., and R. D. Bollman, "Understanding Rural Canada: Implications for Rural Development Policy and Rural Planning Policy" in: Douglas, D. (ed.), Rural Planning and Development in Canada, Nelson Education Ltd., Toronto, 2010.

112   World Bank, Reshaping Economic Geography, Word Bank Report 2009, Washington, D. C., 2010.

113   See, in particular: Stabler, J. C., and R. Olfert, Restructuring Rural Saskatchewan: The Challenge of the 1990's, University of Regina Press. Also see: Statistics Canada, "Rural and Small Town Canada: Analysis Bulletin," various issues.

114   A Census Metropolitan Area (CMA) is defined as having an urban core of 50,000 and a total population of 100,000 or more people.

115   Rostow, W. W., The Stages of Economic Growth, Cambridge University Press, 1960.

116   Schumpeter, J., Capitalism, Socialism, and Democracy, 1942.

117	Bollman, R. D., Factors Driving Canada's Rural Economy, 1914–2006, Research Paper #083, Statistics Canada/Agriculture Division, Ottawa, 2007.
118	Canada Council on Learning, The Rural-Urban Gap in Education, Ottawa, 2006.
119	Immigrants, overwhelmingly, go to metro centres. In predominately urban areas, about 30% of residents have been born outside of Canada whereas in rural areas the number is about 6 percent.
120	Pew Research Center, "Canada's Changing Religious Landscape," June 27, 2013.
121	Bollman, R. D., Factors Driving Canada's Rural Economy, 1914-2006, Research Paper #083, Statistics Canada/Agriculture Division, Ottawa, 2007.
122	For a rigorous interpretation and application of any legal issue identified herein, the reader is strongly encouraged to consult a qualified lawyer.
123	Reclamation is estimated to cost about $300,000 per well. See "Orphan Wells: Alberta's $47 Billion Problem," The Western Producer, March 22, 2018.
124	Two recent articles: "Frustrations Over Rural Crime in the Spotlight," The Western Producer, March 1, 2018; and "Stand Your Ground," The Western Producer, March 15, 2018.
125	Arnason, R., "What Canadian Law Says About Self-Defence," The Western Producer, March 15, 2018.
126	In Saskatchewan, nearly 100 of the province's 266 RMs and some rural communities are currently establishing or reviving Rural Crime Watch efforts. (Western Producer, November 1, 2018.) A Rural Crime Watch app on rural resident cell phones might be very effective.
127	Western Producer, December 6, 2018.
128	Graney, J., "Fighting Crime with Tax Credits," The Edmonton Sun, November 19, 2018.
129	Cloutis, E., et. al., Socio-Economic Vulnerability of Prairie Communities to Climate Change, PARC Project No. 48, University of Winnipeg, June 2001.
130	Also see: Bedard, A. F., J. Belanger, E. Guimond, and C. Penny, Measuring Remoteness and Accessibility – A Set of Indices for Canadian Communities, Catalogue #18-001, Statistics Canada, Ottawa, 2017.
131	Lauzon, A., R. Bollman, and W. Ashton, State of Rural Canada, CRRF, Brandon University, 2016. Also see: Federation of Canadian Municipalities, Rural Challenges, National Opportunity, Winnipeg, 2018.
132	Also see Chapter 9, "Role of Government."
133	World Bank, Reshaping Economic Geography, World Bank Report 2009, Washington, D. C.
134	Speer, S., "Natural Resources are a Win for Rural Communities," Toronto Sun, August 5, 2017.
135	Ibid.
136	This is now a well-worn thesis. See, in particular: Putman, R., Bowling Alone: The Collapse and Revival of American Community, 2000. Also see Bloom, A., The Closing of the American Mind, 1987.
137	The "average" level of support indicated is for truly rural societies. Much of the total Alberta funding available goes to the Calgary Stampede, Edmonton Exhibition, and other predominately "urban" societies.
138	As reported in the Western Producer, January 24, 2019, p. 17.
139	Dunbar's number specifically refers to the cognitive limit to the number of people with whom one can maintain stable social relationships. A number larger than this generally requires more restrictive rules, laws, and enforced norms to maintain a stable, cohesive group (Wikipedia). This often applies to the critical number of people required to successfully execute a community initiative.

140   Harper, Stephen J., Right Here, Right Now, Penguin Random House Canada, Toronto, 2018.

CHAPTER 8

141   Incomplete. The 2015 Alberta Directory alone lists 530 agricultural associations and organizations.

142   Photo of 1949 AFU Convention delegates, District #4, from the personal collection of Don & Marilyn Macyk.

143   The United Farmers of Alberta (UFA) temporarily morphed into a political movement, governing Alberta from 1921 until 1935.Two subsequent political parties on the prairies, the Social Credit Party (Alberta) and the Cooperative Commonwealth Federation (Saskatchewan) were both imbued with this same cooperative philosophy. See: Irving, J. A., The Social Credit Movement in Alberta, U of T Press, 1959; and Archer, J. A., Saskatchewan: A History, Western Producer Prairie Books, 1980.

144   Ruguly, E., "Viterra another example of Canadian short-sightedness," Globe and Mail, March 23, 2012.

145   This section borrows from: Jaques, Carrol, UNIFARM, A Story of Conflict and Change, University of Calgary Press, 2001.

146   Much of this information is drawn from: Pilger, G., "Are Canada's Farm Organizations Actually Listening to Their Members?" Country Guide, June 2015.

147   Ibid.

148   The Alberta Federation of Agriculture, for example, has about 250 members representing about 1,000 farm families. That is less than 3% of all farm families in Alberta.

149   Aside from farm organizations, members include: six grain companies (Cargill Canada, G3 Canada, Parrish & Heimbecker, Richardson International, Viterra, and Louis Dreyfus Canada), two processors (Canada Bread and Warburtons), and seven seed and/or life science companies (BASF, Bayer, Canterra Seed, Dow-Dupont AgroSciences, FP Genetics, SeCan and Syngenta). The Saskatchewan Wheat Development Commission is not a member. The Western Wheat Growers Association is an affiliate member.

150   CIGI promotes Canadian crops to domestic and international processors. Besides overseas missions, CIGI runs seminars in Winnipeg showing domestic and foreign millers and bakers how to get the most out of Canadian wheat. Members include Alberta Wheat Commission, Manitoba Wheat and Barley Growers Association, Saskatchewan Wheat Development Commission, and seven grain companies (Cargill Canada, G3 Canada, Parrish & Heimbecker, Paterson Grain, Richardson International, Viterra, and the Inland Terminal Association of Canada. (See Alberta Farm Express, June 4, 2018.) As of mid-2019, a Cereals Canada–CIGI merger is still being considered. (The Western Producer, July 11, 2019.)

151   CropLife Canada represents the Canadian manufacturers, developers, and distributors of pest control and modern plant breeding products. It is part of the CropLife International consortium, a global federation with members across 91 countries, that champions agricultural innovations in crop protection and biotechnology. The big-four, Bayer-Monsanto, Syngenta-ChemChina, Dow-Dupont, and BASF, are major players.

152   Seed Synergy Collaboration Project, White Paper on the Next Generation Seed System, 2018.

153   The options being considered are end-point royalties (when the grain is sold) or trailing royalties (an on-going annual royalty to plant breeders) for the use of farm-saved seed, with either option only being applicable to newly-developed varieties. A third hybrid option is now being considered. (See Alberta Farmer, July 15, 2019, p. 7.) The Canadian Food Inspection Agency enforces all health and safety standards under the food and drug regulations and is also responsible for the administration of non-health and safety regulations concerning food advertising, labelling, and advertising. Nominally under AAFC (Section 9.5), it now reports directly to the Minister of Health (Health Canada).

154 See, especially: Jaques, Carrol, UNIFARM, A Story of Conflict and Change, U of C Press, 2001. (Chapter 14 "Showdown")

155 McKenna, B., "Canada's Dairy Industry is a Rich, Closed Club," Globe and Mail, June 25, 2015.

156 See, in particular, extensive analyses by Josling, Barichello, Arcus, Schmidt, Veeman, and Mussell. See, especially: Mussell, A., "Economists Suggest Overhaul of Dairy System," Western Producer, November 1, 2018, p. 13.

157 See especially: Skogstad, Grace, The Politics of Agricultural Policy-Making in Canada, University of Toronto Press, 1987.

158 The judgement of Kevin Hursh, "Commodity Groups Abound, but System Works," Western Producer, January 25, 2018.

159 Western Producer, December 20, 2018.

160 Western Producer, December 13, 2018.

161 See http://gfo.ca/Marketing/Dealers.

162 Western Producer, November 15, 2017.

CHAPTER 9

163 This chapter borrows heavily from the work of Dr. D. D. Hedley. See especially: Hedley, D. D., "Governance in Canadian Agriculture," Canadian Journal of Agricultural Economics, Vol. 65 (2017), pp. 523-541, and D. D. Hedley, The Evolution of Agricultural Support Policy in Canada, CAES Fellows Paper 2015-1. Also see: Skogstad, Grace, The Politics of Agricultural Policy-Making in Canada, U of T Press, 1987.

164 Dick, L., "History of Agriculture to the Second World War," Canadian Encyclopaedia, 2015.

165 Thus, crop Iinsurance is under provincial control. The federally sponsored PFRA was only supported by the provinces because of the feds' fiscal capacity. Under Section 92, the establishment of the CWB was also an invasion of provincial powers.

166 The prairie region also included the Dawson Creek area of B.C. and a sliver of Western Ontario. For direct sales, licenses for the export or import of wheat, oats, and barley were granted to western grain elevator companies (pools, UGG, Pioneer, etc.), as well as the Ontario Wheat Board. Companies manufacturing a product with 25% or more of these grains (e.g., Nabisco Shredded Wheat) also had to have an export license.

167 With respect to the dairy industry, the federal ASA (Aggregate Measure of Support) subsidies only covered industrial milk, not fluid milk. Subsidies (federal) were set and paid annually during 1958–1974 until quotas and the $2.66/cwt (later $6.03/hl metric equivalent) subsidy on industrial milk was established. (Terminated in 1995/96.)

168 A good discussion of the issues surrounding this freight change can be found in Skogstad, G., The Politics of Agricultural Policy-Making in Canada, U of T, 1987, pp. 121-156.

169 See, especially: Gray, et al., "A New Safety Net Program for Canadian Agriculture: GRIP," Choices, 3rd Quarter, 1991, pp. 34-35.

170 AIDA was created specifically because of the collapse of hog prices in 1997 and 1998, and to circumvent the ongoing countervailing duties on hogs imposed by the US starting in the mid-1980s. Arrangements were made with indigenous farmers for access to the program other than through tax filing since they do not file or pay income tax.

171 Descriptions of on-going programs may be inaccurate or incomplete. The reader is encouraged to consult the official documents.

172 See especially: Canadian Transportation Agency, At the Heart of Transportation: A Moving History, Ottawa, 2017.

173   Ibid, p. 62.

174   This was the conclusion of a long-standing controversy. See, in particular: Skogstad. G., The Politics of Agricultural Policy-Making in Canada, U of T, 1987, Chapter 6. Also see: MAA, The Crow Rate Debate: Issues and Impacts, Alberta Economic Development, 1982.

175   For an entertaining and informative account of CN and CP in recent years see: Green, H., Railroader, CEO Hunter Harrison, Page Two Books, 2018.

176   This section relies heavily upon: AIC, An Overview of the Canadian Agricultural Innovation System, Ottawa, July 2017.

177   A similar graphic for the 1990–2013 period can be found in AAFC, An Overview of the Canadian Agriculture and Agri-Food System, 2015, Ottawa, 2016, Chart D.1, p. 41.

178   AAFC, op. cit., Chart D.4, p. 42.

179   White, E., "Canadian Ag Falls Behind in Research and Development," Western Producer, December 13, 2018.

180   The 2002-2013 average was $523 million. See Chapter 4, Section 4.4.

CHAPTER 10

181   Canada's land area is 8,886,356 km2 (2.1 B acres).

182   Saskatchewan is currently (late 2018) re-assessing how to more effectively address this issue. Verbal or written permission by the land owner is being advocated by Saskatchewan municipalities.

183   See especially: PRISM Environmental Consultants, The Economics of Grazing Systems for Waterfowl Habitat Conservation: An Evaluation & Guidelines for Implementation, Recreation, Parks & Wildlife Foundation, Edmonton, May 1987.

184   See, in particular: Prairie Pothole Region, Wikipedia and the Ducks Unlimited website.

185   Pedology Consultants/Marv Anderson & Assoc., Farmland Drainage in Central and Northern Alberta, Farming for the Future 82-0070, Edmonton, 1984.

186   In the USA, the estimate is 50 percent. See: Wiebe, K. D., and R. E. Heimlich, "The Evolution of Federal Wetlands Policy," Choices, 1995.

187   "Economists put a Price Tag on Nature," Western Producer, January 8, 2009.

188   For an estimated distribution of benefits, see: Pattison, W. S., Alberta Waterfowl Habitat Enhancement Projects: Benefit Distribution, MAA/Alberta Environment, Edmonton, March 1984.

189   This draws heavily upon Knapik et al. and the references therein. See: Knapik, L., et al., Agricultural Land Degradation in Western Canada: A Physical and Economic Overview, Agriculture Canada, Edmonton, September 1984.

190   A fifth would be land fragmentation, especially the isolation of small parcels around urban areas.

191   Henry, L, Henry's Handbook of Soil and Water, Henry Perspectives, Saskatoon, SK.

192   Henry, L., B. Harron, and D. Flaten, The Nature and Management of Salt-Affected Land in Saskatchewan, Agdex 518, Saskatchewan Agriculture, 1987.

193   These estimates vary. Another 1984 source says that the extent of salinized soils in the dryland regions of Canada was already 5.4 million acres and that it was growing at a rate of 10% per annum. See: Sparrow, H. O., Soil at Risk, GOC, Ottawa, 1984.

194   Canadians for a Sustainable Society, Why You Should Care About Farmland Loss, June 2016.

195	This may be an underestimate. In Ontario, it was already 18% in 2001; 6% in Alberta. See: Hoffmann, N., "Urban Consumption of Agricultural Land," Rural and Small Town Canada, Statistics Canada, September 2001.
196	Immen, W., "Urban Sprawl Threatens Prime Farmland," Globe and Mail, Updated April, 2018.
197	Stan, K., "The Edmonton-Calgary Corridor: Simulating Future Land Cover Change Under Potential Government Intervention," PhD dissertation, University of Alberta, 2018.
198	Animals + perennial crops; animals + fish ponds, etc.
199	A combination of trees, grasses, and livestock.
200	See comparative budgets from Manitoba Agriculture, 2019. Also see: Gilmour, G., "Organic production an economic winner," Alberta Farm Express, February 11, 2019.
201	Rodale Institute, Regenerative Organic Agriculture and Climate Change: A Down-to-Earth Solution to Global Warming, 2014. Also see: Gabe Brown, Dirt to Soil, Chelsea Green Publishing, undated.
202	According to Laura Reiley, Washington Post, June 12, 2019.
203	Examples in Alberta include integrated resource plans for Jean D'Or Prairie, Big Bend, Lakeland, and Castle River. (Studies by Stewart-Weir & Co. for Alberta Energy & Natural Resources, Edmonton, 1984–85.) Also see: PLAN:NET/GOA, Alberta Integrated Resource Planning Program: Evaluation, Edmonton, April 1992.
204	A counter-argument to the success of the PFRA pasture management program is that it has become subsidized pasture for ranchers who are obliged to keep it in pasture to get their pasture at a very low cost. And, much similar land just across the Alberta-Montana border has remained in cropland, but seems to respond well to continuous cropping technology, which was of course unknown at the time these lands were put into government hands in the 1930s. Some of the PFRA work on cropping types (row crops, shelter belts, etc.) should also be mentioned. Much of that was probably instrumental in helping farmers survive in some of the arid areas that were not taken over by the PFRA/government during the 1930s and later. Also see: MAA, Valuation of Secondary Benefits and Externalities Associated with Community Pasture Programs, PFRA, Regina, 1978.
205	Canada ranks #8 in the world with approximately 10X as much, on a per capita basis, as the United States (ranked #55).
206	Wikipedia.
207	A high threat to water availability means that more than 40% of the water in rivers is withdrawn for human use.
208	A weighted average rate/hectare for the 13 Districts in Alberta in 2017. Basic data from: AAF, Alberta Irrigation Information, 2017, Lethbridge, July 2018.
209	Approximate average rehabilitation per irrigated acre/yr. (i.e., ($30million/year * 25%)/1.42 million acres).
210	Average, 13 Alberta Irrigation Districts. Basic data from: AECOM Canada Ltd., Irrigation Sector: Conservation, Efficiency, and Planning Report Appendices, Alberta Agriculture, 2010.
211	Alberta Agriculture, Irrigation in Alberta, 2016. The crop and system composition is similar in Saskatchewan.
212	Paterson, B., Economic Value of Irrigation in Alberta, AIPA, Lethbridge, 2015, and K. D. Russell/UMA, Irrigation Development in Alberta: The Economic Impact, AIPA, Lethbridge, 1984. Also see: MAA, The Benefits of Irrigation in Southern Alberta in the Year 2000 and Beyond, Alberta Agriculture, Lethbridge, 1999.
213	Alberta Agriculture & Forestry, Irrigation in Alberta, website.
214	The current composition of Alberta's electrical energy is: coal and coke 47%, natural gas 40%, wind 7%, biogas and geothermal 3%, and HYDRO 3 percent. Contrast this to the percentage of hydro-electricity in other provinces: Manitoba 97%, Quebec 95%, B.C. 88%, and Saskatchewan 13 percent. See National Energy Board (NEB), Provincial and Territorial Energy Profiles, 2016, Ottawa.

215 The comparative current costs of alternative energy sources, including the external (social) cost, is approximately as follows (US cents/kwh): existing coal 8.8 cents, natural gas 5.2 cents, wind with NG backup 8.3 cents, solar PV with NG backup 11.8 cents, and HYDRO 6.4 cents. See: Greenstone, M., The True Costs of Alternative Energy Sources, Brookings, Washington, 2012. Also see Wikipedia.

216 Since water storage projects often have negative direct benefits at or near the project location but most benefits accrue elsewhere, the NIMBY mentality will most likely persist.

217 See especially: Dinar, A., and A. Subramanian (eds.), Water Pricing Experiences: An International Perspective, World Bank Technical Paper #386, Washington, D. C., 1997.

218 Water has many of the characteristics of a public good; defined as those goods: a) which violate the exclusion principle such that you get "free riders"; b) for which there are inherent externalities (spillovers, side-effects, or third-party costs); and/or c) for which there is imperfect competition. Perfect competition requires, in particular, numerous buyers and sellers as well as no "transaction costs" (i.e., perfect information).

219 There is, however, some public support for the introduction of tradable water entitlements which protect the interests of established water users. See: Dinar and Subramanian, op. cit., p. 39.

220 Authorized sales, as per Sections 81 and 82. If the transfer is permanent, Section 83 of the Act also allows the AE Regional Director to withhold 10% of the transfer when the new license is issued.

221 Pedology Consultants/MAA, Farmland Drainage in Central and Northern Alberta, FF Research Project No. 82-0070, Edmonton, 1984.

222 WER Engineering, On-Farm Water Management Study: An Evaluation of On-Farm Water Management Options in Alberta, Alberta Agriculture, 1993.

223 "Muskeg" (a Cree derivative) is the standard term in Western Canada and Alaska, while "bog" is the common term elsewhere. Muskeg consists of dead plants in various states of decomposition (as peat) and tends to have a water table near the surface. They are ideal habitat for beaver, muskrat, and a wide variety of other organisms. The vegetation sometimes includes tamarack (larch) and poplar.

224 A B/C ratio <1 is not feasible. The analyses also assumed that the social value of wildlife was equal to the cost of wildlife mitigation; a questionable assumption. See: GOA/Interdepartmental Steering Committee on Drainage, Inventory of Alberta's Drainage Requirements, Edmonton, 1985.

225 The annual flooding in Bangladesh is Exhibit #1. See: SNC Lavalin. Kalni-Kushiyara River Improvement Project and the Dampara Water Management Project, CIDA, Dhaka, 1997. Also see: WER Engineering-AFH/CIDA, Guidelines and Criteria for Planning and Design of River Flood Control (Indonesia), Jakarta, 1993.

226 There are various measurement techniques: Contingency Valuation Method (CVM), sometimes called the Willingness to Pay Method (WTP), as well as the Travel Cost Method (TCM). Numerous websites provide detailed explanations of each.

227 Robert Costanza, University of Vermont. See: Western Producer, January 8, 2009.

228 Anderson, M. S., and M. L. Lerohl, "Ecological Economics and Agriculture in Alberta: Status Report," unpublished, 1992.

229 Repetto, R., "Nature's Resources as Productive Assets," Challenge, Fall 1989.

230 Cullen, M., "Go Green, You May Live Longer," Edmonton Sun, February 25, 2018.

231 See especially: Smith, K. R., "Public Payments for Environmental Services from Agriculture: Precedents and Possibilities," American Journal of Agricultural Economics, Vol. 88, 2006, pp. 1167–1173. Also see: Baylis, K., et. al., "Payments for Environmental Services: A Comparison of US and EU Agri-Environmental Policies," Ecological Economics, 2008.

232 "A New Furrow," The Economist, September 1, 2018.

233 Based on 2.1 million US farmers and 900 million acres of cropland.

234 Wilson, B., "KAP Proposal Pays Farmers for Environmental Care," The Western Producer, January 31, 2002.

235 Primarily based on information from: Arnason, R., Western Producer, December 1, 2017.

CHAPTER 11

236 This is data from the U.S. Environmental Protection Agency (EPA) for the USA.

237 Bjorn Lomborg, The Skeptical Environmentalist, Cambridge University Press, Cambridge, 1998.

238 Nordhaus, W. D., "Reflections on the Economics of Climate Change," Journal of Economic Perspectives, Vol. 7 (Fall 1993), pp. 11–25.

239 This and subsequent Sections of this chapter draw heavily upon: Toma-Bouma. Marginal Abatement Cost Assessment of Greenhouse Gas Emissions in Alberta's Agricultural Sector, AESRD/AAFRD, June 2014, and: MAA, Policy Implications of Climate Change on Alberta Agriculture, AAFRD/Policy Secretariat. Edmonton, 2008.

240 Note that this only considers fossil fuels and cement manufacturing.

241 Approximate. Agriculture estimate from Table 11.1 is for 2011. Data for the oil and gas sector, transportation sector, and the electricity sector are for 2015 from GOC, "Greenhouse gas emissions by Canadian economic sector." Remaining estimates are imputed from 2005 data.

242 A megatonne is a million tonnes. Our source: Arnason, R., The Western Producer, October 13, 2016.

243 Ibid.

244 Martin, L., "Canada may already be carbon neutral, so why are we keeping it a secret?," Financial Post, March 2, 2016. (Figure 11.1 indicates the Big Four polluters are responsible for only 55% of global CO_2 emissions.)

245 The estimate in 2011 was 19.4 Mt. See: Worth, D.E., et al., "Agricultural Greenhouse Gases," in: Clearwater, R. I., et al. (eds.), Environmental Sustainability of Canadian Agriculture: Agri-environmental Indicator Report Series – Report #4, Agriculture and Agri-Food Canada, Ottawa, 2016.

246 This also ignores potential increases over time. The 2015 estimate is reportedly 57 Mt.

247 Smith, P., e. al., "Agriculture, Forestry, and Other Land Use," in: Edenhofer, O., et al.(eds.), Climate Change 2014: Mitigation of Climate Change – Contribution of Working Group III to the Fifth Assessment Report, Intergovernmental Panel on Climate Change (IPCC), Cambridge University Press, Cambridge, 2014, pp. 811–922.

248 Lamb, A., et. al., "The Potential for Land Sparing to Offset Greenhouse Gas Emissions from Agriculture," Nature Climate Change, Vol. 6 (2016), pp. 488–492.

249 Brazilian and Indonesian forests, CO_2 "sinks," which analysts consider essential to maintaining a sustainable eco-system are generally assigned CO_2e sink values of 4-8 tonnes/acre/year (i.e., 10-20 tonnes/ha./year). See: Toma, D., and M. Anderson, Emissions in Alberta's Agricultural Sector, Alberta Environment, June 2014.

250 A mega-tonne is a million tonnes. This is also sometimes expressed in terms of teragrams (Tg) which is equal to 10^{12} grams. One Tg = one trillion grams = one mega tonne = one million tonnes. Data estimate from: NIR, "National Inventory Report 1990-2014: Greenhouse Gas Sources and Sinks," in: Canada's Submission to the United Nations Framework Convention on Climate Change, Environment and Climate Change Canada, 2017.

251 Fan, J., et al., "Increasing Crop Yields and Root Input Make Canadian Farmland a Large Carbon Sink," Geoderma, Vol. 336 (2019), pp. 49–58.

252 Lamb, A., et. al., "The Potential for Land Sparing to Offset Greenhouse Gas Emissions from Agriculture," Nature Climate Change, Vol. 6 (2016), pp. 488–492.

253 The Rodale Institute is a US-based non-profit organization that supports research into organic farming and promotes a holistic, whole-systems approach to agriculture and biologically sustainable solutions to climate change. An undercurrent to the Institute's work is the need for productive agricultural alternatives to pesticides, herbicides, and (chemical) fertilizers.

254 Also see: Intergovernmental Panel on Climate Change (IPCC) – UNEP-World Meteorological Organization, Climate Change 2007, IPCC Assessment Report, Geneva, 2014. See: www.ipcc.ch/

255 Ibid.

256 See, especially, research by the Prairie Adaptation Research Collaborative (PARC), Regina.

257 Plants `breath in carbon dioxide and `breath out oxygen.

258 McGinn, S., and A. Shepherd, "Impact of Climate Change Scenarios on the Agri-Climate of the Canadian Prairies," Canadian Journal of Soil Science, Vol. 83 (2003), pp. 623–630.

259 Smith, W. N., et. al., Assessing the effects of climate change on crop production and GHG emissions in Canadian agriculture," Ecosystems and Environment, August 2013.

260 Kimball, B., et. al., "Data from the Arizona FACE Experiments on Wheat at Ample and Limiting Levels of Water and Nitrogen," Open Data Journal for Agricultural Research, USDA/ARS, 2017.

261 Williams, G. D., et. al., *Estimating Effects of Climate Change on Agriculture in Saskatchewan, Canada*, Environment Canada, Downsville, 1988. As reported in Stewart, D. B., "Climate Change and Agricultural Production in the Canadian Prairies," in: Wall, G. (ed.), Symposium on the Impacts of Climate Change and Variability on the Great Plains, University of Waterloo, Waterloo, 1990, p 135.

262 Carter, T.R., and R. A. Saarikko, "Estimating Regional Crop Potential in Finland under a Changing Climate, Agricultural and Forest Meteorology, Vol. 79 (1996), 301–313.

263 As reported in: Parry, M., and Z. Jiachen, "The Potential Effect of Climate Changes on Agriculture," in: Jager, J., and H. L. Fergusson (eds.), Climate Change: Science, Impacts and Policy, Proceedings of the Second World Climate Conference, Cambridge University Press, Cambridge, 1991, p. 281.

264 MAA, Policy Implications of Climate Change on Alberta Agriculture, AAFRD/Policy Secretariat. Edmonton, 2008.

265 As reported in the Edmonton Sun, January 22, 2018.

266 Alberta Report, June 1987.

267 See, for example, Alberta Agriculture (2008), Alberta Research Council-ARC (various), and the Prairie Adaptive Research Collaborative-PARC (various).

268 Pietroniro, A., B. Toth, and J. Toyra, "Water Availability in the SSRB under Climate Change," Climate Change and Water in the Prairies Conference, Saskatoon, Saskatchewan, June 22, 2006. (SSRB = South Saskatchewan River Basin)

269 Research in this area is extensive. See, for example:

Bootsma, A., S. Gameda, and D. W. McKenney, Adaptation of Agricultural Production to Climate Change in Atlantic Canada, AAFC/CFS, Halifax, June 2001.

French, A. Most Recent Scenarios of Future Climate Change for Canada and Alberta: Implications for Agriculture, AIA Conference, Banff, April 2014.

Herrington, R., B. Johnson, and F. Hunter, Responding to Global Climate Change in the Prairies, Climate Impacts and Adaptation, Canada Country Study Vol. III, Environment Canada, Ottawa, 1997.

Koshida, G., and W. Avis (eds.), The Country Canada Study: Climate Impacts and Adaptation – National Sectoral Volumes, Environment Canada, Ottawa, 1998.

Nyirfa, W., and W. Heron, Assessment of Climate Change on the Agricultural Resource of the Canadian Prairies, PARC and AAFC/PFRA, Regina, February 2004. See: www.parc.ca/research_pub_scenarios.htn.

Parry, M. L., T. R. Carter, and N. T. Konijn (editors), The Impact of Climatic Variations on Agriculture, Volume 1: Assessments in Cool Temperate and Cold Regions, Kluwer Academic Publishers, Boston, 1988.

270   MAA, An Analysis of Agricultural Research and Productivity in Alberta, Environment Council of Alberta, Edmonton, January 1983.

271   Statistics Canada, Aggregate Productivity Measures, 1981, Catalogue No. 14-201, Ottawa.

272   Also see: Grise, J., and S. N. Kulshreshtha, "Farmers Choice of Crops in Canadian Prairies under Climate Change: An Econometric Analysis," Journal of Earth Science & Climate Change, February 2016.

CHAPTER 12

273   This Section draws upon numerous sources. A good primer is "The Effects of Pollution" in the Encyclopaedia Britannica.

274   Soares, J. R., et al., Scientific Reports 6, No. 30349, 2016.

275   The Earth Institute, "Global Study Shows How Agriculture Interacts with Industry," Columbia University, 5-16.

276   Based on data from the National Centre for Scientific Research (CNRS), France, March 2018. (Professor Juilliard; reported by Josh Gabbatiss.)

277   Farm & Food Care Ontario, Survey of Pesticide Use in Ontario, 2013/14, November 2015. (Imperial equivalencies = 1.1 lbs/acre, 25.1 lbs/acre, and 3.2 lbs/acre, respectively.)

278   Properly called neonicotinoids, these are a class of neuro-active insecticides chemically similar to nicotine. They are rapid action broad-spectrum systemic insecticides, applied as sprays, drenches, seed and soil treatments. In Ontario, in recent years, nearly all corn seed and 60 percent of soybeans have been treated with a neonic. Soybeans in Manitoba, as well as an increasing amount of canola seed on the prairies has similarly been treated with a neonic. In the USA, about 95% of corn and canola crops, most cotton, sorghum, and sugar beets, and about half of all soybeans are treated with a neonic. They are also used on most US fruits and vegetables.

279   The Canadian data is from: Farm & Food Care Ontario, op. cit. The U.S. data is from: USDA/ERS, Pesticide Use in U.S. Agriculture: 21 Selected Crops, 1960-2008, Washington, DC, 2009.

280   Farm & Food Care Ontario, op. cit.

281   Farm & Food Care Ontario, op. cit.

282   As reported in the Western Producer, January 24, 2019, p. 43.

283   Arnason, R., ""U.S. releases clear stance on glyphosate". Western Producer, May 30, 2019.

284   Kniss, A. R., "Long-term trends in the intensity and relative toxicity of herbicide use," Nature Communications 8, Article No. 14865, April 2017. An exception is copper sulfate, which is an eternally persistent heavy metal that can be toxic to certain organisms (e.g., sheep).

285   A diamide is a chemical compound containing two amide groups. An amide is a metallic derivative of ammonia.

286   Also see: "Twelve Tips to Clubroot Management," Grainews, January 8, 2019.

287   GOA, Alberta Clubroot Management Plan, Agdex 140/638-2, 2014. Also see: AARD, Agri-Facts, Clubroot Disease of Canola and Mustard, May 2007.

288   Farmers often inaccurately refer to this as alkali. It is, instead, a salt.

289   Alberta Agriculture & Forestry, Natural/Irrigation Salinity, 2017, and Irrigation Canal Seepage Salinity, 2017.

290   Shapley, P. S., "Pollution from Industry and Agriculture: Nitrogen Oxides in the Atmosphere," UIUC, undated.

291   The author does not either endorse or authenticate any claims made by any products identified herein.

292   Also see: MAA, Socio-Economic Assessment of Low Phytate Barley, AAFRD, Edmonton, September 2002.

293   As reported in New Holland News, winter 2018.

294   This is extracted almost verbatim from: Toma, D., and M. Anderson, Emissions in Alberta's Agricultural Sector, Alberta Environment, June 2014.

295   Basarab, J., "Residual feed intake (RFI): An indirect approach for reducing GHG emissions," Alberta Beef Magazine, 2013.

CHAPTER 13

296   This section draws very heavily upon "Organic Food" in Wikipedia and "Regulating organic products in Canada," documentation available through the Canadian Food Inspection Agency.

297   Chemical ripening (e.g., banana) generally uses either ethylene or propylene gas. Food irradiation (with gamma rays, X-rays, or electron beams) is a technology that improves the safety and extends the shelf-life of foods by reducing or eliminating microorganisms and insects.

298   According to the Canada Organic Trade Association. As reported in the Western Producer, December 30, 2018.

299   This somewhat over-estimates the producing acreage since green manure crops are generally planted in a 2/3 or 3/4 rotation. Data from the Canadian Organic Trade Association.

300   Statistics Canada, Canada at a Glance, 2017.

301   Statistics Canada, Canada's Dairy Industry at a Glance, 2017.

302   Western Producer, November 22, 2018, p. 5. The estimate for milk is from the Washington Post, May 4, 2017.

303   KPMG, Global Organic Milk Production Market Report, June 2018, p. 3.

304   Product-specific and still subject to further research. KPMG finds organic dairy farming costs are 1.3 to 1.6 times higher than conventional dairy costs. Also see Chapter 10, Section 10.5.

305   A meta-analysis is basically an empirical analysis of numerous (sometimes 100s) of independent scientific studies to try to derive a statistically significant consensus.

306   NDTV, 2016.

307   Unless otherwise noted, all information in this Section is derived from Agriculture and Agri-Food Canada websites.

308   See "Nutraceutical," Wikipedia.

309   See especially: Hasler, C. M., "Functional Foods: Benefits, Concerns, and Challenges," The Journal of Nutrition, Vol. 132, December 2002.

310   See, especially, Wikipedia and "Restrictions on Genetically Modified Organisms: Canada," U.S. Library of Congress.

311   Bio-engineering or biotechnology is a general term used to describe the amalgamation of science, technique and

art that has as its focus the manipulation and study of life forms and processes at the level of molecular information and regulation. Encompassed in this definition are molecular, cellular, organism, and whole system replication, transformation, and interaction. Also sometimes referred to as nanotechnology engineering, gene technology, recombinant DNA technology, or genetic engineering.

312   Data from the Western Producer, July 5, 2018. Arable land estimate includes crops, fallow, and improved pasture, Chapter 3, Table 3.2.

313   Stuart Smyth (U. of Saskatchewan), "New York Times ignored GM crop benefits," Western Producer, November 10, 2016.

314   Obituary, Calestous Juma, The Economist, January 2018.

CHAPTER 14

315   See, in particular, Fagan, J. B., "Restrictions should be placed on the genetic engineering of food," in: Dudley, W. (ed.), Opposing Viewpoints in Social Issues, Greenhaven Press, San Diego, 2000, pp. 268–275.

316   The insights of Mr. Don Macyk are gratefully acknowledged. Also see: Michael Boehlje, "Megatrends Impacting Agriculture," Professor Emeritus, Purdue University, 1988 and numerous, more recent, presentations.

317   Technology Quarterly – The Future of Agriculture, The Economist, June 2016, pp. 7-8.

318   The Innovation Issue, The Western Producer, December 27, 2018, pp. 19, 46-49.

319   So-called "information technologies" embrace a myriad of related technologies, including: a) GPS "precision" agriculture; b) artificial intelligence (AI); c) monitoring devices; and d) robotics.

320   Western Producer, December 27, 2018, p. 43.

321   And as we write this, Bunge and ADM are still rumoured to be courting each other.

Index

Please note that the italicized "*t*" after the page number refers to a table; "*f*" refers to a figure and "*n*" to endnotes.

2015 *Alberta Directory, The,* 137*n*141, 264*n*141
3-D printers, ... 252
4-H, ... 133–134, 137

A
"A New Furrow" *(The Economist),* 203*n*232, 269*n*232
AAFC. *See* Agriculture and Agri-Food Canada (AAFC)
Aboriginal peoples, ... 8, 13, 125
    see also First Nations peoples; individually-named peoples
    in British Columbia, ... 126, 126*t*7.2
    history of agricultural development and, ... 1
    in Manitoba, ... 126, 126*t*7.2
    in Ontario, ... 126, 126*t*7.2
    in Saskatchewan, ... 126, 126*t*7.2
    unemployment rate for, ... 126
absolute trade advantage, ... 88
Acadian peoples, ... 8
Adaptation of Agricultural Production to Climate Change in Atlantic Canada (Bootsma, Gameda & McKenney), ... 219*n*269, 270*n*269
Advance Payments for Crops Act (APCA, 1977), ... 157*t*9.2, 158
AEPs. *See* agri-environmental policies (AEPs)
AESs. *See* agri-environmental Schemes (AESs)
Ag Advisor, ... 236*t*13.2
Aggregate Measure of Support (AMS), ... 62*n*39, 156*n*167, 259*n*39, 265*n*167
Aggregate Productivity Measures, 1981 (Statistics Canada), ... 219*n*271, 271*n*271
Agreement on Internal Trade (AIT), ... 163–164
agri-business,
    see also dairy product manufacturing industry; food and beverage processing industry;
        grain and oilseed milling industry; meat product manufacturing industry
    employment in, ... 93, 93*t*6.1

grain and oilseed milling sector, ...96–104
implications of consolidations in, ... 252
linkages and primary agriculture make up, ...116–117
mergers and consolidations in, ... 252, 252*n*321, 273*n*321
size of, ...93
vertical integration in, ... 254, 255
agri-environmental policies (AEPs), ... 203
in Canada, ...204–205
in European Union (EU), ... 203
in United States (US), ...203–204
agri-environmental schemes (AESs), ... 203
Agri-Facts, ... 227*n*287, 272*n*287
Agri-Food Canada, ...3, 163
agri-food industries, ...2
backward linkages to producers, ... 93–95
forward linkages to producers, ... 93–95
Agri-Food Innovation Council (AIC), ... 166*n*176, 169*f*9.4, 266*n*176
agri-food products,
export sales of, ...75, 76*t*5.3
import sales of, ...75, 76*t*5.3
net trade surplus in, ... 77
agricultural adaptation (to climate change),
efficient use of fertilizer, ... 220
land use (cropping pattern) adjustments, ...219–220
minimum tillage, ... 220
technological changes, ... 219
Agricultural Agreement (WTO), ... 74, 159
Agricultural and Rural Development Act, The (*ARDA,* 1962), ... 156
Agricultural Degradation in Western Canada (Knapik), ...182*t*10.2
Agricultural Economics, ...203*n*231, 268*n*231
Agricultural Forest Meteorology, ...218*n*262, 270*n*262
"Agricultural Greenhouse Gases" (Worth), ... 214*n*245, 269*n*245
Agricultural Implements (Hambre & Asseltine), ... 113*n*98, 262*n*98
Agricultural Income Disaster Assistance (AIDA), ... 158–159, 159*n*170, 160, 265*n*170
Agricultural Land Degradation in Western Canada: A Physical and Economic Overview
(Knapik), ... 178*n*189, 266*n*189
Agricultural Land Reserve (ALR), ... 187
in British Columbia, ...18
Agricultural Marketing Programs Act (1997), ... 156, 158
agricultural organizations, ...137–152
commodity groups, ...143–145
evolution of, ... 152
growth of, ... 152

    national marketing boards, ... 145–146
    provincial marketing boards and commissions, ... 147–150
    recent decline in farmer participation in, ... 152
    roles of, ... 152
    rural cooperatives, ... 137–139
    supplier and processor organizations, ... 152
    umbrella (general farm) organizations, ... 139–142
Agricultural Policy Framework (APF), ... 159, 170
Agricultural Producers of Saskatchewan (APAS), ... 140
Agricultural Products Cooperative Marketing Act (1985), ... 158
Agricultural Products Marketing Board Act (1947), ... 143, 154, 158
agricultural robots, ... 248
agricultural societies, ... 125, 133
Agricultural Stabilization Act (ASA, 1958), ... 155–156, 156n167, 157, 157t9.2, 265n167
agriculture,
    agri-business structure variable in future of, ... 243, 252–256
    consumer-driven variable in future of, ... 243, 245
    future of, ... 256
    global position system (GPS) precision, ... 246–247, 248n319, 273n319
    internationalization variable in future of, ... 243–245
    mitigation of GHG measures, ... 216, 216t11.6
    multi-spectral sensors in, ... 247–248
    as part of complex biological system, ... 256
    regulation variable in future of, ... 243, 245–246
    robots in, ... 246, 248, 248n319, 273n319
    science as bedrock of, ... 256
    state supported, ... 153
    technological change variable in future of, ... 243, 246–251
    unique regions for, ... 153
    water availability variable in future of, ... 243, 251–252
"Agriculture, Forestry, and Other Land Use" (P. Smith), ... 214n247, 269n247
Agriculture and Agri-Food Canada (AAFC), ...39f3.18, 40f3.19, 71n48,
    72n49, 72n50, 73n51, 76t5.3, 99n74, 102n84, 106t6.7, 107t6.8, 118n108, 142, 161, 166f9.2,
    192n208, 213f11.8, 213t11.3, 218f11.11, 219n269, 236n307, 237, 259n48, 259n49, 259n50,
        259n51, 261n74, 261n84, 262n108, 267n208, 270n269, 271n269, 272n307
    organizational structures of, ... 161–162, 162f9.1
    Rural Secretariat, ... 132
AgriInsurance (prev. All-Risk Insurance; prev. *Crop Insurance Act* (1959)), ... 160
AgriInvest, ...31, 158, 160, 173
AgriMarketing, ...112f6.10
AgriRecovery, ... 160
AgriStability, ... 31, 159–160, 173
AIA. *See* Alberta Institute of Agrologist (AIA)

AIC. *See* Agri-Food Innovation Council (AIC)
AIDA. *See* Agricultural Income Disaster Assistance (AIDA)
air pollution, ... 221, 221*t*12.1
AIT. *See* Agreement on Internal Trade (AIT)
Alberta,
    agricultural societies in, ... 133
    ALUS in, ... 205
    barley farming in, ... 49
    beef production in, ... 14, 56
    canola farming in, ... 46
    climate change transposition in, ... 218, 218*f*11.11
    feed grain in, ... 105
    historical view of agriculture in, ... 13–18
    Hutterites in, ... 126
    importance of irrigation to, ... 22, 22*f*3.4
    integrated resource plans, ... 190*n*203, 267*n*203
    irrigation in, ... 192–193, 193*f*10.7, 194, 194*f*10.8
    meat product manufacturing industry in, ... 94
    potato farming in, ... 101, 102
    provincial crown land in, ... 175
    pulse farming in, ... 55
    resource-based programs in, ... 170–171
    sheep farming in, ... 66
    Social Credit party in, ... 16–17
    soybean farming in, ... 53
    sugar beet farming in, ... 56
    sugar processing in, ... 103
    urban sprawl initiatives in, ... 186
    wheat farming in, ... 48
Alberta Agriculture, ... 22*f*3.4, 193*f*10.7, 193*n*210, 199*n*222, 218*n*267,
                                              267*n*210, 267*n*210, 268*n*222, 270*n*267
Alberta Agriculture and Forestry, ... 196*n*213, 227*n*289, 267*n*213, 272*n*289
Alberta Beef Magazine, ... 231*n*295, 272*n*295
Alberta Economic Development, ... 165*n*174, 266*n*174
Alberta Environment, ... 170, 231*n*294, 272*n*294
Alberta Environment & Parks, ... 194
Alberta Farm Express, ... 142*n*150, 188*n*200, 264*n*150, 267*n*200
Alberta Farmer, ... 60*n*35, 142*n*153, 259*n*35, 264*n*153
Alberta Federation of Agriculture, The, ... 140*n*148, 264*n*148
Alberta Institute of Agrologist (AIA), ... 219*n*269, 270*n*269
Alberta Irrigation Projects Association, ... 192
Alberta Report, ... 218*n*266, 270*n*266
Alberta Research Council, ... 218*n*267, 270*n*267

Alberta Waterfowl Habitat Enhancement Projects: Benefit Distribution (Pattison), ... 178*n*188, 202*t*10.9, 266*n*188
Alberta Wheat and Alberta Barley Newsletter, ... 83*n*63, 85*n*65, 260*n*63, 260*n*65
Alberta Wheat Commission, ... 142*n*150, 264*n*150
Alberta Wheat Pool, ... 138
All-Risk Insurance. *See* AgriInsurance (prev. All-Risk Insurance; prev. prev. *Crop Insurance Act* (1959))
ALR. *See* Agricultural Land Reserve (ALR)
alternative ecological systems, ... 249
    hydroponics, ... 249
    inland aquaculture, ... 249
    vertical farming, ... 249
alternative energy, ... 197*n*215, 268*n*215
alternative food sources, ... 249–250
alternative food uses, ... 250
alternative land use (ALU) model, ... 190, 255–256
alternative land use services (ALUS), ... 204, 206, 255–256
    farmers could be recruited as, ... 202
    in Manitoba, ... 205
    in PEI, Ontario, Saskatchewan and Alberta, ... 204–205
ALU. *See* alternative land use (ALU) model
ALUS. *See* alternative land use services (ALUS)
Amazon.com, ... 234
American Revolutionary War, ... 11
AMS. *See* Aggregate Measure of Support (AMS)
Analysis of Agricultural Research and Productivity in Alberta, An (MAA), ... 219*n*270, 271*n*270
Anderson, M., ... 215*n*249, 231*n*294, 269*n*249, 272*n*294
Anderson, M.S., ... 202*n*228, 268*n*228
Animal Diseases and Protection Act (1970), ... 157*t*9.2
animal food manufacturing industry, ... 94
animal husbandry, ... 13
Animal Nutrition Association of Canada, ... 114*n*103, 262*n*103
APAS. *See* Agricultural Producers of Saskatchewan (APAS)
APCA. See Advance Payments for Crops Act (*APCA,* 1977)
APF. *See* Agricultural Policy Framework (APF)
Archer, J.A., ... 264*n*143
ARDA. *See* Agricultural and Rural Development Act, The (*ARDA,* 1962)
"Are Canada's Farm Organizations Actually Listening to Their Members?" (Pilger), .. 140*n*146, 140*n*147, 264*n*146, 264*n*147
Arnason, R., .. 127*n*125, 205*n*235, 213*n*242, 225*n*283, 263*n*125, 269*n*235, 269*n*242, 271*n*283
artificial intelligence (AI), ... 248, 248*n*319, 249, 273*n*319
ASA. *See* Agricultural Stabilization Act (*ASA,* 1958)
Ashton, William (Bill), ... 119*n*109, 131*n*131, 262*n*109, 263*n*131

Asian Development Bank, ... 202t10.9
Asseltine, A., ... 113n98, 262n98
"Assessing the effects of climate change on crop production and GHG emissions in Canadian agriculture" (W.N. Smith), ... 218n259, 270n259
Assessment of Climate Change on the Agricultural Resources of the Canadian Prairies (Nyirfa & Heron), ... 218n267, 219n269, 270n267, 271n269
At the Heart of Transportation: A Moving History (Canadian Transportation Agency), . 164n172, 164n173, 265n172, 265n173
Atlantic Canada Opportunities Agency, ... 132
Atlantic provinces,
 farm statistics in, ... 19–21, 19f3.1, 20t3.1, 21f3.2
 historical view of agriculture in, ... 8–9
 importance of retail cooperatives in, ... 138
Avis, W., ... 219n269, 271n269

B

Back, J., ... 50n27, 258n27
backward linkages, ... 93, 96–104, 107–116
 in food and beverage processing industry, ... 93, 96
 impact on Canadian economy, ... 116–117
 importance to economy of, ... 107
 input suppliers for primary producers, ... 93
Bank of Canada, ... 78
barley farming, ... 2, 9, 10, 14, 16, 48–50
 barley malting plants, ... 99
 CPTPP and, ... 85
 exports of, ... 72, 100
 international prices for, ... 77
 as major crop, ... 45, 45t4.1, 46f4.1, 46t4.2
 production process, ... 49
 reasons for production drop in last ten years, ... 49–50
 as third most important cereal crop in Canada, ... 49
 uses of, ... 49
"Barley Production and Research Trends in Canada" (Brophy), ... 48n26, 258n26
Barratt, R.F., ... 102n85, 261n85
Basarab, J., ... 231n295, 272n295
Baylis, K., ... 203n231, 268n231
Bear, F.E., ... 236t13.2
Bedard, A.F., ... 128n130, 263n130
Bedard, M., ... 116n106, 262n106
beef production (ranching), ... 3, 8, 9, 10, 11, 56–57
 in Alberta and Saskatchewan, ... 14, 56
 animal diseases and, ... 227
 auction markets, ... 104

INDEX 281

backgrounding process in, ... 57
in British Columbia, ... 56
Canada as sixth world exporter of, ... 73
CETA and, ... 87
cow-calf operations component of, ... 56–57
CUSMA and, ... 83
exports of, ... 72–73
feed grain, ... 105
feed lots, ... 104–195
feeder (slaughter cattle) operations component of, ... 56, 57
finishing process in, ... 57
GHG and, ... 228
international prices for, ... 77
in Manitoba, ... 56
in Ontario, ... 56
as principal source of GHG, ... 211, 211*f*11.5
processing plants, ... 105, 105*t*6.6
residual feed intake (RFI) and, ... 231
size of, ... 56
slaughterhouses in, ... 57
US exports, ... 72, 72*f*5.4
Belanger, J., ... 128*n*130, 263*n*130
Bells, Robert A., ... 100*n*80, 261*n*80
Beothuk peoples, ... 8
beverage manufacturing industry, ... 94
distilling industry, ... 100
major brewing companies, ... 100
wine industry, ... 100
Bible, The (King James version), ... 69
Bill C-34. See *Transportation Appeal Tribunal of Canada Act* (2002) (prev. *Bill C-34*)
bio-engineering, ... 238*n*311, 272–273*n*311
biofortification, ... 248
biofuels, ... 52
Bioproducts Discovery and Development Centre, University of Guelph, ... 250
biotechnology, ... 238*n*311, 272–273*n*311
Bloom, A., ... 133*n*136, 263*n*136
BNA Act. See *Constitution Act* (prev. *British North America Act (BNA Act)*), ... 1867
Boehjie, Michael, ... 243*n*316, 273*n*316
bog (muskeg in Cree), ... 200*n*223, 268*n*223
Bollman, Ray D., ... 121*f*7.3, 119*f*7.1, 119*n*109, 119*n*111, 123*n*117, 124*f*7.5, 126*n*121, 131*f*7.8,
    131*n*131, 262*n*109, 262*n*111, 263*n*117, 263*n*121, 263*n*131
Bootsma, A., ... 219*n*269, 270*n*269
bottom-up rural policies, ... 133–136

4-H and, ... 133–134
    agricultural societies, ... 133
    boy scouts, ... 135
    churches, ... 135
    community-based organizations, ... 133
    recreational activities, ... 135
    service clubs, ... 134–135
    Women's Institute, ... 134
Bowling Alone: The Collapse and Revival of American Community (Putman), 133*n*136, 263*n*136
bread and bakery product manufacturing industry, ... 94
    fragmentation of, ... 99
Brisson, Y., ... 105*n*90, 261*n*90
British Colombia,
    Aboriginal peoples in, ... 126, 126*t*7.2
    ALR in, ... 187
    beef production in, ... 56
    canola farming in, ... 46
    farm statistics in, ... 19–21, 19*f*3.1, 20*t*3.1, 21*f*3.2
    food and beverage processing industry in, ... 94
    GHG from agriculture in, ... 211*t*11.2
    historical view of agriculture in, ... 18
    irrigation in, ... 22, 22*f*3.4, 192
    meat product manufacturing industry in, ... 94
    net farm incomes in, ... 34, 44
    provincial crown land in, ... 175
    sales of organic foods in, ... 234
    urban sprawl initiatives in, ... 186–187
British North America Act (BNA Act). See *Constitution Act* (prev. *British North America Act
(BNA Act)*), ... 1867
Brockman, J.L., ... 198*t*10.8
Bronfman, C., ... 100*n*79, 261*n*79
Brophy, M., ... 48*n*26, 258*n*26
BSE. *See* diseases, agricultural, bovine spongiform encephalopathy (BSE)
Bucks for Wildlife, ... 190

C

Cabot, John. *See* Caboto, Giovanni (*a.k.a.* John Cabot)
Caboto, Giovanni (*a.k.a.* John Cabot), ... 8
CAIS. *See* Canadian Agricultural Income Stabilization (CAIS) Program
Caisse Populaire, ... 137
Canada,
    amount of GHGs from, ... 208–210, 208*f*11.2, 209*t*11.1
    beef production in, ... 56
    dependence of agriculture on a few major crops, ... 68

INDEX   283

    different regional evolutions of agriculture in, ... 7–8
    dominance of some provinces' farm receipts, ... 31, 31*f*3.13
    educational levels of farmers in, ... 28–29, 29*f*3.10
    farm ownership in, ... 25, 26*t*3.5
    as fifth largest agricultural exporter, ... 70
    future of agriculture in, ... 4
    GDP percentage of international trade, ... 69, 69*f*5.1
    GHG from agriculture in, ... 211*t*11.2
    historic agricultural role in, ... 7–8
    importance of international trade to, ... 69, 69*f*5.1
    international competitiveness rank, ... 81, 81*t*5.4
    land area, ... 175*n*181, 266*n*181
    major exporters and importers, ... 70, 70*f*5.2
    mineral rights in, ... 23
    national profile of farms and farmers in, ... 43–44
    net exports, ... 70, 70*t*5.1
    net farm incomes across, ... 34
    non-indigenous immigration during 1895-1914 in, ... 15
    number of people involved in agriculture today in, ... 1
    temperate climate limits crop production in, ... 67
    vastness of agriculture in, ... 67
    world commodity production compared to, ... 46*t*4.2
Canada Agricultural Review Tribunal, The (CART, 1983), ... 163
Canada at a Glance (Statistics Canada), ... 234*n*300, 272*n*300
Canada Border Services Agency, ... 163
Canada Council on Learning, ... 124, 124*n*118, 263*n*118
Canada Country Study (Environment Canada), ... 219*n*269, 270*n*269
Canada Environmental Protection Act (1999), ... 163
Canada Grain Commission, ... 97*t*6.4
"Canada may already be carbon neutral, so why are we keeping it a secret?" (Martin), ...
. . . . 214*n*244, 269*n*244
Canada Organic Trade Association, ... 234*n*298, 234*n*299, 272*n*298, 272*n*299
Canada-South Korea Free Trade Agreement (2015, CKFTA), ... 82
Canada-United States-Mexico Agreement (CUSMA, *a.k.a.* NAFTA 2.0; USMCA; T-MEC), . .
. . . . 73, 82–84, 147, 244
Canada West Foundation, ... 101*f*6.3
"Canada's Changing Religious Landscape" (Pew Research Centre), ... 125*n*120, 263*n*120
"Canada's Dairy Industries a Rich, Closed Club" (McKenna), ... 61*n*36, 146*n*155,
. . . . 259*n*36, 265*n*155
Canada's Dairy Industry at a Glance (Statistics Canada), ... 234*n*301, 272*n*301
Canada's Supply Management System (Heminthavong), ... 59*n*32, 258*n*32
"Canadian Ag Falls Behind in Research and Development" (White), ... 168*n*179, 266*n*179
Canadian Agricultural Income Stabilization (CAIS) Program, ... 159, 160

Canadian Agricultural Partnership (CAP), ... 161
Canadian Atlas, The (Royal Ontario Museum), ...7*n*1, 257
Canadian Beef Cattle Market Development and Promotion Agency, ... 148
Canadian Beef Grading Agency, ... 148*t*8.5
Canadian Cancer Society, ... 225
Canadian Canola Growers Association, ... 117*f*6.13
Canadian Cattlemen's Association (CCA), ...144, 144*t*8.4, 148
    Canfax market analysis division of, ... 57, 104
Canadian Chamber of Agriculture, ... 139
Canadian Chicken Marketing Agency (CCMA), ...64
Canadian Commonwealth Federation (CCF), ... 140
Canadian Council of Agriculture. *See* Canadian Federation of Agriculture (CFA, prev. Canadian Council of Agriculture)
Canadian Council of Food Processors, ...93*n*72, 260*n*72
Canadian Dairy Commission Act, The (1967), ... 156, 157*t*9.2
Canadian Dairy Commission (CDC, 1966), ... 61, 146, 156, 161
Canadian Dairy Information Centre, ...59*t*4.4, 60*f*4.6, 60*f*4.7, 74*f*5.7, 106*f*6.5, 106*n*91,
                                                                                235*f*13.1, 261*n*91
Canadian Egg Marketing Agency (CEMA), ... 64, 65
Canadian Encyclopaedia, The, ...7*n*1, 8*n*5, 8*n*7, 93*t*6.1, 94*t*6.2, 153*n*164, 257, 265*n*164
Canadian Farm Income Program (CFIP, 2000), ... 158–159, 160
Canadian Federation of Agriculture (CFA, prev. Canadian Council of Agriculture), . .139–140,
                                                                                140*t*8.1
Canadian Food Inspection Agency (CFIA), . . 142, 142*n*153, 162, 163, 233*n*296, 234, 236, 237,
                                                                239, 264*n*153, 272*n*296
Canadian Food Processors Association. *See* Food Processors of Canada (prev. Canadian Food Processors Association) . . . .
Canadian Frozen Food Association, ... 103
Canadian Grain Commission (CGC), ... 162, 163
Canadian Grains Council, ... 141, 141*t*8.3, 142
Canadian Hatching Egg Producers (CHEP), ... 107*t*6.8, 146, 161
Canadian Horticultural Council, ... 139
Canadian Industry Statistics - Agricultural Implement Manufacturing, ... 114*n*101, 262*n*101
Canadian Institute of Actuaries, ... 218
Canadian International Development Agency (CIDA), ... 195, 230
Canadian International Grains Institute (CIGI), ... 142, 142*n*150, 264*n*150
Canadian Journal of Agricultural Economics, ... 81*n*58, 153*n*163, 260*n*58, 265*n*163
Canadian Journal of Economics, ...62*n*38, 259*n*38
Canadian Journal of Soil Science, ... 217–218, 217*n*258, 270*n*258
Canadian Meat Council, ... 152
Canadian Milk Supply Management Committee (CMSMC), ... 61, 156
Canadian National (CN) Railway (prev. CNR), . . . 15, 150, 164, 165, 166, 166*n*175, 266*n*175
Canadian National Millers Association, ...98*n*73, 260*n*73

Canadian Northern Railway, ... 164
Canadian Oilseed Processors Association (COPA), ... 101*f*6.2
Canadian Organic Standards,
    Organic Production Systems - General Principles and Management Standards, ... 233
    Organic Production Systems - Permitted Substance Lists, ... 233
Canadian Pacific (CP) Railway, ... 15, 18, 150, 164, 165, 166*n*175, 266*n*175
Canadian Para-Mutual Agency, ... 162, 163
Canadian Plant Technology Agency (CPTA), ... 142
Canadian Pork Council (CPC), ... 149
Canadian Revenue Agency (CRA), ... 160
Canadian Seed Growers' Association (CSGA), ... 142, 152
Canadian Seed Institute (CSI), ... 142
Canadian Seed Trade Association (CSTA), ... 142, 152
Canadian Socio-Economic Information Management System (CANSIM), ... 73*f*5.5,
    93*t*6.1, 192*f*10.6
Canadian Sugar Institute, ... 103*n*87, 261*n*87
Canadian Transportation Agency, ... 164*n*172, 164*n*173, 165, 265*n*172, 265*n*173
Canadian Turkey Marketing Agency (CTMA), ... 64
Canadian Veterinary Medical Association, The (CVMA), ... 114*n*104, 262*n*104
Canadian Wheat Board (CWB), ... 17, 90, 142, 153*n*165, 155, 158, 164, 165, 265*n*165
    ceased operations in 2011, ... 48, 158
    establishment of in 1935, ... 154
Canadians for a Sustainable Society, ... 184*n*194, 266*n*194
Canola Council of Canada (CCC), ... 144, 144*t*8.4, 149
canola farming, ... 2
    benefits of GMO-produced, ... 240
    breaking wheat-canola rotation, ... 229
    Canada as largest exporter of, ... 71
    contribution of to Canadian economy, ... 99
    domination of, ... 30, 30*f*3.11, 46
    economic benefits of, ... 117*f*6.13
    as GMO, ... 46–47
    herbicide use on GMO, ... 224
    linkages exceed value of, ... 116
    as major crop, ... 45, 45*t*4.1, 46*f*4.1, 46*t*4.2
    NAFTA and, ... 83
    oilseed crushing plants for, ... 99
    production process, ... 47
    single most important crop, ... 46
    uses of, ... 47
Canola Guide (Entz), ... 229*t*12.2
Canola Performance Trials (CPT), ... 149
CANSIM. *See* Canadian Socio-Economic Information Management System (CANSIM)

CAP. *See* Canadian Agricultural Partnership (CAP)
capital-intensive farming, ...36
Capitalism, Socialism, and Democracy (Schumpeter), ... 122*n*116, 262*n*116
carbon sinks, ...3
Carrol, Jaques, ... 139*n*145, 143*n*154, 264*n*145, 265*n*154
Carson, Rachel, ... 221
CART. *See* Canada Agricultural Review Tribunal, The (CART, 1983)
Carter, C., ...62*n*38, 259*n*38
Carter, T.R., ... 218*n*262, 219*n*269, 270*n*262, 271*n*269
CCA. *See* Canadian Cattlemen's Association (CCA)
CCF. *See* Canadian Commonwealth Federation (CCF)
CCMA. *See* Canadian Chicken Marketing Agency (CCMA)
CDC. *See* Canadian Dairy Commission (CDC, 1966)
CEMA. *See* Canadian Egg Marketing Agency (CEMA)
cement manufacturing, ... 209*n*240, 269*n*240
Census Metropolitan Area (CMA), ... 121*n*114, 262*n*114
Census of Agriculture (2005, Statistics Canada), ... 67*t*4.7
Census of Agriculture (2007, Statistics Canada), ... 54*f*4.2
Census of Agriculture (2011, Statistics Canada), ... 58*t*4.3
Census of Agriculture (2016, Statistics Canada), ... 2, 26*n*19, 34*n*20, 38*t*3.10, 41*t*3.12, 43,
56*n*31, 67*t*4.7, 258, 258*n*31
    age of farm operators, ...26–27, 27*f*3.8
    agricultural land base, ... 21, 21*f*3.3
    farm definition, ...20
    farm structures, ... 26*f*3.7
    farm types, ... 25*t*3.4
    farmland area by tenure, ... 24*f*3.3
    gender of farm operators, ... 28*f*3.9
    land use and principal crops, ... 22*f*3.2
    number of farms and sizes of, ... 20*t*3.1
Census of Agriculture (1981-2011, Statistics Canada), ... 55*f*4.3, 58*f*4.4
Central Intelligence Agency (CIA), ... 70*t*5.1
Cereals Canada, ... 142
CETA. *See* Comprehensive and Economic Trade Agreement (CETA, *a.k.a.* Canada-EU Trade Agreement)
CFA. *See* Canadian Federation of Agriculture (CFA, prev. Canadian Council of Agriculture)
CFC. *See* Chicken Farmers of Canada (CFC)
CFF. *See* Cooperative Commonwealth Federation (CFF)
CFIA. *See* Canadian Food Inspection Agency (CFIA)
CFIP. *See* Canadian Farm Income Program (CFIP, 2000)
CGC. *See* Canadian Grain Commission (CGC)
CGIAR. *See* Consultative Group on International Agricultural Research (CGIAR)
Challenge, ... 202*n*229, 268*n*229

Champlain, Samuel de, ... 9, 11
"Changing Face of the Canadian Hog Industry, The" (Brisson), ... 105*n*90, 261*n*90
chemical ripening, ... 233*n*297, 272*n*297
CHEP. *See* Canadian Hatching Egg Producers (CHEP)
chicken, egg and turkey production industry, ...146–147
    as supply management industry, ...146–147
Chicken Farmers of Canada (CFC), ... 63*t*4.5, 107*t*6.8, 146, 161
Chicken Producers Marketing Board, ... 145, 146
China, ... 52, 71, 72, 73, 75, 77, 81, 95, 101
    Free Trade Agreement (FTA) with, ...82
    as major trading partner, ... 70, 70*f*5.2
Choices, ... 158*n*169, 202*t*10.9, 265*n*169
Christensen, Naomi, ... 101*n*82, 261*n*82
Church of Latter-Day Saints. *See* Mormons (*a.k.a.* Church of Latter-Day Saints)
CIA. *See* Central Intelligence Agency (CIA)
CIDA. *See* Canadian International Development Agency (CIDA)
CIGI. *See* Canadian International Grains Institute (CIGI)
Cision PR Newswire, ... 237
CKFTA. *See* Canada-South Korea Free Trade Agreement (2015, CKFTA)
Clearwater, R.I., ... 214*n*245, 269*n*245
Climate Change: Science, Impacts and Policy, Proceedings of the Second World Climate Conference
(Jager & Ferguson, eds.), ... 218*n*263, 270*n*263
*Climate Change 2014: Mitigation of Climate Change - Contribution of Working Group III to the Fifth
Assessment Report, Intergovernmental Panel on Climate Change (IPCC)* (Edenhofer), ...
  214*n*247, 269*n*247
"Climate Change and Agricultural Production in the Canadian Prairies" (Stewart), ...
  218*n*261, 270*n*261
Climate Change and Water in the Prairies Conference, ... 219*n*268, 270*n*268
climate change (global warming), ... 3
    see also agricultural adaptation (to climate change); CO_2 fertilization effect; greenhouse
        gases (GHG)
        exacerbation of supply-demand imbalance from, ... 256
        extended droughts from, ...218–219
        extreme weather effects from, ...218–219
        models for, ...207–208
        net positive impact of on Canadian agriculture, ... 217
        pests and diseases increase from, ... 219
        regenerative organic culture (ROA) and, ... 189
        responses to, ... 245
        social challenge of in rural areas, ... 124, 128
        strains on production in Canada by, ... 244
        transposition of weather from, ... 218
climate modelling, ... 3, 207–208, 218

Closing of the American Mind, The (Bloom), ... 133*n*136, 263*n*136
Cloutis, E., ... 128*n*129, 263*n*129
Clubroot Disease of Canola and Mustard (Agri-Facts), ... 227*n*287, 272*n*287
Clubroot Management Plan (GOA), ... 227*n*287, 272*n*287
CMA. *See* Census Metropolitan Area (CMA)
CMSMC. *See* Canadian Milk Supply Management Committee (CMSMC)
CN. *See* Canadian National (CN) Railway (prev. CNR)
CNR. *See* Canadian National (CN) Railway (prev. CNR)
CNRS. *See* National Centre for Scientific Research (CNRS)
CO_2 fertilization effect, ... 217, 217*f*11.10
Co-op fédérée, La, ... 137
Codex Alimentarius, ... 79
Colby-Saliba, B., ... 198*t*10.8
commercial farms, ... 38–40, 38*t*3.10
    farm type financial structures, ... 39, 39*t*3.11
    Gini Coefficient and, ... 40, 40*f*3.20
    number of, ... 20, 21
    statistics, ... 38–40, 38*t*3.10, 39*f*3.18, 39*t*3.11, 40*f*3.19
Commercial Seed Analysts Association of Canada (CSAAC), ... 142
Commission de protection du territoire agricole du Québec, ... 185
commodity groups, ... 143–145
    criticisms of non-supply-managed, ... 150
    finances for, ... 143
    friction with other value-chain members, ... 143
    list of, ... 144–145*t*8.4
    more powerful than umbrella organizations, ... 143
    value-chain perspective in most, ... 144
"Community Groups Abound, but System Works": (Hursh), ... 147*n*158, 265*n*158
comparative advantage, ... 88, 88*t*5.7
Comprehensive and Economic Trade Agreement (CETA, *a.k.a.* Canada-EU Trade Agreement),
    . . . . 73, 82, 87–88, 87*n*66, 147, 244, 260*n*66
Comprehensive and Progressive Agreement for Trans-Pacific Partnership (CPTPP, prev. Trans-Pacific Partnership (TPP)), ... 73, 82, 84–85, 86*t*5.5, 91, 147, 244
Confederation, ... 23
    transcontinental railroad and, ... 164
Conference Board of Canada, ... 62*n*38, 259*n*38
Constitution Act (prev. *British North America Act (BNA Act)*), 1867, ... 153
Consultative Group on International Agricultural Research (CGIAR), ... 168, 244
consumer subsidy equivalents (CSEs), ... 80*n*56, 260*n*56
contingency valuation method (CVM), ... 202*n*226, 268*n*226
COOL. *See* US Country of Origin Labelling (COOL)
Cooperative Commonwealth Federation (CFF), ... 16, 17, 137*n*143, 264*n*143
Cooperative Elevators Companies, ... 137

cooperatives, rural, ...137–139, 256
    challenges for, ...138–139
    East Coast fishing, ... 138
    first Canadian formed in the Maritimes, ... 137
    formation of on prairies between 1910-1930, ... 154
    grain elevators, ... 138
    growth of between 1940-1990, ... 138
    importance of in Quebec, ... 138
    importance of retail in prairies and Atlantic Canada, ... 138
    many formed before World War I, ... 137
    Rural Electrification Associations (REAs), ... 139
    in 21st century, ... 138
    wheat pools, ... 138
COPA. *See* Canadian Oilseed Processors Association (COPA)
Corn Laws (Britain), ...12
corn (maize) farming, ...2, 8, 9, 12, 13, 51–52
    as fourth most important crop in Canada, ...51
    as GMO, ...51
    for grain type of, ... 51, 52
    grown with soybeans, ...52
    importance of, ... 30, 30*f*3.11
    international prices for, ...77
    as major crop, ... 45, 45*t*4.1, 46*f*4.1, 46*f*4.2
    organic sweet corn, ... 234
    production process, ... 51–52
    for silage type of, ... 51, 52
    for sweet corn type of, ... 51, 52
    varieties of, ...51
cost-price squeeze, ...36–37
    unit cost reduction method to counteract, ... 36–37
    value-added increase to counteract, ... 36–37
Costanza, Robert, ... 202*n*227, 268*n*227
Country Canada Study, The: Climate Impacts and Adaptation (Koshida and Avis), ... 219*n*269,
                                                                                                                                            271*n*269
Country Guide, ... 140*n*146, 140*n*147, 264*n*146, 264*n*147
Countryside Stewardship and Organic Farming Scheme (UK), ... 203
Cows and Fish Program, ... 177
CP. *See* Canadian Pacific (CP) Railway
CPC. *See* Canadian Pork Council (CPC)
CPSR. *See* wheat farming, Canada prairie spring red (CPSR) class
CPT. *See* Canola Performance Trials (CPT)
CPTA. *See* Canadian Plant Technology Agency (CPTA)

CPTPP. *See* Comprehensive and Progressive Agreement for Trans-Pacific Partnership (CPTPP, prev. Trans-Pacific Partnership (TPP))
CRA. *See* Canadian Revenue Agency (CRA)
Crime Stoppers Tips Line, ... 127
Criminal Code of Canada, ... 127, 128
CRISP-Cas9 technology, ... 248
Crop Development Centre, University of Saskatchewan, ... 151
crop insurance, ... 153*n*165, 265*n*165
Crop Insurance Act (1959) (became AgriInsurance), ... 156, 157, 157*t*9.2
Crop Insurance Program, ... 158
CropLife Canada, ... 142, 142*n*151, 264*n*151
Crow Benefit, ... 157
Crow Rate. *See* railways, Crow's Nest Pass freight rates (*a.k.a.* Crow Rate)
Crow Rate Debate, The: Issues and Impacts (Alberta Economic Development), ... 156*n*174,
                                                                                                                                      266*n*174

Crown corporations,
    Canadian Dairy Commission (CDC), ... 61, 156
    Farm Credit Corporation (FCC), ... 161
    National Farm Products Board, ... 156
Crown land,
    CCF and, ...17
    federal, ... 175
    Natural Resources Act (1932) and, ... 153–154
    policies until 1825, ...12
    provincial, ... 175
    resource management of, ... 175–185
Crow's Nest Pass Agreement (1897), ... 15, 154, 157, 164, 165
CRP. *See* US Conservation Reserve Program (CRP)
CSAAC. *See* Commercial Seed Analysts Association of Canada (CSAAC)
CSEs. *See* consumer subsidy equivalents (CSEs)
CSGA. *See* Canadian Seed Growers' Association (CSGA)
CSI. *See* Canadian Seed Institute (CSI)
CSTA. *See* Canadian Seed Trade Association (CSTA)
CTMA. *See* Canadian Turkey Marketing Agency (CTMA)
Cullen, M., ... 203*n*230, 268*n*230
cultural diversity, ... 125–126
    challenges of in rural areas, ... 125–126
CUSMA. *See* North American Free Trade Agreement (NAFTA), rev. Canada-United States-Mexico Agreement (CUSMA, *a.k.a.* NAFTA 2.0; USMCA; T-MEC)
CVM. *See* contingency valuation method (CVM)
CVMA. *See* Canadian Veterinary Medical Association, The (CVMA)
CWAD. *See* wheat farming, Canada western amber durum (CWAD) class
CWB. *See* Canadian Wheat Board (CWB)

CWRS. *See* wheat farming, Canadian western red spring (CWRS) class
CWSWS. *See* wheat farming, Canada western soft white spring (CWSWS) class

D

Dade, Carlo, ... 101n82, 261n82
Dairy Farmers of Canada, ...139, 144t8.4
Dairy Farmers of Ontario, ...146, 148t8.5
Dairy Farmers of Quebec, ... 11, 62n38, 89t5.8, 259n38
dairy farming, ...9, 10, 11, 12–13, 16, 18, 59–62
    see also milk production industry
    Canada compared to US, ...60
    CETA and, ... 87–88, 89, 89t5.8
    cheeses, ...13
    CPTPP and, ... 85, 89, 89t5.8
    CUSMA and, ... 83, 89, 89t5.8
    farmer profiles, ... 59t4.4
    federal subsidies for, ... 156n167, 265n167
    genetics in, ... 115
    greenhouse gases and, ...228
    guaranteed minimum prices, ...62
    import controls, ... 62–63
    management of, ...61
    national quotas, ... 61, 62f4.8
    production controls, ...61
    statistics, ...59
    supply management in, ...37
    trade limited by quotas and import tariffs, ...73
Dairy Farming in Canada, ... 11n11, 257
dairy product manufacturing industry, ...94
    organic milk, ... 234, 235f13.1
    processing plants, ... 106–107, 107f6.5
    size of, ... 106
"Data from the Arizona FACE Experiments on Wheat at Ample and Limiting Levels of Water and Nitrogen" (Kimball), ... 218n260, 270n260
debt-asset ratio (farms), ... 41–42
decoupling, ...79
density definitions, ... 119–120
deoxyribonucleic acid (DNA), ... 225, 230, 232, 233, 238, 240
development, agricultural,
    domestic consumer and international marketplace presently drive, ... 173–174
    government driven until after WWII, ... 173
diamide, ... 225, 225n285, 271n285
Dick, Lyle, ... 7n1, 8n7, 153n164, 257, 265n164
Dinar, A., ... 197n217, 198n219, 268n217, 268n219

direct and indirect industry effect, ... 117
Directory of Canadian Agricultural Associations, ... 137
diseases, agricultural, ...221, 226–227
    animal, ... 227
    bovine spongiform encephalopathy (BSE), ...72
    Clubroot, ... 226–227, 226*f*12.3
    as environmental issue, ...226–227
    genomics and, ... 248
distance definitions, ... 120
Distilled (Bronfman), ... 100*n*79, 261*n*79
distilling industry, ... 100
DNA. *See* deoxyribonucleic acid (DNA)
"Do Water Market Prices Appropriately Measure Water Values?" (Colby-Saliba), ... 198*t*10.8
Dominion Lands Act/Homestead Act, 1872, ...15
donkey farming, ...67
Douglas, D., ... 119*n*111, 262*n*111
duck farming, ...66
Ducks Unlimited, ... 177n184, 178, 266*n*184
Dudley, W., ... 241*n*315, 273*n*315
Dunbar's number, ... 136, 136*n*139, 263*n*139

E
Earth Institute, The, ... 222*n*275, 271*n*275
EC. *See* European Union (EU) (prev. European Economic Community (EEC)); (prev. European Community (EC))
Ecological Economics, ... 203*n*231, 268*n*231
"Ecological Economics and Agriculture in Alberta: Status Report" (Anderson & Lerohl), ...
. . . . 202*n*288, 268*n*228
Economic Assessment of the Value of Wildlife Resources in Alberta, An (Phillips), ... 202*t*10.9
economic changes,
    indirect ripple effects (spin-offs) of, ... 117, 118
    induced income effect of, ... 117
    initial direct industry impact effect of, ... 117
    multipliers as means to measure, ... 117, 118
Economic Valuation of Environmental Impacts: A Workbook (Asian Development Bank), ... 202*t*10.9
Economics – Introductory Analysis (Samuelson), ... 88*n*69, 260n69
Economics of Grazing Systems for Waterfowl Habitat Conservation, The: An Evaluation & Guidelines for Implementation, Recreation, Parks & Wildlife Foundation (Prism Environmental Consultants), ... 177*n*183, 266*n*183
"Economics Suggest Overhaul of Dairy System" (Mussell), ... 147*n*156, 265*n*156
economies of scale, ... 37, 37*f*3.17
Economist, The, ... 203*n*232, 241*n*314, 246*n*317, 269*n*232, 273*n*314, 273*n*317
"Economists Put a Price on Nature" *(The Western Producer),* 178*n*187, 266*n*187

INDEX 293

"Economists Suggest Overhaul of Dairy System" (White), ...60n34, 259n34
Ecosystems and Environment, ... 218n259, 270n259
Edenhofer, O., ... 214n247, 269n247
"Edmonton-Calgary Corridor, The: Simulating Future Land Cover Change Under Potential Government Intervention" (Stan), ... 186n197, 267n197
Edmonton Sun, The, ... 128n128, 203n230, 218n265, 263n128, 268n230, 270n265
education,
    levels of farmers', ... 124
    levels of in rural Canada, ... 124
EEC. *See* European Union (EU) (prev. European Economic Community (EEC)); (prev. European Community (EC))
EFC. *See* Egg Farmers of Canada (EFC)
"Effects of Pollution, The" *(Encyclopaedia Britannica)*, ... 221n273, 271n273
EFP. *See* Environmental Farm Planning (EFP)
Egg Farmers of Canada (EFC), ... 63t4.5, 107t6.8, 146, 161
Egg Producers Marketing Board, ... 145, 146
Emissions in Alberta's Agricultural Sector (Toma & Anderson), ... 215n249, 231n294,
                                                                                                                                                                                     269n249, 272n294
employment in agri-business, ... 93, 93t6.1
    GDP and, ... 118f6.14
employment in rural communities, ... 129, 130f7.7, 131
Encyclopaedia Britannica, ... 70f5.2, 221n273, 271n273
end-point royalties, ... 142n153, 264n153
Entz, P., ... 229t12.2
environment, ...3
   see also climate change (global warming); greenhouse gases (GHG),
    ethanol biofuels and, ... 104
    sustainability of, ... 161, 245
Environment and Climate Change Canada, ... 215n250, 269n250
Environment Canada, . . 211t11.2, 215t11.5, 218n261, 219n269, 270n261, 270n269, 271n269
Environment Council of Alberta, ... 219n270, 271n270
environmental amenities,
    accounts of stocks of, ...202–203
    advantages of public payments to farmers for, ...205–206
    existence value externality for, ... 201
    GDP measure of, ... 202
    GNP measure of, ... 202
    non-market indicative values for, ...202, 202t10.9
    opportunity cost of, ... 202
    option value externality for, ... 201
    request value externality for, ... 201
    social problem of, ... 201
Environmental Farm Planning (EFP), ... 170

Environmental Performance Reviews - Canada 2017 (OECD), ... 210*f*11.4
Environmental Stewardship and Climate Change program, ... 161
Environmental Sustainability of Canadian Agriculture: Agri-environmental Indicator Report Series (Clearwater), ... 214*n*245, 269*n*245
Enviropig, ... 231
EPA. *See* U.S. Environmental Protection Agency (EPA)
Equestrian Canada (prev. Equine Canada), ...67*n*47, 259*n*47
Equine Canada. *See* Equestrian Canada (prev. Equine Canada)
Estimates of Producer and Consumer Subsidy Equivalents: Government Intervention in Agriculture, 1982-87 (Webb), ...80*n*56, 260*n*56
Estimating Effects of Climate Change on Agriculture in Saskatchewan, Canada (Williams), ... 218*n*261, 270*n*261
"Estimating Regional Crop Potential in Finland under a Changing Climate" (Carter & Saarikko), ... 218*n*262, 270*n*262
ethanol and biodiesel production industry, ... 103
    environmental benefits of, ... 104
    percentage required in gasoline, ... 103
    plants for, ...103, 104*t*6.4
EU. *See* European Union (EU) (prev. European Economic Community (EEC)); (prev. European Community (EC))
European Commission, ...213*t*11.3
European immigration, ... 8, 12, 14
    during 1895-1914, ...15
    history of agricultural development and, ...1
European Union (EU) (prev. European Economic Community (EEC)); (prev. European Community (EC)), ... 80, 87, 87*n*66, 260*n*66
    GMOs restrictions in, ... 240
eutrophication, ... 222
"Evolution of Agricultural Support Policy in Canada, The" (Hedley), ... 153*n*163, 265*n*163
"Evolution of Federal Wetlands Policy" (Wiebe & Heimlich), . . . 178*n*186, 202*t*10.9, 266*n*186
exchange rates, ... 2, 244
    Canada *vs.* US, ...78–79, 78*f*5.10
    fluctuations in and impacts of, ...78
exports, agricultural, ...2
    agri-food products, ... 75, 77
    barley, ... 72, 100
    beef, ... 72–73
    Canada as fifth largest in world, ... 70, 75
    Canada as major, ...68
    canola, ...71
    to CPTTP countries, ... 84, 84*f*5.12
    dairy quotas, ...73
    exchange rates and, ...78

INDEX   295

flower milling, ...13
of food and beverage products, ...95
major destinations of Canadian, ... 70, 70*f*5.2
Nairobi Declaration (WTO, 2015) and subsidies, ... 73*n*52, 259–260*n*52
pork, ... 73, 73*f*5.5
of pork, ...59
of poultry and eggs, ...74
pulses, ...72
soybeans, ...53, 72
top ten Canadian net, ... 70, 70*t*5.1
total sales figures, ...75
wheat, ... 15, 71, 71*f*5.3

F

FACE. *See* free-air CO2 enrichment (FACE) experiments
Factors Driving Canada's Rural Economy (Bollman), ... 123*n*117, 263*n*117, 263*n*121
Fafard, M., ... 102*n*85, 261*n*85
Fagan, J.B., ... 241*n*315, 273*n*315
Fair Rail Freight Services Act (2013), ... 165
family farm, ...44
Fan, J., ...215*n*251, 216*f*11.9, 269*n*251
FAO. *See* Food and Agricultural Organization (FAO) of the United Nations
Farm and Food Care (FFC) Ontario, ... 66*n*45, 223*n*277, 223*n*279, 223*n*280, 224*f*12.2,
                                            225*n*281, 259*n*45, 271*n*277, 271*n*279, 271*n*280, 271*n*281
farm bankruptcies, ... 157
Farm Credit Act, The (1959), ... 156
Farm Credit Corporation (FCC, 1959), ... 156, 161, 162
farm implements, ... 113–114
Farm Income Protection Act (*FIPA,* 1991), ... 157, 158, 159
Farm Journal, ... 113*f*6.11
Farm Loan Program, ... 171
Farm Operating Expenses and Depreciation Charges (Statistics Canada), ... 32*f*3.7
Farm Product Council of Canada (FPCC), ... 107*t*6.8, 161
Farm Products Marketing Agencies Act (1972), ...64
farm supplies (inputs), ... 107–108
    chemical fertilizers, ... 107
    depreciation costs, ... 108
    farm implements, ... 113
    feed for livestock purchases, ... 107
    livestock costs, ... 107
    machinery operating costs, ... 108
    off-shore sourcing of as variable in future of, ... 243, 244
    other providers of, ... 115–116
    pesticide costs, ... 108

  seed costs, ... 108
farmer-controlled input supply, ... 36–37
farmer-controlled product purchases, ... 37
farmers,
  age of operators, ... 26–27, 27*f*3.8
  benefits of drainage of wetlands, ... 199
  dairy statistics, ...59–60, 59*t*4.4
  decline in participation of in agricultural organizations, ... 152
  educational levels of, ... 28–29, 29*f*3.10, 124
  employment in agri-business and, ... 93, 93*t*6.1
  functional foods and, ... 241
  gender of, ... 27–28, 28*f*3.9
  GMOs and, ... 241
  impact of climate change on, ... 128
  as land custodians, ... 3
  long-term credit for, ... 161
  national profile of, ... 43–44
  organic food production and, ... 241
  purchase of machinery and implements by, ... 113
  regulatory regime for, ... 246
  statistics on, ... 19–21, 19*f*3.1, 20*t*3.1, 21*f*3.2, 24, 25*f*3.6, 25*t*3.4
  ways to counteract cost-price pressures by, ... 36–37
"Farmers Choice of Crops in Canadian Prairies under Climate Change: An Econometric
Analysis" (Grise & Kulshreshtha), ... 219*n*272, 271*n*272
Farmland Drainage in Central and Northern Alberta, Farming for the Future (Pedology
Consultants/Marv Anderson & Associates), ...178*n*185, 199*n*221, 266*n*185, 268*n*221
farms,
  alternative fuels for machinery on, ... 231
  bi-modality in, ... 253
  capital value of, ... 44
  Census of Agriculture and, ... 2
  characteristics of, ... 1–2
  costs and prices compared, ... 36–37, 36*f*3.16
  debt increases, ... 41–42
  definition of, ... 20
  diversity in, ... 253
  future structure of, ... 253, 253*f*14.1
  highly-efficient operating, ... 254
  impact of climate change on, ... 128
  income components, ... 34–35, 35*t*3.9, 44
  increase in capital-intensive, ... 36
  increase in self-sufficiency n, ... 255
  increase in size of, ... 20–21, 20*f*3.2, 20*t*3.1

innovative finances for, ... 254
legal structure of, ... 26f3.7
narrower margins in, ... 254–255
national profile of, ... 43–44
net farm income *vs.* disposable income, ... 33
net income, ... 33–35, 33t3.8
number of commercial, ... 20, 21
operational expenses, ...31–32, 32t3.7
organization, ... 25–26, 26t3.5
as percentage of rural population, ... 43, 43t3.13
revenue sources of, ... 29–31, 30f3.11, 30f3.12, 30t3.6
safety regulations for, ... 246
statistics on, ... 19–21, 19f3.1, 20t3.1, 21f3.2
synergies and efficiencies in, ... 254
uniqueness of, ... 1–2, 19
wireless security systems for, ... 251
FCC. *See* Farm Credit Corporation (FCC, 1959)
federal government's role (in agriculture), ...3
    see also provincial government's role (in agriculture)
    1870-1950 initiatives, ...3
    1950-1980 initiatives, ...3
    1980-2000 initiatives, ...3
    Agricultural Policy Framework (APF), ... 179
    establishment of CWB in 1935, ...17
    to moderate price and cost fluctuations, ...37
    post-2000 initiatives, ...3
    regulatory environment, ...3
federal legislation (1860-1950), ...155t9.1
federal legislation (1951-1980), ...157t9.2
federal legislation (1981-2000), ...159t9.3
federal legislation (2001-2018), ...171t9.4
federal regulators,
    Agreement on Internal Trade (AIT), ... 163–164
    Canada Agricultural Review Tribunal, The (CART, 1983), ... 163
    Canadian Food Inspection Agency (CFIA), ... 142, 163
    Canadian Grain Commission (CGC), ... 162, 163
    Canadian Para-Mutual Agency, ... 162, 163
    Pest Management Regulatory Agency (PMRA), ... 163
    Public Health Agency of Canada, ... 163
    Transport Canada, ... 163
Federal Regulatory Framework for Biotechnology, ... 239
Federated Cooperative of Saskatchewan, ... 138
fee simple, ...23

Feed Freight Assistance Act (1941), ... 154
Feed Freight Assistance Program (1941), ... 158
Feed Grain Freight Assistance Program, ... 164
Ferguson, H.L., ... 218*n*263, 270*n*263
fertilizers,
    on barley, ...49
    costs of chemical, ... 107
    costs of in grain and oilseed industry, ...108–109
    creation of aerosols, ... 222
    impact of unused nitrogen and phosphorous, ... 222
    increase in production and usage of, ... 222
    negative impacts of, ... 222
    nitrogen-based chemical, ... 228, 230
    pollution by synthetic, ... 221, 221*t*12.1, 222
    potash and, ... 108
    production companies, ... 108–109, 108*f*6.6
    production process, ... 109
    synthetic as principal source of GHG, ...211, 211*f*11.5
    on wheat, ...47
FFC Ontario. *See* Farm and Food Care (FFC) Ontario
"Fighting Crime with Tax Credits" (Graney), ... 128*n*128, 263*n*128
Final Offer Arbitration (FOA), ... 103
Financial Post, ... 214*n*244, 269*n*244
FIPA. See Farm Income Protection Act *(FIPA,* 1991)
First Nations peoples, ...14
    see also Aboriginal peoples; individually-named peoples
fishing, ...8, 9
    cod, ...8, 9
    inland aquaculture, ... 249
    packing, ...18
    in Quebec, ...11
Flaten, D., ... 179*n*192, 266*n*192
flax farming, ...9
FOA. *See* Final Offer Arbitration (FOA)
Food and Agricultural Organization (FAO) of the United Nations, . .46*t*4.2, 73*n*51, 79, 260*n*51
food and beverage processing industry, ...93
    as Canada's second largest manufacturing industry, ...93
    employment in, ...93
    GDP of, ...93
    largest manufacturing industry in most provinces, ...94
    manufacturing companies for, ... 95, 95*f*6.1
    number of processing establishments for, ... 94, 94*t*6.2
    profiled companies in, ... 95, 96*t*6.3

statistics for, ... 94–95
food and drug regulations, ... 142n153, 264n153
Food Directorate, Health Canada, ... 237
food irradiation, ... 233n297, 272n297
Food Processing and Food Processing's Top 100 (foodprocessing.com), ... 95f6.1
food processing industries,
    challenges to, ...76
food production,
    self-sufficiency in developing countries, ... 244
food safety, ... 161, 246
Food Processors of Canada (prev. Canadian Food Processors Association), ... 103
foodprocessing.com, ... 95f6.1
forward linkages, ... 93–96
    in grain and oilseed milling industry, ... 96–104
    impact on Canadian economy, ... 116–117
    in livestock and poultry industry, ...96
    processors, retailers and intermediaries, ...93
fossil fuels, ... 209n240, 269n240
FPCC. *See* Farm Product Council of Canada (FPCC)
Fraser Institute, ... 62n38, 259n38
Free Trade Agreement (FTA), ...82
free trade initiatives, ...2
French, A., ... 219n269, 270n269
fruit and food preserving industry, ...94
"Fruit and Vegetable Industry" (Barratt & Fafard), ... 102n85, 261n85
fruit farming, ...12
    apples, ...18
    grapes, ... 13, 18
    robot fruit pickers, ... 248
"Frustrations Over Rural Crime in the Spotlight" *(The Western Producer),* . . 127n124, 263n124
FTA. *See* Free Trade Agreement (FTA)
functional foods, ... 4, 236–237, 245
    compared to natural health products, ... 237
    components and sources, ... 238t13.3
    definitions of, ... 237
    farmers and, ... 241
    health outcomes from, ... 237, 238t13.3
    market for, ... 237
"Functional Foods: Benefits, Concerns, and Challenges" (Hasler), ... 237n309, 272n309
fungicides, ... 222, 223
    on barley, ...49
    on wheat, ...48
fur trade, ... 7, 13, 18

"Future of Agriculture, The" *(The Economist)*, ... 246*n*317, 273*n*317

G
Gabbatiss, Josh, ... 223*n*276, 271*n*276
Galbraith, John Kenneth, ... 243
Gameda, S., ... 219*n*269, 270*n*269
Gaston, C., ... 116*n*106, 262*n*106
GATT. *See* General Agreement on Tariffs and Trade (GATT)
GCMs. *See* global climate models (GCMs)
GDP. *See* gross domestic product (GDP)
GEDS. *See* Government Electronic Directory Services (GEDS)
gene technology, ... 238*n*311, 273*n*311
General Agreement on Tariffs and Trade (GATT), ... 79, 80, 81, 244
general farm organizations. *See* umbrella (general farm) organizations
genetic engineering, ... 238, 238*n*311, 273*n*311
genetic modification (GM), ... 238, 245
genetically modified organisms (GMOs), ... 4, 84*n*64, 87, 90, 233, 237, 238–241,
245, 246, 260*n*64
    canola as, ... 46–47
    commercial crops from, ... 239, 239*t*13.4
    corn (maize) as, ... 51
    countries that restrict, ... 240
    definitions of, ... 238
    farmers and, ... 241
    financial benefits of, ... 240–241
    genome editing and, ... 230
    labelling of, ... 239
    not permitted in organic food production, ... 233
    production of, ... 240
    regulatory process for, ... 239
    seed companies and, ... 111
    soybeans as, ... 52
genetics,
    artificial insemination and, ... 114
    companies, ... 115, 115*f*6.12
genome editing, ... 229, 230, 241, 248–249
    possible benefits of, ... 248–249
genomic splicing, ... 226
Geoderma, ... 215*n*251, 216*f*11.9, 269*n*251
geographic indicators, ... 87*n*68, 260*n*68
GHG. *See* greenhouse gases (GHG)
Gilmour, G., ... 188*n*200, 267*n*200
Gini Coefficient, ... 40*n*24, 258
    commercial farms and, ... 40, 40*f*3.20

Global Affairs Canada, ... 63*n*40, 65*n*42, 259*n*40, 259*n*42
global climate models (GCMs), ... 218
Global Organic Milk Production Market Report (KPMG), ... 235*n*303, 235*n*304,
272*n*303, 272*n*304
global positioning system (GPS), ... 248
"Global Study Shows How Agriculture Interacts with Industry" (The Earth Institute), ...
222*n*275, 271*n*275
global warming. *See* climate change (global warming)
globalization, ...76
    of financial markets as variable in future of agriculture, ... 244
    importance of, ... 244
    new income sources for rural communities, ... 130
Globe and Mail, . . 61*n*36, 139*n*144, 146*n*155, 185*n*196, 259*n*36, 264*n*144, 265*n*155, 267*n*196
glucosinolates, ... 46, 46*n*25, 258*n*25
glyphosates, ...3, 224–225
    glyophosate-resistant crops, ... 230
GM. *See* genetic modification (GM)
GMOs. *See* genetically modified organisms (GMOs)
GNP. *See* gross national product (GNP)
"Go Green, You May Live Longer" (Cullen), ... 203*n*230, 268*n*230
GOA. *See* Government of Alberta (GOA)
"Goat Fact Sheet" (Farm and Food Care Ontario), ...66*n*45, 259*n*45
goat farming, ...66
GOC. *See* Government of Canada (GOC)
"Governance in Canadian Agriculture" (Hedley), ... 81*n*58, 153*n*163, 260*n*58, 265*n*163
Government Electronic Directory Services (GEDS), ... 162f9.1
Government of Alberta (GOA), ... 227*n*287, 272*n*287
Government of Canada (GOC), ... 74*f*5.6, 75*f*5.8, 84*f*5.12, 86*t*5.5, 87*t*5.6, 106*n*91, 208*f*11.2,
209*f*11.3, 211*t*11.2, 212*f*11.7, 261*n*91
    Population Census, 1851-2016, ... 121*f*7.2
GPS. *See* global positioning system (GPS)
grain and oilseed milling industry, ...94
    cereal and oilseed elevators, ... 97–98, 97*t*6.4, 138
    fertilizer costs, ...108–109
    oilseed crushing plants, ...99
"Grain and Rail in Western Canada" (Hall Commission, 1977), ... 164–165
grain elevator system, ... 97–98, 138
    cooperatively owned, ... 138
    main grain companies, ... 98*t*6.5
    ports for, ...97
    privately owned, ... 138
Grain Elevators in Canada Licensed by the Canada Grain Commission (Canada Grain
Commission), ... 97*t*6.4

Grain Farmers of Ontario, ...148*t*8.5, 149, 150
Grain Growers' Grain Company, ... 17, 137
Grain Growers of Canada, ... 141, 142, 149
Grainews, ... 226*n*286, 271*n*286
Grand Trunk Pacific railway, ... 164
Graney, J., ... 128*n*128, 263*n*128
Gray, Richard, ... 158*n*169, 265*n*169
Great Depression of 1930s, ...9, 16, 17, 18, 154
    farm numbers during, ... 19, 20*t*31
Green, H., ... 166*n*175, 266*n*175
Greencover Canada, ... 170
"Greenhouse gas emissions by Canadian economic sector" (Government of Canada), . .209*n*241,
  269n241
greenhouse gases (GHG),
    aerobic digestion to reduce, ... 232
    agricultural land as major sink for, ...213, 214, 214*t*11.4, 216*f*11.9
    agricultural sector's share of, ...209, 209*f*11.3
    agricultural source by province, ...211*t*11.2
    agricultural sources of, ... 210–211, 211*f*11.5, 211*t*11.2, 228
    beef cattle as principal source of, ...211, 211*f*11.5, 228
    benefits of mitigation of, ... 217
    Canada's contribution to, ... 208–210, 208*f*11.2, 209*t*11.1
    Canada's total increasing since 1990, ... 210
    constituents of, ... 208
    crop mix and yield changes as mitigation potential for, ... 214, 215
    Crown lands as major sink for, ... 213, 214, 214*t*11.4
    emissions from livestock, ... 211–212, 212*f*11.6, 228
    human activity and, ... 207
    indirect soil emissions as principal source of, ...211, 211*f*11.5, 212
    land management changes as mitigation potential for, ... 214, 215, 215*t*11.5
    land use changes as mitigation potential for, ... 214, 215
    passive decomposing to reduce, ... 232
    sink potential in Canada for, ... 213–214, 216*f*11.9
    synthetic fertilizers as principal source of, ...211, 211*f*11.5, 222
Greenstone, M., ... 197*n*215, 268*n*215
Grier, K., ... 103*n*86, 261*n*86
GRIP. *See* Gross Revenue Insurance Program (GRIP)
Grise, J., ... 219*n*272, 271*n*272
gross domestic product (GDP), ... 10, 118, 196
    of agriculture in 1920s Quebec, ...10
    Canada's international trade as percentage of, ... 69, 69*f*5.1
    density as value-added in, ... 120
    environmental amenities and, ... 202

food and beverage processing industry and, ... 93
by industry, ... 42, 42f3.22
percentage of rural in, ... 131
gross national product (GNP), ... 7, 202
Gross Revenue Insurance Program (GRIP), ... 157, 158, 158n169, 159, 160, 265n169
Growing Forward 2 programs, ... 161
growth poles. *See* regional rural service centers (*a.k.a.* growth poles)
Guebert, A., ... 60n35, 259n35
Guimond, E., ... 128n130, 263n130
Guy, John, ... 8

H
Haber-Bosch process, ... 53n59, 258n29
Haldimand, Frederick, ... 12
Hambre, B.M., ... 113n98, 262n98
Handbook of International Food and Agricultural Policies, ... 76f5.9
Harari, Yuval N., ... 7n2, 257
Harper, S.J., ... 82n60, 136n140, 260n60, 264n140
Harron, B., ... 179n192, 266n192
Hasler, C.M., ... 237n309, 272n309
Hatching Egg Producers Marketing Board, ... 145, 146
HBC. *See* Hudson Bay Company (HBC)
health and safety standards, ... 142n153, 264n153
Health Canada, ... 142n153, 158, 162, 225, 239, 264n153
Hebert, Louis, ... 9
Hedley, D.D., ... 76f5.9, 81n58, 172f9.6, 260n58, 265n163
Heimlich, R.E., ... 178n186, 202t10.9, 266n186
Heminthavong, K., ... 59n32, 258n32
Henry, L., ... 179n191, 179n192, 266n191, 266n192
Henry's Handbook of Soil and Water (Henry), ... 179n191, 266n191
herbicides, ... 222, 223
on barley, ... 49
genomics and, ... 248
glyphosate-based, ... 3
total usage of, ... 223, 224, 224f12.2
on wheat, ... 48
Heron, W., ... 219n269, 271n269
Herrington, R., ... 219n269, 270n269
"Hidden Costs of Supply Management in a Small Market" (Carter), ... 62n38, 259n38
Highlights of the 1974 Interprovincial Input-Output Model (Gaston & Bedard), ... 116n106,
262n106
Historic Statistics of Canada (GOC), ... 9n10, 257
history of agricultural development, ... 1
Aboriginal peoples and, ... 1

European immigration and, ...1
    in Maritimes, ...1
    in Ontario, ...1
    in Quebec, ...1
History of Agriculture in Ontario, A (Reaman), ... 8n4, 10n10, 12n11, 257
"History of Agriculture to the Second World War" (Dick), ... 153n164, 265n164
Hoffman, M., ... 184n195, 267n195
Holm, H.M., ...180t10.1
homesteaders, ... 11, 15–16, 16
horse farming, ...67
How Agriculture Made Canada: Farming in the Nineteenth Century (Russell), ... 13n14, 257
"How Lessons from Canola Can Help Unleash Pulse Fractionation on the Prairies" (Canada
West Foundation), ... 101n82, 261n82
Hudson Bay Company (HBC), ...14
Hunter, F., ... 219n269, 270n269
hunter gatherers, ... 7–8
Huron peoples, ...11
Hursh, Kevin, ... 147n158, 265n158
Hutterites, ...18, 25, 125, 126, 254
    colonies of, ... 126
    location of, ...26
    manufacturing expertise, ... 131
    as successful farmers and ranchers, ... 126
hydroponics, ... 249

I

IDRC. *See* International Development Research Centre (IDRC)
Immen, W., ... 185n196, 267n196
immigration, ... 125n119, 263n119
"Impact of Climate Change Scenarios on the Agri-Climate of the Canadian Prairies" (McGinn
& Shepherd), ... 217–218, 217n258, 270n258
*Impact of Climatic Variations on Agriculture, The, Volume 1: Assessments in Cool Temperate and
Cold Regions* (Parry, T.R. Carter & Konijn), ... 219n269, 271n269
Implications for Agriculture (AIA Conference, 2014), ... 219n269, 270n269
imports, agricultural, ...2
    agri-food products, ... 75, 77
    Canada as sixth largest in world, ...75
    dairy custom tariffs, ... 73, 74f5.6
    dairy supply controls, ... 62–63
    exchange rates and, ...78
    major destinations of to Canada, ... 70, 70f5.2
    major farm implements, ... 113
    of poultry and eggs, ... 74–75, 75f5.8
    poultry farming controls on, ...65

incentives, ... 132
income elasticity of demand, ... 91*n*71, 260*n*71
"Increasing Crop Yields and Root Input Make Canadian Farmland a Large Carbon Sink" (Fan), 215*n*251, 216*f*11.9, 269*n*251
induced income effect, ... 117
information technologies, ... 248*n*319, 273*n*319
    genomics, ... 229, 230, 241, 248–249
    GPS precision agriculture, ... 248*n*319, 273*n*319
    monitoring devices, ... 248*n*319, 273*n*319
    multi-spectral sensors, ... 247–248
    robotics, ... 248, 248*n*319, 273*n*319
infrastructure, ... 132
inland aquaculture, ... 249
Inland Terminal Association of Canada, ... 142*n*150, 264*n*150
Innovation Issue (The Western Producer), ... 248*n*318, 273*n*318
Inquiry on Federal Water Policy (1985, GOC), ... 198*t*10.8
insecticides, ... 222, 223
    on canola, ...47
    total usage of, ... 224
Intergovernmental Panel on Climate Change (IPCC), ... 216*n*254, 216*t*11.6, 217, 217*n*255, 270*n*254, 270*n*255
International Agency for Research on Cancer, ... 225
International Development Research Centre (IDRC), ... 168
International Food Information Council, ... 238*t*13.3
international marketing and trade, ... 161
international trade (in agriculture), ... 2
    see also exchange rates; exports; imports; trade balance, agricultural; trading prospects
    advantages of freer, ... 88–89
    amber category of distortion in, ...79
    Canada's competitiveness rank in, ... 81, 81*t*5.4
    commodity and income support measures, ...79
    comparative advantage in, ... 88–89, 88*t*5.7
    composition of over time, ...69
    future prospects, ... 89–91
    green category of distortion in, ...79
    liberalization of, ...79
    protectionism levels, ... 80–81
    red category of distortion in, ...79
    rules and regulations for, ... 79–80
internationalization, ... 243–245
    of R&D as variable in future of farming, ... 243, 244
Inuit peoples, ...8
IPCC. *See* Intergovernmental Panel on Climate Change (IPCC)

IRPs. *See* land management, integrated resource plans (IRPs)
irrigation, ...22, 227–228
    in Alberta, ... 192–193, 193*f*10.7, 194, 194*f*10.8
    in British Columbia, ... 192
    on-stream reservoirs, ...194–197
    in Ontario, ... 22, 192
    robotic, ... 248
    in Saskatchewan, ... 22, 192, 193
    soil salinization from, ...227–228
    technology to help, ...251–252
Irrigation in Alberta (Alberta Agriculture), ... 194*n*211, 267*n*211
Irving, J.A., ... 17*n*16, 137*n*143, 257, 264*n*143

J

Jager, J., ... 218*n*263, 270*n*263
Jiachen, Z., ... 218*n*263, 270*n*263
Johnson, B., ... 219*n*269, 270*n*269
Josling, Tim, ... 116*n*106, 262*n*106
Journal of Earth Sciences & Climate Change, ... 219*n*272, 271*n*272
Journal of Economic Perspectives, ... 208*n*238, 269*n*238
Journal of Nutrition, ... 237, 237*n*309, 272*n*309
Juma, Calestous, ... 241, 241*n*314, 273*n*314

K

KAP. *See* Keystone Agricultural Producers (KAP)
"KAP Proposal Pays Farmers for Environmental Care" (B. Wilson), ... 205*n*234, 269*n*234
Keystone Agricultural Producers (KAP), ... 205
Kimball, B., ... 218*n*260, 270*n*260
kitchen gardens, ... 10
Knapik, L., ... 178*n*189, 182*t*10.2, 183*t*10.3, 266*n*189
Kniss, A.R., ... 225*n*284, 271*n*284
Konijn, N.T., ... 219*n*269, 271*n*269
Koshida, G., ... 219*n*269, 271*n*269
Kroeger, Arthur, ... 165
Kroeger Report (1999, on grain handling and transportation), ... 165
Kulshreshtha, S.N., ... 198*t*10.8, 219*n*272, 271*n*272
Kyoto Protocol (1997), ... 210

L

Labour Force Survey (Statistics Canada), ... 131*f*7.8
Lamb, A., ...215*n*248, 215*n*252, 269*n*248, 270*n*252
land clearing, ... 177
land degradation, ...178–184
    loss of organic matter (OM) and nutrients, ...178, 183–184

soil acidification, ...178, 182–183
soil erosion by wind and water, ...178, 180–182
soil salinization, ... 178, 179–180, 180t10.1, 227–228
land fragmentation, ... 175, 178n190, 266n190
land management,
    changes in as potential for mitigation of GHG, ... 214, 215, 215t11.5
    community pasture program, ... 190
    conservation authorities, ... 190
    Environmental Farm Planning (EFP) programs, ... 190
    integrated resource plans (IRPs), ...189–190
    regional land use plans, ... 189
    shelterbelt program, ... 190
land ownership, ... 23–24, 175, 175f10.1
land resources, characteristics of, ... 3
land tenure, ... 23–24, 24t3.3
land types, ... 3
land use, ... 21–22, 21f3.3
    acreage of urban areas by turn of century, ... 184
    agricultural, ... 184, 184t10.4
    alternative land use (ALU) model, ... 190
    alternative land use services (ALUS), ...255–256
    appreciation farmland, ...40–41
    changes in as potential for mitigation of GHG, ... 214, 215
    competition for, ... 256
    crop production and fallow, ... 176–177, 1176f10.2
    pasture, ...176, 176f10.2, 177
    principal crops and, ... 23, 23t3.2
    woodlots and wetlands, ... 17–178, 176, 176f10.2
    zoning for, ...189–190
land values, ... 40–42
    capital in land and buildings, ... 41, 41t3.12
Laurier, Wilfred, ... 9n8, 15, 257
Lauzon, A., ... 119n109, 131n131, 262n109, 263n131
legumes,
    in kitchen garden, ... 10
    as nitrogen-fixing plants, ... 53, 229, 229t12.2, 248, 249
    perennial and forage, ... 229
Lerohl, M.L., ... 202n228, 268n228
Leslie, L., ... 179f10.3, 180f01.4
LIFT. *See* Lower Inventories for Tomorrow Program (LIFT, 1970)
Lincoln, Abraham, ... 3, 133
linkages. *See* backward linkages; forward linkages
livestock and poultry industry, ... 2

auction markets via video, ... 251
beef auction markets, ... 104
beef feedlots, ...104–105
beef processing plants, ... 105, 105*t*6.6
costs of, ... 107
electric monitoring of, ... 247–248
feed costs, ... 107
feed mill costs in, ... 114
forward linkages in, ...96
increase in artificial insemination costs in, ... 114
pork processing plants, ... 105, 106*t*6.7
statistics, ... 30, 30*f*3.12
veterinary costs, ... 114
Lomborg, Bjorn, ... 208*n*237, 269*n*237
"Long-term trends in the intensity and relative toxicity of herbicide use" (Kniss), ... 225*n*284, 271*n*284
Lower Canada. *See* New France (*a.k.a.* Lower Canada)
Lower Inventories for Tomorrow Program (LIFT, 1970), ... 157*t*9.2
lumber, ...18

M

MAA. *See* Marv Anderson & Associates (MAA)
machinery operating costs, ...108, 111–112
production companies, ... 111–113, 112*f*6.10
Macyk, Don, ...137*n*142, 243*n*316, 264*n*142, 273*n*316
Macyk, Marilyn, ...137*n*142, 264*n*142
maize. *See* corn (maize) farming
Malthus, Thomas Robert, ...36
malting barley processing industry, ...99
managed agricultural trade, ...80–81
Manitoba,
Aboriginal peoples in, ... 126, 126*t*7.2
agricultural societies in, ... 133
ALUS in, ... 205
beef production in, ...56
canola farming in, ...46
corn (maize) farming in, ...51
grain and oilseed milling industry in, ...94
historical view of agriculture in, ... 13–18
Hutterites in, ... 126
irrigation in, ... 192
pork production in, ... 58, 59, 59*f*4.5
potato farming in, ... 101, 102
pulse farming in, ...55

soybean farming in,	53, 229
wheat farming in,	48
Manitoba Agriculture,	188*n*200, 188*t*10.5, 267*n*200
Manitoba Wheat and Barley Growers Association,	142*n*150, 264*n*150
Manitoba Wheat Pool,	138
maple syrup,	11
Marginal Abatement Cost Assessment of Greenhouse Gas Emissions in Alberta's Agricultural Sector (Toma & Bouma Management Consultants),	208*n*239, 269*n*239
Maritimes,	
first cooperative in Canada formed in,	137
food and beverage processing industry in,	94
GHG from agriculture in,	211*t*11.2
in history of agricultural development,	1
income supplement programs in,	170
net farm incomes in,	34, 44
rural growth *vs.* urban growth,	121
youth unemployment in,	123
market price support (MPS),	62*n*39, 259*n*39
marketing, domestic,	2
marketing boards,	3
about 70 of them by 1970,	154
marketing commissions,	3
Marketing Freedom for Grain Farmers Act (2011),	165
Marquis wheat,	15, 47
Martin, L.,	214*n*244, 269*n*244
Marv Anderson & Associates (MAA),	218*f*11.11, 218*n*264, 219*n*270, 231*n*292, 270*n*264, 271*n*270, 272*n*292
McEachern, Gordon,	116*n*106, 262*n*106
McGinn, S.,	217–218, 217*n*258, 270*n*258
McKenna, B.,	61*n*36, 146*n*155, 259*n*36, 265*n*155
McKenney, D.W.,	219*n*269, 270*n*269
Measuring Remoteness and Accessibility – A Set of Indices for Canadian Communities (A.F. Bedard, Belanger, Guimond & Penny),	128*n*130, 263*n*130
meat product manufacturing industry,	94
mechanization,	37
Medicare,	17
"Megatrends Impacting Agriculture" (Boehjie),	243*n*316, 273*n*316
Mennonites,	18
manufacturing expertise,	131
Merchants of Grain (Morgan),	98*t*6.5
meta-analysis,	235*n*305, 272*n*305
Métis peoples,	14
MFN. *See* Most Favored Nation (MFN)	

Mi'kmaq (MicMac) peoples, ...8
Milk Producers Marketing Board, ... 145, 146
milk production industry, ... 146
    see also dairy farming
    as supply management industry, ... 146
Miller, Virginia P., ... 8*n*6, 257
Ministry of Agriculture, ... 132
monoculture agriculture, ... 68, 176
Montreal Economic Institute, ...62*n*38, 259*n*38
Morgan, Dan, ... 98*t*6.5
Mormons (a.k.a. Church of Latter-Day Saints), ...14
Morrison, R. Bruce, ... 8*n*3, 257
Most Favored Nation (MFN), ...79
MPS. *See* market price support (MPS)
Mueller, R.A., ...198*t*10.8
mule farming, ...67
multi-cultural pioneers in agriculture, ...18
multinationals, ...2
    seed/chemical companies consolidate into, ... 110
Murray, M., ...82*n*61, 260*n*61
Mussell, A., ... 103*n*86, 147*n*156, 261*n*86, 265*n*156

N

NAFTA. *See* North American Free Trade Agreement (NAFTA), rev. Canada-United States-Mexico Agreement (CUSMA, *a.k.a.* NAFTA 2.0; USMCA; T-MEC)
NAFTA 2.0. *See* North American Free Trade Agreement (NAFTA), rev. Canada-United States-Mexico Agreement (CUSMA, *a.k.a.* NAFTA 2.0; USMCA; T-MEC)
NAFTA and Agriculture (Murray), ...82*n*61, 260*n*61
Nairobi Declaration (WTO, 2015), ...73*n*52, 79, 259–260*n*52
nanotechnology engineering, ... 238*n*311, 273*n*311
NASS. *See* United States Department of Agriculture (USDA), National Agricultural Statistics Service (NASS)
National Academy of Sciences, ... 104*n*89, 261*n*89
national agricultural policy, ... 137
    amalgam of agricultural organizations, ... 137
National Centre for Scientific Research (CNRS), ... 223*n*276, 271*n*276
National Energy Board (NEB), ... 197*n*214, 267*n*214
National Farm Products Marketing Act (1972), ... 143, 147, 156, 157*t*9.2
National Farm Stewardship Program (NFSP), ... 170, 204
National Farmers Union (NFU), ... 140–141, 141*t*8.2
National Inventory Report: Greenhouse Gas Sources and Sinks in Canada, 1990-2011 (GOC), ...211*t*11.2
National Inventory Report: Greenhouse Gas Sources and Sinks in Canada, 2013 (GOC), ...215*t*11.5

"National Inventory Report 1990-2014: Greenhouse Gas Sources and Sinks" (Environment and Climate Change Canada), ... 215*n*250, 269*n*250
national marketing boards, ...145–146
    criticisms of supply management and, ... 147
National Milk Marketing Plan (NMMP), ...61
National Transportation Act (*NTA*, 1987), ... 165
National Water Supply Expansion Program (NWSEP), ... 170
Native People: The Canadian Experience (Morrison & Wilson), ... 8*n*6, 257
Natural Health Foods Directorate, ... 237
natural health products (nutraceuticals), ... 237
    compared to functional foods, ... 237
Natural/Irrigation Salinity, 2017 (Alberta Agriculture and Forestry), ... 227*n*289, 272*n*289
Natural Resource Journal, ...198*t*10.8
natural resources, ...202–203
Natural Resources Act (1930; incorporated into *British North America Act* (1865-1965)), . . 153
"Natural Resources are a Win for Rural Communities" (Speer), ... 132*n*134, 263*n*134
Nature and Management of Salt-Affected Land in Saskatchewan, The (Henry, Harron & Flaten),
                                                                                             179*n*192, 266*n*192
Nature Climate Change, ...215*n*248, 215*n*252, 269*n*248, 270*n*252
Nature Communications, ... 225*n*284, 271*n*284
"Nature's Resources as Productive Assets" (Repetto), ... 202*n*229, 268*n*229
NB. *See* New Brunswick (NB)
NDTV. *See* New Delhi Television Limited (NDTV)
NEB. *See* National Energy Board (NEB)
neonicotinoids, ... 223*n*278, 226, 271*n*278
    regulations for, ... 223
neonics, ... 223
net income stabilization account (NISA), ... 158, 159, 160
Neutral/Petun peoples, ... 11
New Brunswick (NB),
    historical view of agriculture in, ... 8–9
    potato farming in, ... 101
    provincial crown land in, ... 175
    seafood product manufacturing industry in, ... 94
New Delhi Television Limited (NDTV), ... 236*n*306, 272*n*306
New France (*a.k.a.* Lower Canada),
    historical view of agriculture in, ... 9–11
New Holland News, ... 231n293, 272*n*293
"New Safety Net Program for Canadian Agriculture, A: GRIP" (Gray), . . . 158*n*169, 265*n*169
"New York Times ignored GM crop benefits" (Smyth), ... 240*n*313, 273*n*313
Newfoundland and Labrador (NL),
    historical view of agriculture in, ... 8
    provincial crown land in, ... 175

seafood product manufacturing industry in, ...94
Next Generation Seed System (Seed Synergy Collaboration Project), ... 142*n*152, 264*n*152
"Next generational change in farming is already underway, The" (Guebert), ...60*n*35, 259*n*35
NFSP. *See* National Farm Stewardship Program (NFSP)
NFU. *See* National Farmers Union (NFU)
NGOs. *See* non-governmental organizations (NGOs)
NHL. *See* non-Hodgkin's lymphoma (NHL)
NIMBY. *See* Not in My Backyard (NIMBY)
NISA. *See* net income stabilization account (NISA)
nitrogen fertilizer, ...53*n*29, 258*n*29
NL. *See* Newfoundland and Labrador (NL)
NMMP. *See* National Milk Marketing Plan (NMMP)
nominal producer protection coefficient (NPPC), ... 80, 80*f*5.11, 81, 81*f*5.4
non-GMO crops, ... 226
non-governmental organizations (NGOs), ... 166, 178
    agricultural R&D by, ... 168
non-Hodgkin's lymphoma (NHL), ... 225
non-tariff barriers (NTBs), ... 79, 80, 88
Nordhaus, W.D., ... 208*n*238, 269*n*238
North American Free Trade Agreement (NAFTA, rev. Canada-United States-Mexico
Agreement (CUSMA, *a.k.a.* NAFTA 2.0; USMCA; T-MEC)), ... 82–83
Northwest Territories (NWT),
    Crown land in, ... 175
Not in My Backyard (NIMBY), ... 197*n*216, 268*n*216
Nova Scotia (NS),
    historical view of agriculture in, ... 8–9
    seafood product manufacturing industry in, ...94
NPPC. *See* nominal producer protection coefficient (NPPC)
NS. *See* Nova Scotia (NS)
NTA. See National Transportation Act (*NTA,* 1987)
NTBs. *See* non-tariff barriers (NTBs)
Nunavut,
    Crown land in, ... 175
"Nutraceutical" *(Wikipedia),* 237*n*308, 272*n*308
nutraceuticals. *See* natural health products (nutraceuticals)
NWSEP. *See* National Water Supply Expansion Program (NWSEP)
NWT. *See* Northwest Territories (NWT)
Nyirfa, W., ... 219*n*269, 271*n*269

O

oats farming, ... 50–51
    falls in world production of, ...51
    flour mills for, ...98
    as major crop, ... 45, 45*t*4.1, 46*f*4.1, 46*t*4.2

performance (pony) oats in, ...50
production process, ...50
types of, ...50
uses of, ... 50–51
OECD. *See* Organization for Economic Cooperation and Development (OECD)
OFA. *See* Ontario Federation of Agriculture (OFA)
O'Grady, K.L., ... 198*t*10.8
Olfert, R., ... 120*n*113, 262*n*113
O&M. *See* operation and maintenance (O&M)
OM. *See* organic matter (OM) loss (loss of nutrients)
OMAFRA. *See* Ontario Ministry of Agriculture, Food and Rural Affairs (OMAFRA)
OMERS. *See* Ontario Municipal Employees Retirement System (OMERS)
On Farm Water Management Study: An Evaluation of On-Farm Water Management Options in Alberta (Alberta Agriculture), ... 199*n*222, 268*n*222
on-stream reservoirs, ...194–195
   economics of, ... 195–196, 195*t*10.7
   functions of, ...194–195
   problems with, ... 195
Ontario Federation of Agriculture (OFA), ... 139, 140*t*8.1
Ontario Grain Farmer, ...7
Ontario Marketing Board, ...13
Ontario Ministry of Agriculture, Food and Rural Affairs (OMAFRA), ... 103*n*86, 261*n*86
Ontario Municipal Employees Retirement System (OMERS), ...25
Ontario Processing Vegetables: An Economic Analysis (Mussell & Grier), ... 103*n*86, 261*n*86
Ontario Teachers' Pension Plan (OTPP), ...25
Ontario (Upper Canada),
   Aboriginal peoples in, ... 126, 126*t*7.2
   agricultural societies in, ... 133
   ALUS in, ... 205
   beef production in, ...56
   canola farming in, ...46
   chicken production in, ...63
   corn (maize) farming in, ...51
   dairy farming in, ...60, 60*f*4.6, 60*f*4.7, 61
   duck farming in, ...66
   farm statistics in, ...19–21, 19*f*3.1, 20*t*3.1, 21*f*3.2
   flour mills in, ...48
   food and beverage processing industry in, ...94
   GHG from agriculture in, ... 211*t*11.2
   goat milk production in, ...66
   greenbelts in, ... 186
   historical view of agriculture in, ... 11–13
   in history of agricultural development, ...1

  irrigation in, ... 22, 22*f*3.4, 192
  licensed vegetable processors in, ... 102
  meat product manufacturing industry in, ... 94
  net farm incomes in, ... 34, 44
  number of farmers today in, ... 13
  pork production in, ... 58, 59, 59*f*4.5
  potato farming in, ... 101, 102
  pulse farming in, ... 55
  sheep farming in, ... 66
  soybean farming in, ... 52, 53, 229
  urban sprawl initiatives in, ... 185–186
  wheat farming in, ... 48
Ontario Wheat Board, ... 17, 48, 154*n*166, 265*n*166
Open Data Journal for Agricultural Research, ... 218*n*260, 270*n*260
operation and maintenance (O&M), ... 139
Opposing Viewpoints in Social Issues (Dudley, ed.), ... 241*n*315, 273*n*315
organic agriculture. *See* regenerative organic agriculture (ROA)
Organic Alberta, ... 235
organic food production, ... 233, 237, 245
  artificial food additives prohibited in, ... 233
  compared to conventional, ... 235–236, 235*t*13.1, 236*t*13.2
  farmers and, ... 241
  GMO not permitted in, ... 233
  as holistic system, ... 233
  increase in sales of, ... 234
  labelling of, ... 236
  legal definition of, ... 233–234
  meat production and, ... 233
  non-synthetic pesticides allowed in, ... 233
  organic logo on, ... 234
  pesticide residues on, ... 236
  prohibition of molecular DNA manipulation in, ... 233
  regulated by CFIA, ... 234
Organic Food (Wikipedia), ... 233*n*296, 272*n*296
organic matter (OM) loss (loss of nutrients), ... 178, 183–184
"Organic production an economic winner" (Gilmour), ... 188*n*200, 266*n*200
Organization for Economic Cooperation and Development (OECD), 80, 80*f*5.11, 81*t*5.4, 124,
                 169*f*9.4, 172*f*9.5, 202, 210*f*11.4
  definition of agricultural support, ... 171–172
"Orphan Wells: Alberta's $47 Billion Problem" *(The Western Producer)*, ... 127*n*123, 127*n*124,
                      263*n*123, 263*n*124
OTPP. *See* Ontario Teachers' Pension Plan (OTPP)

Overview of the Canadian Agricultural Innovation System, An (AIC), ...166*n*176,
    169*f*9.4, 266*n*176
Overview of the Canadian Agriculture and Agri-Food System, An, 2015 (AAFC), ...39*f*3.18,
    40*f*3.19, 71*n*48, 75*n*53, 118*f*6.14, 167*n*177, 167*n*178, 259*n*48, 260*n*53, 266*n*177, 266*n*178

P

Palliser Triangle, Saskatchewan and Alberta, ... 154
PARC. *See* Prairie Adaptation Research Collaborative (PARC)
parliamentary structure of Canada,
    farm gate as separator of power, ... 153
    health-care responsibilities shared, ... 153
    House of Commons, ... 154
    justice system, ... 153
    power shared over immigration and agriculture, ... 153
    provincial power over civil and contract law and education, ... 153
    Senate, ... 153
    taxation system, ... 153
Parry, M.L., ...218*n*263, 219*n*269, 270*n*263, 271*n*269
Paterson, B., ...196*n*212, 267*n*212
Pattison, W.S., ... 178*n*188, 202*t*10.9, 266*n*188
"Payments for Environmental Services: A Comparison of US and EU Agri-Environmental
Policies" (Baylis), ... 203*n*231, 268*n*231
PCPA. See Pest Control Products Act (PCPA, 1995)
Pearce, R., ... 111*n*97, 262*n*97
Pedology Consultants/Marv Anderson & Associates, . . .178*n*185, 199*n*221, 266*n*185, 268*n*221
PEI. *See* Prince Edward Island (PEI)
Penny, C., ... 128*n*130, 263*n*130
Pest Control Products Act (*PCPA*, 1995), ... 158, 163
Pest Management Control Agency, Health Canada, ... 223
Pest Management Regulatory Agency (PMRA), ... 163, 225
pesticides,
    advantages of biological, ... 230
    biological, ... 228, 230
    costs of, ... 108
    environmental concerns about, ...222–223
    fungicides, ...48, 49, 222, 223
    genomics and, ... 248
    glyphosate health concerns, ...224–225
    health issues from, ...224–225
    herbicides, ... 3, 38, 49, 222, 223, 224, 224*f*12.2
    insecticides, ...47, 132, 222, 223, 224
    new technologies and, ... 226
    non-synthetic allowed in organic food production, ... 233
    pollution by, ...221, 221*t*12.1, 222–223

residue of, ... 236
total usage of, ... 223, 224, 224*f*12.1, 244
usage of less toxic, ... 225–226
Pew Research Centre, ... 125*n*120, 263*n*120
PFRA. *See* Prairie Farm Rehabilitation Administration (PFRA)
Phillips, W.E., ... 202*t*10.9
Pietroniro, A., ... 219*n*268, 270*n*268
Pilger, G., ... 140*n*146, 140*n*147, 264*n*146, 264*n*147
Pitt, Robert D., ... 8*n*5, 257
PMRA. *See* Pest Management Regulatory Agency (PMRA)
Policy Implications of Climate Change on Alberta Agriculture (MAA), ... 208*n*239, 218*f*11.11,
218*n*264, 269*n*239, 270*n*264
"Political Economy of Agricultural Policy in Canada, The" (Hedley), ... 76*f*5.9, 172*f*9.5
Politics of Agricultural Policy-Making in Canada, The (Skogstad), .. 77*n*54, 147*n*157, 153*n*163,
157*n*168, 165*n*174, 260*n*54, 265*n*157, 265*n*163, 265*n*168, 265*n*174
pollution, agriculture-related, ... 4
abiotic, ... 221*t*12.1
biological pest control as mitigative measure of, ... 228, 230
biotic, ... 221*t*12.1
genome editing as mitigative measure of, ... 229, 230
irrigation as source of, ... 221, 221*t*12.1, 227–228
machinery emissions and discharges as sources of, ... 221, 221*t*12.1, 228
nitrogen-based fertilizers as mitigative measure of, ... 228, 230
nitrogen-fixing plants as mitigative measure of, ... 228, 229
pesticides and disease as sources of, ... 221, 221*t*12.1, 226–227
spore-infested soils and, ... 226–227
synthetic fertilizer as source of, ... 221, 221*t*12.1
waste processing as source of, ... 221, 221*t*12.1, 228
"Pollution from Industry and Agriculture: Nitrogen Oxide in the Atmosphere" (Shapley), ...
228*n*290, 272*n*290
Pond, Peter, ... 13
population,
growth, ... 121, 121*f*7.3, 185
growth of as variable in future of agriculture, ... 243–244
higher densities of and socio-economic blending, ... 129
rural Canada, ... 2
rural *vs.* urban, ... 122*t*7.1
world growth in, ... 251
Population Census of Canada, 1851-2016 (Government of Canada), ... 121*f*7.2
Population Census of Canada, 1981-2016 (Government of Canada), ... 121*f*7.3
Population Census of Canada, 2016 (Statistics Canada), ... 126*t*7.2, 129*f*7.6
pork production, ... 57–59
AIDA for, ... 159*n*170, 265*n*170

farrow-finish operations type of, ... 58, 58*t*4.3
farrowing nurseries operations type of, ...58, 58*t*4.3
finishing operations type of, ... 58, 58*t*4.3
hog numbers and farms, ... 57, 58*f*4.4
intensification of farms, ...57
international prices for, ...77
in Manitoba, ... 58, 59, 59*f*4.5
in Ontario, ...58, 59, 59*f*4.5
processing plants, ...105, 106*t*6.7
potash, ... 11, 108, 244
potato farming, ... 8, 9, 11, 12, 13
in Alberta, ... 101, 102
frozen products, ... 102
in Manitoba, ... 101, 102
in New Brunswick, ... 101
in Ontario, ... 101, 102
in Prince Edward Island, ... 101
processing, ...101–102
Potato Market Information Review, ... 102n84, 261n84
"Potential Effect of Climate Change on Agriculture, The" (Parry & Jiaschen), ...218*n*263,
270*n*263
"Potential for Land Sparing to Offset Greenhouse Gas Emissions from Agriculture, The"
(Lamb), ...215*n*248, 215*n*252, 269*n*248, 270*n*252
potholes (temporary wetlands), ... 178
poultry farming, ... 2, 63–65
see also chicken, egg and turkey production industry; duck farming; goat farming;
sheep farming
CETA and, ...87
chicken production methods, ...63
components of, ...63
CUSMA and, ...83
egg production methods, ...64
exports and imports of poultry and eggs, ... 74–75
import controls on, ...65
management of, ...64
marketing, ...64
minimum prices for, ...65
poultry and egg processing plants, ...107, 107*t*6.8
producer numbers, ... 63, 63*t*4.5
production controls, ... 64–65
production quotas for, ...64–65, 65*t*4.6
supply management in, ...37
turkey production methods, ... 63–64

Prairie Adaptation Research Collaborative (PARC), .. 217*n*256, 218*n*267, 219*n*269, 270*n*256, 270*n*267, 271*n*269
Prairie Farm Rehabilitation Act (PRFA), 1935, ...17
Prairie Farm Rehabilitation Administration (PFRA), ...153*n*165, 190, 190*n*204, 265*n*165, 267*n*204
Prairie Grain Advance Payments Act (1959), ...157*t*9.2, 158
Prairie Oat Growers Association, ... 147
Prairie Oat Growers Manual, ...50*n*27, 258*n*27
Prairie Pothole Region (Wikipedia), ... 177*n*184, 266*n*184
Prairie provinces,
    1871-land survey of, ... 14–15
    Clubroot in, ... 226–227, 226*f*12.3
    cultural diversity challenges in rural, ...125–126
    farm statistics in, ... 19–21, 19*f*3.1, 20*t*3.1, 21*f*3.2
    food and beverage processing industry in, ...94
    GHG from agriculture in, ...211*t*11.2
    historical view of agriculture in, ... 13–18
    importance of retail cooperatives in, ... 138
    net farm income in, ...34
    number of farmers today in, ...18
    organic farmers on, ... 234
    as problem children, ... 153
    rural growth *vs.* urban growth, ... 121
    salinization in, ...179, 227–228
    soil acidification on, ... 182
    summer fallow in, ... 21–22
    two-thirds of agricultural GHG from, ... 212
precision agriculture, ... 246–247, 248*n*319, 273*n*319
PRFA. *See Prairie Farm Rehabilitation Act* (PRFA), 1935
price-takers, ...77
primary agriculture, ...2
Prince Edward Island (PEI), ... 184
    ALUS in, ...204–205
    historical view of agriculture in, ... 8–9
    potato farming in, ... 101
    seafood product manufacturing industry in, ...94
producer subsidy equivalents (PSEs), ...80*n*56, 260*n*56
producer support estimate (PSE), ...80
production, domestic, ...2
protectionism, ... 80–81
Provincial and Territorial Energy Profiles (NEB), ... 197*n*214, 267*n*214
provincial government's role (in agriculture), ...169–171
    ad hoc programs, ... 169–170, 170*t*9.6

Agricultural Policy Framework (APF),	170
applied research facilities,	171
cost-sharing formula with federal government,	169
credit programs,	171
resource-based programs,	170–171
provincial marketing boards and commissions,	147–150
for beef production,	148
for pork production,	149
for soybeans and corn production,	150
for wheat, barley and oats production,	149
provincial role (in agriculture),	3
PSE. *See* producer support estimate (PSE)	
PSEs. *See* producer subsidy equivalents (PSEs)	
Public Health Agency of Canada,	163
public land management,	3
Public Payments for Environmental Services from Agriculture: Precedents and Possibilities" (K.R. Smith),	203*n*231, 268*n*231
pulse farming,	2, 54–56
attractiveness of,	54
Canada as largest exporter of,	71
elevators for,	97–98
exports of,	72, 100
increases in production,	55–56
as major crop,	45, 45*t*4.1, 46*f*4.1, 46*t*4.2
NAFTA and,	83
prairie climate and soils good for,	54
processing,	100–101, 101*f*6.3
protein fractionation,	101
types of,	54, 55*f*4.3
uses of,	55
varieties of,	55
"Pulse Fractionation a Huge Opportunity" (Christensen & Dade),	101*n*82, 261*n*82
Putman, R.,	133*n*136, 263*n*136

Q

Quebec,	
canola farming in,	46
chicken production in,	63
corn (maize) farming in,	51
dairy farming in,	60, 60*f*4.6, 60*f*4.7, 61
duck farming in,	66
farm statistics in,	19–21, 19*f*3.1, 20*t*3.1, 21*f*3.2
flour mills in,	48
food and beverage processing industry in,	94

GDP of agriculture in 1920s, ...10
GHG from agriculture in, ...211*t*11.2
historical view of agriculture in, ... 9–11
in history of agricultural development, ...1
importance of cooperatives in, ... 138
meat product manufacturing industry in, ...94
net farm incomes in, ... 34, 44
number of farms today in, ...11
pork production in, ... 58, 59, 59*f*4.5
pulse farming in, ...55
sheep farming in, ...66
urban sprawl initiatives in, ... 185

R

Railroader (Green), ... 166*n*175, 266*n*175
Railway Association of Canada, ... 165
railways, ...3, 164–166
    see also Canadian National (CN) Railway; Canadian Pacific (CP) Railway; specific transport legislation
    completion of transcontinental railroad, ... 164
    Crow's Nest Pass freight rates (*a.k.a.* Crow Rate), ... 164, 165
    relationship to grain transportation, ... 166
    subsidies for grain transportation, ... 164
    third largest network in world, ... 166
    transcontinental, ... 15, 16
    Transport Canada and, ... 165
ranching. *See* beef production (ranching)
R&D. *See* research and development (R&D), agricultural
Reaman, G. Elmore, ... 8*n*4, 10*n*10, 12*n*11, 257
REAs. *See* Rural Electrification Associations (REAs)
"Recent Scenarios of Future Climate Change for Canada and Alberta: Implications for Agriculture" (French), ... 219*n*269, 270*n*269
recombinant DNA technology, ... 238*n*311, 273*n*311
Red River Colony, ...14
"Reflections on the Economics of Climate, Change" (Nordhaus), ... 208*n*238, 269*n*238
regenerative organic agriculture (ROA), ... 3, 4, 187–189, 206, 216, 232
    biodiversity increases from, ... 188
    compared to conventional agriculture, ... 188–189, 188*f*10.5, 188*t*10.5
    components of, ... 188
    crop rotations, ... 226
    environmental sustainability from, ... 255
    as integral to climate solution, ... 189
    multi-spectral sensors for, ... 247
    on-farm soil sequestration, ... 188–189

practicalities of, 187–188
reduction in greenhouse gases from, 188, 188t10.5
self-sufficiency from, 255
Regenerative Organic Culture and Climate Change: A Down-to-Earth Solution to Global Warming (Rodale Institute), 189n201, 266n201
Regional Niagara Policy Plan (1981), 185
regional rural service centers (*a.k.a.* growth poles), 107, 129
"Regulating organic products in Canada" (Canadian Food Inspection Agency), 233n296, 272n296
Reguly, E., 139n144, 264n144
Reiley, Laura, 189n202, 267n202
Reimer, B., 119n111, 262n111
relative efficiency, 88
Renewable Industries Canada, 104f6.4
Repetto, R., 202n229, 268n229
research and development (R&D), agricultural, 3, 37, 166–168
    federal, 166–167, 166f9.2
    funding compared to other countries, 168–169, 169f9.4
    industry groups, 168
    international, 168
    internationalization of as variable in future, 243, 244
    NGOs, 168
    percentage of GDP, 168
    private sector, 167, 167f9.3
    provincial, 167, 167t9.5
    universities, and colleges, 168
residual feed intake (RFI),
    beef animals and, 231
"Residual feed intake (RFI): An indirect approach for reducing GHG emissions" (Basarab), 231n295, 272n295
resource-based provincial programs, 170–171
    Agricultural Policy Framework (APF) and, 170
    in Alberta, 170–171
    land drainage in Ontario and Quebec, 171
resource management (in agriculture), 175–205
    crop land and other users, 177
    Crown land and, 175–185
    extension of to include private land, 176
    land clearing issue, 177
    soil erosion and soil loss equation, 181
    wetland drainage issue, 177
"Responding to Global Climate Change in the Prairies, Climate Impacts and Adaptation" (Herrington, Johnson & Hunter), 219n269, 270n269

"Restrictions on Genetically Modified Organisms: Canada" (U.S. Library of Congress), ... 238*n*310, 272*n*310
"Restrictions should be placed on the genetic engineering of food" (Fagan), 241*n*315, 273*n*315
Restructuring Rural Saskatchewan: The Challenge of the 1990s (Stabler & Olfert), ... 120*n*113, 262*n*113
RFI. *See* residual feed intake (RFI)
Right Here, Right Now (Harper), ... 82*n*60, 136*n*140, 260*n*60, 264*n*140
RM. *See* rural municipality (RM)
ROA. *See* regenerative organic agriculture (ROA)
Rodale Institute, ... 188*f*10.5, 189*n*201, 216*n*253, 267*n*201, 270*n*253
Rostow, W.W., ... 122*n*115, 262*n*115
Rubik's Cube, ... 137
rural,
    climate change adaptation social challenge in, ... 124, 128
    cohort, ... 121
    communities, ... 107
    consolidation of public infrastructure trend in, ... 122
    cultural diversity challenges in, ... 125–126
    as defined by Statistics Canada, ... 120
    definitions of, ... 119
    density concept in, ... 119–120
    distinct community social challenges in, ... 124, 125–126
    dribble back trend in, ... 128
    educational issues in, ... 124
    falling prices of goods and services transportation trend in, ... 122
    as geographical concept, ... 119
    growth of compared to urban growth, ... 121
    hollowing out of, ... 120, 123
    importance of construction to, ... 129
    importance of educational services to, ... 129
    importance of health care to, ... 129
    importance of manufacturing to, ... 129, 130–131, 131*f*7.8
    labor constraint social challenges in, ... 124, 125
    labor-market shortage in, ... 125
    lack of religious leadership in, ... 125
    law and order preservation social challenge in, ... 124, 127–128
    long distance to density characteristic of, ... 119–120, 119*f*7.1
    low population density characteristic of, ... 119–120, 119*f*7.1
    multiple resource use on private land in, ... 126–127
    out-migration of youth from, ... 123, 124*f*7.5
    percentage of Canada's population in, ... 131
    primary producers as percentage of, ... 129
    reduction in information transference costs trend in, ... 122

relative increase in value of human time trend in, ... 122
self-defense legislation and, ... 127
socio-economic blending trend in, ... 129
socio-economic characteristics of, ... 122–123
socio-economic trends in, ... 120–121
technology and physical proximity trends in, ... 130
vitality in growth-pole communities' trend in, ... 129
rural and regional planners,
anti-rural development policy, ... 132
institutions, infrastructure and incentives policy framework, ... 132
Rural and Small Town Canada, Analysis Bulletin (Statistics Canada), ... 119*n*110, 262*n*110
Rural Canada 2013: An Update: A Statement of the Current Structure and Trends in Rural Canada, Federation of Canadian Municipalities (Bollman), ... 119*n*109, 262*n*109
Rural Crime Watch, ... 127
Rural Electrification Associations (REAs), ... 139
rural municipality (RM), ... 130, 140
Rural Planning and Development in Canada (D. Douglas, ed.), ... 119*n*111, 262*n*111
Rural-Urban Gap in Education, The (Canada Council on Learning), ... 124*n*118, 263*n*118
Russell, Peter A., ... 13*n*14, 257
Rutgers University, ... 236*t*13.2
rye farming, ... 9

S

Saarikko, R.A., ... 218*n*262, 270*n*262
SAASE. *See* Saskatchewan Association of Agricultural Societies and Exhibitions (SAASE)
Samuelson, P.A., ... 88*n*69, 260n69
Sapiens: A Brief History of Humankind (Harai, Y.N.), ... 7*n*2, 257
SAR. *See* sodium adsorption ratio (SAR)
Saskatchewan,
Aboriginal population in, ... 126, 126*t*7.2
agricultural societies in, ... 133
ALUS in, ... 205
barley farming in, ...49
beef production in, ... 14, 56
canola farming in, ...46
CCF in, ... 16, 17
dealing with other users of crop land, ... 177*n*182, 266*n*182
grain and oilseed milling industry in, ...94
historical view of agriculture in, ... 13–18
Hutterites in, ... 126
irrigation needs in, ... 22, 22*f*3.4, 192, 193
lentil and chickpea production in, ... 229
out-migration from rural, ... 124*f*7.5
rural Crime Watch in, ... 127*n*126, 263*n*126

    soybean farming in, ...53
    trespass law in, ...128
    wheat farming in, ...48
Saskatchewan: A History (Archer), ... 137*n*143, 264*n*143
Saskatchewan Agricultural Report No. 25M (1983), ... 180*t*10.1
Saskatchewan Association of Agricultural Societies and Exhibitions (SAASE), ... 133
Saskatchewan Wheat Development Commission, ... 142*n*149, 142*n*150, 264*n*149, 264*n*150
Saskatchewan Wheat Pool, ... 138
Saunders, Charles, ... 15
Schumpeter, J., ... 122*n*116, 262*n*116
Scientific Reports, ... 222*n*274, 271*n*274
seafood product industry, ... 94
Secan at 40 (Pearce), ... 111*n*97, 262*n*97
See Canadian Pacific (CP) railway, ... 166
seed-chemical industry,
    certification of, ... 111
    consolidation of, ... 109, 109*f*6.7, 110–111
    costs of for crop producers, ... 109
    GMOs and, ... 111
    production companies, ...110–111, 110*f*6.8, 110*f*6.9
Seed Synergy Collaboration Project, ... 142, 142*n*152, 264*n*152
seigneuries (lordships), ... 10
service clubs, ... 1334–135
    Chambers of Commerce, ... 134
    Elks, ...125, 132, 134, 135
    Hospital Auxiliary, ... 134, 135
    Kiwanis, ... 134
    Lions, ... 134, 135
    Rotary, ... 125, 134
    Royal Canadian Legion, ... 134, 135
    Royal Purple, ... 134, 135
Shapley, P.S., ... 228*n*290, 272*n*290
sheep farming, ... 8, 66
Shepherd, A., ... 217–218, 217*n*258, 270*n*258
Sifton, Clifford, ... 15
Silent Spring (Carson), ... 221
Simcoe, John Graves, ... 12
Six Nations Indians, ... 11
Skeptical Environmentalist, The (Lomborg), ... 208*n*237, 269*n*237
Skogstad, Grace, ... 77*n*54, 147*n*157, 153*n*163, 157*n*168, 165*n*174, 260n54,
                                        265*n*157, 265*n*163, 265*n*168, 266*n*174
Smith, K.R., ... 203*n*231, 268*n*231
Smith, P., ... 214*n*247, 269*n*247

INDEX   325

Smith, W.N., ... 218, 218*n*259, 270*n*259
Smyth, Stuart, ... 240*n*313, 273*n*313
Soares, J.R., ... 222*n*274, 271*n*274
SOC. *See* soil organic carbon (SOC) sequestration
Social Credit Movement in Alberta, The (Irving), ... 17*n*16, 137*n*143, 257, 264*n*143
Social Credit Party, ... 16, 17, 264*n*143
Socio-Economic Assessment of Low Phytate Barley (MAA), ... 231*n*292, 272*n*292
socio-economic profile, rural, ... 2, 44
    gender, ...27–28, 28*f*3.9
"Socio-Economic Value of Water in Canada, The" (Mueller), ... 198*t*10.8
Socio-Economic Vulnerability of Prairie Communities to Climate Change (Cloutis), ...128*n*129,
  263*n*129
sodium adsorption ratio (SAR), ... 179, 180*f*10.4
soil acidification, ...178, 182–183
    effects of on crops, ...183*t*10.3
    natural acidity and addition of acid-forming materials, ... 182
Soil at Risk (Sparrow), ... 179*n*193, 266*n*193
soil erosion (by water), ... 178
    effects of on crop yields, ... 182*t*10.2
    soil degradation from, ... 180
soil erosion (by wind), ...178, 181–182
    effects of on crop yields, ... 182*f*10.2
soil organic carbon (SOC) sequestration, ... 214
soil pollution, ... 221, 221*t*12.1
    manure treatments to prevent, ...231–232
Soil Salinity. A Study in Crop Tolerance and Cropping Practices (Holm), ... 180*t*10.1
soil salinization, ... 178, 179–180, 180*t*10.1, 227–228
    effects of on crops, ...180*t*10.1
    on the prairies, ... 179
    primary or secondary, ... 179
    saline seep mechanism, ... 179, 179*f*10.3
    self-affected soil classes, ... 180*f*10.4
South Saskatchewan River Basin (SSRB), ... 218*n*268, 270*n*268
soybean farming, ...2, 13, 52–54
    export of, ... 53, 72
    grown with corn (maize), ...52
    importance of, ... 30, 30*f*3.11
    importance of oil from, ...53
    international prices for, ...77
    legume characteristics, ...53
    as major crop, ... 45, 45*t*4.1, 46*f*4.1, 46*t*4.2
    in Manitoba, ... 229
    in Ontario, ... 229

versatility of, ... 53
Sparling, David, ... 39n22, 258
Sparrow, H.O., ... 179n193, 266n193
Speer, S., ... 132n134, 263n134
spin-offs. *See* economic changes, indirect ripple effects (spin-offs) of
SSRB. *See* South Saskatchewan River Basin (SSRB)
Stabler, J.C., ... 120n113, 262n113
Stages of Economic Growth, The (Rostow), ... 122n115, 262n115
Stan, K., ... 186n197, 267n197
"Stand Your Ground" *(The Western Producer),* 127n124, 263n124
State of Rural Canada; Introduction (Lauzon, Bollman & Aston), ... 119n109, 131n131,
                                                                                                                    262n109, 263n131
Statistics Canada, ...2, 11, 13, 19, 19f3.1, 20, 20f3.1, 20t3.1, 21f3.3, 21n17,
   21t3.2, 22f3.2, 24t3.3, 25f3.4, 26t3.5, 27f3.8, 28f3.9, 29f3.10, 31f3.13, 32f3.7, 35t3.9, 36f3.16,
   38n21, 38t3.10, 40n23, 41t3.12, 42f3.22, 43t3.13, 45t4.1, 46f4.1, 54f4.2, 55f4.3, 55n30, 58f4.4,
   58t4.3, 59f4.5, 62f4.8, 73f5.5, 94, 105n90, 116n106, 118n108, 119n110, 120, 121f7.3, 123f7.4,
   123n117, 126t7.2, 128n130, 129f7.6, 130f7.7, 131f7.8, 167f9.3, 170t9.6, 175f10.1, 176f10.2,
   184t10.4, 192f10.6, 193t9.7, 219n271, 234n300, 234n301, 258, 258n30, 261n90, 262n106,
                   262n108, 262n110, 263n117, 263n130, 271n271, 272n300, 272n301
Statistics of Canada's Animal Genetics, ... 115
Stewart, D.B., ... 218n261, 270n261
Subramanian, A., ... 197n217, 198n219, 268n217, 268n219
subsistence farming, ... 9, 10
sugar and confectionery product manufacturing industry, ... 94
   in Alberta, ... 103
   size of, ... 103
sugar beet farming, ...2
   in Alberta, ...56
supplier and processor organizations, ... 152
supply management, ... 36–37, 62n38, 259n38
   chicken, egg and turkey production industry as, ... 146
   criticisms of national marketing boards and, ... 147
   milk production industry as, ... 146
   national legislation led to, ... 143
support, agricultural, ... 80–81
   direct and indirect producer, ... 172, 172f9.5
   direct government payments to farmers, ... 172–173, 172f9.6
   direct payments to producers, ... 173, 173t9.7
   OECD definition, ... 171–172
Survey of Pesticide Use in Ontario (Farm and Food Care Ontario), ... 223n277, 223n279,
                   223n280, 224f12.2, 225n281, 271n277, 271n279, 271n280, 271n281
Sustainable Watersheds Act (2017, Manitoba), ... 205

Symposium on the Impacts of Climate Change and Variability on the Great Plains (ed. Wall), ... 218*n*261, 270*n*261

T

T-MEC. *See* North American Free Trade Agreement (NAFTA), rev. Canada-United States-Mexico Agreement (CUSMA, *a.k.a.* NAFTA 2.0; USMCA; T-MEC)
Talbot, Thomas, ...12
tariff-rate quotas (TRQs), ... 63, 65, 73, 74, 83*n*62, 87, 260*n*62
Taylor, Jeff, ... 7*n*1, 8*n*7, 257
TCM. *See* travel cost method (TCM)
technology, ...37
   see also information technologies
     emerging, ... 245
     new income sources for rural communities, ... 130
     positive and negative outcomes from, ... 256
Temporary Foreign Worker Program (TFWP), ... 125
TFC. *See* Turkey Farmers of Canada (TFC)
TFWP. *See* Temporary Foreign Worker Program (TFWP)
TILMA. *See* Trade, Investment, and Labour Mobility Agreement (TILMA, Alberta & British Columbia)
timber trade, ...9
Tobacco/Allawandaron peoples, ...11
tobacco farming, ... 8, 12, 13
Toma, D., ... 215*n*249, 231*n*294, 269*n*249, 272*n*294
Toma & Bouma Management Consultants, ... 208*n*239, 269*n*239
Toronto Sun, ... 132*n*134, 263*n*134
Toth, B., ... 219*n*268, 270*n*268
Toyra, J., ... 219*n*268, 270*n*268
TPP. *See* Comprehensive and Progressive Agreement for Trans-Pacific Partnership (CPTPP, prev. Trans-Pacific Partnership (TPP))
Trade, Investment, and Labour Mobility Agreement (TILMA, Alberta & British Columbia), ... 164
trade agreements,
   bilateral, ... 81–82
   CUSMA, ...73
   CETA, ...73
   CPTPP, ...73
   GATT, ... 79, 80, 81
   international as variable in future of agriculture, ... 243, 244
   regional, ... 82–83
trade balance, agricultural, ... 2, 71, 71*t*5.2, 75–76, 76*f*5.9, 76t5.3
   in dairy products, ... 74*f*5.6
trademarks, ... 87*n*68, 260*n*68
trading agreements, bilateral, ...2

trading agreements, regional, ...2
trading prospects, ...2
trailing royalties, ... 142n153, 264n153
Transport Canada, ... 163
Transportation Appeal Tribunal of Canada Act (2002) (prev. *Bill C-34*), ... 165
Transportation Safety Board of Canada, ... 165
travel cost method (TCM), ... 202n226, 228n226
Treaty of Paris, 1763, ...10
Tree Fruit Board, ...18
TRQs. *See* tariff-rate quotas (TRQs)
True Costs of Alternative Energy Sources (Greenstone), ... 197n215, 268n215
Trump, Donald, ...84
Turkey Farmers of Canada (TFC), ... 63t4.5, 107t6.8, 161
Turkey Producers Market Board, ... 145, 146
"Twelve Tips to Clubroot Management" *(Grainews),* 226n286, 271n286

U

UFA. *See* United Farmers of Alberta (UFA)
UGG. *See* United Grain Growers (UGG)
UIUC. *See* University of Illinois Urbana/Champaign (UIUC)
UK. *See* United Kingdom (UK)
umbrella (general farm) organizations, ... 139–142
    Canada Grains Council, ... 141, 141t8.2, 142
    Canadian Federation of Agriculture, ... 139–140, 140t8.1
    Cereals Canada, ... 142
    commodity groups more powerful than, ... 143
    commodity-specific national, ... 140
    funding sources, ... 142
    Grain Growers of Canada, ... 141, 142
    industry organizations, ... 142
    National Farmers Union (NFU), ... 140–141, 141t8.2
UN Environment Program (UNEP), ... 216n254, 217n255, 270n254, 270n255
"Understanding Rural Canada: Implications for Rural Development Policy and Rural Planning
Policy" (Reimer & Bollman), ... 119n111, 262n111
unemployment,
    Aboriginal peoples and, ... 126
    rural youth, ... 123, 123f7.4
UNEP. *See* UN Environment Program (UNEP)
UNESCO Biosphere Reserve, ... 186
UNIFARM, A Story of Conflict and Change (Carrol), . . 139n145, 143n154, 264n145, 265n154
Union Catholique des Cultivateurs, ...11
Union des producteurs agricoles, L' (UPA), ... 11, 139, 152
United Empire Loyalists, ...12
United Farmers' Cooperative, ... 137

United Farmers of Alberta (UFA), ...16, 36–37, 137, 137*n*143, 138, 140, 264*n*143
United Farmers of Ontario, ...13
United Grain Growers (UGG), ... 37, 138, 139
United States Department of Agriculture (USDA), ...71*f*5.3, 109*f*6.7, 223*n*279,
                           224*t*12.1, 271*n*279
 National Agricultural Statistics Service (NASS), ...60*n*33, 258*n*33
United States (US),
 beef exports to, ... 72, 72*f*5.4
 as major trading partner, ... 70, 70*f*5.2
universal soil loss equation, ... 181
University of Guelph, ... 150, 250
University of Illinois Urbana/Champaign (UIUC), ... 228*n*290, 272*n*290
University of Saskatchewan, ... 240*n*313, 273*n*313
UPA. *See* Union des producteurs agricoles, L' (UPA)
Upper Canada. *See* Ontario (Upper Canada)
"Urban Consumption of Agricultural Land" (Hoffman), ... 184*n*195, 267*n*195
"Urban Sprawl Threatens Prime Farmland" (Immen), ... 185*n*196, 267*n*196
urbanization, ...3, 175, 184–187
 initiatives in Alberta to slow, ... 186
 initiatives in British Columbia to slow, ...186–187
 initiatives in Ontario to slow, ...185–186
 initiatives in Quebec to slow, ... 185
 larger home factor in, ... 185
 population growth factor in, ... 185
Uruguay Round (WTO, 1994), ... 79
U.S. Department of Agriculture, ... 181
U.S. Department of Energy, ... 209*f*11.1
U.S. Environmental Protection Agency (EPA), ... 208*n*236, 225, 269*n*236
"U.S. releases clear stance on glyphosate" (Arnason), ... 225*n*283, 271*n*283
US. *See* United States (US)
US Census Bureau, ... 71*t*5.2
US Conservation Reserve Program (CRP), ... 203–204, 204*f*10.9
US Country of Origin Labelling (COOL), ... 72, 73, 77
US Declaration of Independence (1776), ...11
U.S Library of Congress, ... 238*n*310, 272*n*310
US National Cancer Institute, ... 225
US Oceanic and Atmospheric Administration, ... 218
USDA. *See* United States Department of Agriculture (USDA)
USDA Statistical Bulletin, ...80*n*56, 260*n*56
USMCA. *See* North American Free Trade Agreement (NAFTA), rev. Canada-United States-Mexico Agreement (CUSMA, *a.k.a.* NAFTA 2.0; USMCA; T-MEC)

V

value-added, ... 2, 42, 245

density of in GDP, ... 120
increase of to counteract cost-price squeeze, ... 36–37
leakages and, ... 118
supply management to increase, ... 36–37
Value of Water in Alternative Sources in Saskatchewan, The (Kulshreshtha, Brockman &
O'Grady), ...198*t*10.8
vegetable and fruit processing,
canning plants in Ontario for, ... 102
size of industry, ... 102
vertical farming, ... 249
"Vierra another example of Canadian short-sightedness" (Reguly), ... 139*n*144, 264*n*144

W

Wall, G., ... 218*n*261, 270*n*261
Washington Post, ... 189*n*202, 235*n*302, 235*t*13.1, 267*n*202, 272*n*302
water,
agricultural use of, ... 192, 192*f*10.6
environmental dictates for, ... 246
as global asset, ... 251
hydroponics and, ... 249
increase in demand for safe potable, ... 256
irrigation, ...192–199
marginal value of, ... 198
pollution, ... 221, 221*t*
pricing, ... 198
as public good, ... 197*n*218, 268*n*218
salinization, ...227–228
supply and consumptive uses, ...191–192
technology and development of supply of, ...251–252
tradable entitlements to, ... 198*n*219, 268*n*219
value of, ... 197–198, 198*t*10.8
vertical farming and, ... 249
willingness to pay method (WTP) for, ...197–198
"Water Availability in the SSRB under Climate Change" (Pietoniro, Toth & Toyra), . .219*n*268,
270*n*268
water drainage,
off-farm or public, ...199–200
on-farm or private, ... 199
Ontario support for on-farm, ... 199
Quebec support for on-farm, ... 199
regional development initiatives for, ... 200
socio-economic reasons for public, ... 200
water maintenance,
conservation authorities and, ... 201

    environmental farm plans, ... 201
    flood control, ... 200
    integrated watershed management approach, ... 201
    participatory watershed-based delegated governance models, ... 201
    pollution standards, ... 201
    provincial groundwater registries, ... 201
water management, ...191–201
    legal framework, ... 191
water management units (WMUs), ...199–200
Water Pricing Experiences: An International Perspective (Dinar & Subramanian), ...197*n*217,
                                                               198*n*219, 268*n*217, 268*n*219
water resources, characteristics of, ...3
water rights, ...197–198
    legacy approach to, ... 197
water storage, ... 197*n*216, 268*n*216
Webb, A.J., ...80*n*56, 260*n*56
Western Economic Diversification Fund, ... 132
Western Grain Elevator Association, ... 152
Western Grain Stabilization Act (*WGSA*, 1976), ...156, 157, 157*t*9.2
Western Grains Research Foundation (WGRF), ... 150–151, 150*t*8.6, 151*f*8.5, 151*t*8.7
Western Grains Transition Program (1995), ... 158
Western Grains Transportation Act, The (*WGTA*, 1983), .157, 158, 165
Western Producer, The, ... 60*n*34, 81*n*57, 89*n*70, 102*n*83,
    109*n*92, 110*f*6.8, 111*n*94, 111*n*95, 111*n*96, 127*n*123, 127*n*125, 127*n*126, 128*n*127, 134*n*138,
    142*n*150, 147*n*156, 147*n*158, 149*n*159, 149*n*160, 150*n*162, 168*n*179, 178*n*187, 202*n*227,
    205*n*234, 205*n*235, 213*n*242, 225*n*282, 225*n*283, 226*f*12.3, 234*n*298, 235*n*302, 235*t*13.1,
    240*n*312, 240*n*313, 248*n*318, 248*n*320, 259*n*34, 260*n*57, 260*n*70, 261*n*83, 261*n*92, 261*n*94,
    261*n*95, 261*n*96, 263*n*123, 263*n*125, 263*n*126, 263*n*127, 263*n*138, 264*n*150, 265*n*156,
    265*n*158, 265*n*159, 265*n*160, 265*n*162, 266*n*179, 266*n*187, 268*n*227, 269*n*234, 269*n*235,
    269*n*242, 271*n*282, 271*n*283, 272*n*298, 272*n*302, 273*n*312, 273*n*313, 273*n*318, 273*n*320
Western Wheat Growers Association, The, ... 142*n*149, 264*n*149
wetland drainage, ... 177
WGRF. *See* Western Grains Research Foundation (WGRF)
WGSA. *See Western Grain Stabilization Act* (*WGSA*, 1976)
WGTA. *See Western Grains Transportation Act* (*WGTA*, 1983)
"What Canadian Law Says about Self-Defence" (Arnason), ... 127*n*125, 263*n*125
wheat farming, ...2, 9, 10, 11, 12, 13, 14
    see also Marquis wheat
    breaking wheat-canola rotation, ... 229
    Canada as largest exporter of durum, ...71, 71*f*5.3
    Canada as major exporter, ...71, 71*f*5.3
    Canada as world's seventh largest producer, ... 48
    Canada prairie spring red (CPSR) class, ... 47

Canada western amber durum (CWAD) class, ...47
Canada western soft white spring (CWSWS) class, ...48
Canadian western red spring (CWRS) class, ...47
Canadian Wheat Board, ...17
cooperative grain movements, ... 17
flour mills for, ...98
grain pools, ...17
importance of, ... 30, 30*f*3.11
increase in competition in, ... 245
international prices for, ...77
as major crop, ... 45, 45*t*4.1, 46*f*4.1, 46*t*4.2
non-GMO, ...47
production process, ...48
uses of, ...48
wheat pools, ...17
White, E., ... 60*n*34, 168*n*179, 259*n*34, 266*n*179
WHO. *See* World Health Organization (WHO)
Why You Should Care About Farmland Loss (Canadians for a Sustainable Society), . . .184*n*194, 266*n*194
Wiebe, K.D., ... 178*n*186, 202*t*10.9, 266*n*186
Wikipedia, ...96*t*6.3, 100*n*78, 126*t*7.2, 136*n*139, 177*n*184, 190, 191*n*206, 197*n*215, 233*n*296, 237*n*308, 238*n*310, 239*t*13.4, 257, 261*n*78, 263*n*139, 266*n*184, 267*n*206, 268*n*215, 272*n*296, 272*n*308, 272*n*310
Williams, G.D., ... 218*n*261, 270*n*261
willingness to pay method (WTP), ... 197–198, 202*n*266, 228*n*226
Wilson, B., ... 205*n*234, 269n234
Wilson, C. Roderick, ... 8*n*3, 257
Wine Business Monthly, ... 100*n*80, 261*n*80
wine industry, ... 100
"Wines of Canada" (Bells), ... 100*n*80, 261*n*80
Winnipeg Free Press, ... 101*n*82, 261*n*82
WMUs. *See* water management units (WMUs)
Women's Institute, ... 125, 134
Workmen's Compensation, ... 246
World Bank, ...69*f*5.1, 120*n*112, 132*n*133, 195, 209*t*11.1, 213*t*11.3, 262*n*112, 263*n*133
World Bank, Reshaping Economic Geography (World Bank), ...120*n*112, 132*n*133, 262*n*112, 263*n*133
World Factbook, The: Country Profiles (CIA), ... 70*t*5.1
World Health Organization (WHO), ... 73*n*51, 79, 225, 260*n*51
World Meteorological Organization, ... 216*n*254, 217*n*255, 270*n*254, 270*n*255
World Trade Organization (WTO), ... 62*n*39, 73, 79, 80, 259*n*39
World War I (1914-1918), ... 137
World War II (1939-1945), ...11, 13, 17, 18, 155, 173

Worth, D.E., ... 214*n*245, 269*n*245
WTO. *See* World Trade Organization (WTO)
WTP. *See* willingness to pay method (WTP)

Y
Young Farmers Organization, ... 137
Yukon,
    Crown land in, ... 175